21世纪高等教育计算机规划教材

Android开发与应用

Development and Application of Android

张荣 主编

宋辉 曹小鹏 副主编

人民邮电出版社

北京

图书在版编目（C I P）数据

Android开发与应用 / 张荣主编. -- 北京 : 人民邮电出版社, 2014.2(2018.8重印)
21世纪高等教育计算机规划教材
ISBN 978-7-115-33809-9

Ⅰ. ①A… Ⅱ. ①张… Ⅲ. ①移动终端－应用程序－程序设计－高等学校－教材 Ⅳ. ①TN929.53

中国版本图书馆CIP数据核字(2013)第308140号

内 容 提 要

本书系统讲解 Android 开发的基础知识，全书共有 9 章内容和 8 个实验。内容部分详细地介绍了 Android 的起源和体系特征、Android 开发环境的安装配置、Android 应用程序框架、视图组件的使用、视图界面布局的设计、数据存储与共享、多线程及消息处理、网络通信和多媒体应用等知识点。最后安排的实验部分提供了 8 个典型应用，编者已完成部分功能，另需读者补充完善，达到实验要求提出的效果，并需读者完成遗留的提高部分来增强实践应用能力。为了让读者能够及时地检查自己的学习效果，把握自己的学习进度，每章后面都附有丰富的习题。

本书既可以作为高等院校各专业 Android 开发课程的教材，也可以作为计算机相关培训或技术人员自学的参考资料。

◆ 主　　编　张　荣
副 主 编　宋　辉　曹小鹏
责任编辑　张孟玮
责任印制　彭志环　杨林杰

◆ 人民邮电出版社出版发行　　北京市丰台区成寿寺路 11 号
邮编　100164　　电子邮件　315@ptpress.com.cn
网址　http://www.ptpress.com.cn
固安县铭成印刷有限公司印刷

◆ 开本：787×1092　1/16
印张：17.75　　　　2014 年 2 月第 1 版
字数：465 千字　　　　2018 年 8 月河北第 6 次印刷

定价 45.00 元

读者服务热线：(010)81055256　印装质量热线：(010)81055316
反盗版热线：(010)81055315

前言

Android 是谷歌公司推出的新一代移动设备平台，从诞生之日起就受到了广大程序开发人员的欢迎。然而 Android 系统过于庞大，知识点众多，学习的技术门槛和时间成本都很高，不太适合高校本专科教学。本书抽取出 Android 开发的“精华”和“要点”，剥离了大量琐碎的底层实现细节，进行了高度的概括和总结，将 Android 开发中最基本，应用最广泛的内容进行介绍。

本书以 Android 应用开发能力培养为导向，采用知识讲解和实验训练相结合的方式来组织内容。全书共分为 9 章内容和 8 个实验安排。第 1 章主要介绍 Android 的起源和体系特征；第 2 章介绍 Android 开发环境的安装配置，以及 SDK 中的常用命令；第 3 章介绍 Android 应用程序框架，包含 Android 项目结构、权限和生命周期等；第 4 章介绍视图组件的使用，包含基本视图组件、高级组件和提示框与对话框的用法；第 5 章介绍视图界面布局的设计，主要包含 4 种常见布局管理器的使用和多界面的使用；第 6 章介绍数据存储与共享的 4 种方式；第 7 章介绍多线程及消息处理的用法；第 8 章介绍 HTTP 访问方法、Socket 编程、对 JSON 及 XML 数据的解析和 Web Service 的访问方法；第 9 章介绍音视频的播放和摄像头的使用；最后实验部分安排了 8 个典型应用，来增强读者的实践应用能力。为了让读者能够及时地检查自己的学习效果，把握自己的学习进度，每章后面都附有丰富的习题。

本书条理清楚、语言简练，可帮助读者快速掌握每个知识点。每个部分相互连贯又自成一体，使读者既可以按照本书编排的章节顺序学习，也可以根据实际需要对某一章进行针对性学习。书中不过多地介绍枯燥的理论，注重实用性和可操作性，使用户在掌握相关操作技能的同时，还能学到相应的基础知识。另外开发实例简单实用、针对性强，涵盖了 Android 开发所触及的各个知识点。由于 Android 版本变化很快，本书的实例均适于 API 从 2.2 到 4.2，适用范围广。

本书的参考学时为 48 ~ 64 学时，建议采用理论实践一体化教学模式，各章的参考学时见下面的学时分配表。

学时分配表

章　节	课 程 内 容	学　时
第 1 章	Android 简介	2 ~ 3
第 2 章	Android 开发环境配置和 SDK 命令用法	4 ~ 5
第 3 章	APK 文件结构、程序权限、生命周期、Intent 简介	6 ~ 8
第 4 章	视图组件的使用模式、视图组件用法、提示框与对话框	6 ~ 8
第 5 章	布局管理器的用法、多界面使用	6 ~ 8
第 6 章	数据存储的 4 种方式	6 ~ 8
第 7 章	多线程及消息处理	6 ~ 8
第 8 章	HTTP、Socket 编程、数据解析和 Web Service 的访问方法	6 ~ 8
第 9 章	音视频的播放、摄像头的使用	6 ~ 8
课时总计		48 ~ 64

本书由张荣任主编，宋辉、曹小鹏任副主编。张荣编写了第 2 章、第 4 章、第 5 章、第 9 章和最后的实验部分；宋辉编写了第 3 章、第 6 章、第 7 章、第 8 章；曹小鹏编写了第 1 章。这里非常感谢西安邮电大学计算机学院院长王忠民教授、副院长王曙燕教授、Linux 专家陈莉君教授等提出的宝贵修改意见，同时也非常感谢其他参与编写的人员，如孙家泽、孟伟君、刘永平、王博、王文浪等。由于编者水平和经验有限，书中如有纰漏和不尽如人意之处，恳请读者提出宝贵意见，以便修订时使之更加完善。

编　者

2013 年 12 月

目 录

第 1 章 Android 简介

伴随着信息技术的发展，手机已经从仅具有简单通信功能的设备逐渐发展成为具有强大计算能力的小型计算机系统。近年来手机处理器核心不断增多（目前已经有八核产品上市），其运算速度越来越快，存储设备也在不断扩展，已经接近了几年前微机的水平。在硬件水平不断提高的过程中，基于手机的软件也在迅速发展，逐渐成熟。随着移动互联网和信息技术的发展，手机等手持式设备逐渐取代传统计算机的部分功能已成为现实。在 3G、4G 网络的支持下，尤其是具有小巧、随时随地能够上网、便于携带等特性的手持式设备，对于日常应用如上网浏览网页、购物、收发邮件、观看视频、阅读电子出版物以及游戏娱乐等活动，替代 PC 机已经逐渐成为主流。

1.1 手机操作系统

丰富的手机应用都是需要操作系统的支持，而目前主流的智能手机操作系统有苹果的 iOS、谷歌的 Android 和微软的 Windows Phone 等。这些智能机的操作系统都具有系统平台的功能扩展性，提供良好的开发环境，丰富的硬件驱动，以及多任务支持等特点。

下面对市场上几种主流的智能手机操作系统予以简单的介绍。

1. Symbian

Symbian（塞班）操作系统的前身是英国 Psion 公司推出的 EPOC 操作系统。EPOC 操作系统即 Electronic Piece of Cheese，是专门针对移动信息设备开发的操作系统，坚持“简单”的设计理念。1998 年 6 月，诺基亚、摩托罗拉、爱立信、三菱和 Psion 在英国伦敦共同投资成立 Symbian 公司，将能够运行开放操作系统的移动通信终端产品投入大众消费领域。

Symbian 系统是一个实时性、多任务的 32 位操作系统，具有功耗低、内存占用少等特点，非常适合手机等移动设备使用。经过不断发展完善，Symbian 系统可以支持 GPRS、蓝牙、SyncML 以及 3G 等技术，最重要的是，它是一个标准化的开放式平台，支持任何人为 Symbian 设备开发软件。同时，Symbian 将操作系统的内核与图形用户界面分开，能很好地适应不同方式输入的平台，也可以使各厂商能够为自己的产品制作更加友好的操作界面，符合个性化的潮流。

由于 Symbian 系统的推出主要侧重于电话等通信功能，因此在通话质量上优秀，但对触摸屏、多媒体、新操作界面的支持不足，所支持的各平台之间第三方软件不兼容，同时系统版本之间的兼容性也较差。另外，Symbian 系统结构陈旧、过于复杂，也严重限制了用户开发相关应用。因此，Symbian 系统的没落已经不可避免。

2. iOS

iOS 即 iPhone Operation System，是由苹果公司为 iPhone 开发的操作系统。它主要是给 iPhone、iPod touch 以及 iPad 使用。原本这个系统名为 iPhone OS，直到 2010 年 6 月 7 日 WWDC 大会上宣布改名为 iOS。

iOS 的系统架构分为 4 个层次：核心操作系统层、核心服务层、媒体层和可轻触层。在该系统构架的支持下，iOS 系统提供了华丽的界面，从外观到易用性都拥有直观的用户体验；在与系统互动方面，iOS 中提供了极具创新的多点触控（Multi-Touch）的方式，通过滑动、轻按、挤压及旋转，为用户提供了异常流畅的操作体验；iOS 操作系统还拥有良好的数据安全性，可防止恶意软件和病毒的侵害，也可阻止用户访问敏感信息，iOS 专门设计了针对数据安全的底层硬件和固件功能以及高层的 OS 功能，比如通过获得用户的许可、设置密码锁、对存储的数据加密、支持加密网络通信等方法，避免用户信息的泄漏；另外，iOS 操作系统提供了丰富内置的应用，如使用 Siri 语音和应用进行交互，利用 Facetime 进行视频通话，使用 iMessage 完成短信、照片、视频等的交流，通过 Game Center 进行社交游戏，使用 iCloud 存放 app、照片、通讯录、电子邮件、文档和日历等内容，并可进行信息推送，最后 iOS 平台还拥有世界级庞大的 APP 集合。软硬件的完美搭配使得 iOS 成为优化程度最好、最受欢迎的移动操作系统之一。

3. Windows Phone

2010 年 10 月 11 日，微软公司正式发布了智能手机操作系统 Windows Phone 7，使用 Windows Phone 代替以前的 Windows Mobile。

2012 年 6 月 21 日，微软正式发布全新移动操作系统 Windows Phone 8（以下简称 WP8）。WP8 采用与 Windows 8 相同的内核，意味着可以兼容 Windows 8 应用，让微软的 Windows 系统打通了 PC、平板和手机三大平台。由于内核相同，开发者仅需少量修改就能让应用同时在 WP8 手机和 Windows 8 电脑上运行，除此之外 WP8 还具有以下几个显著的特点：支持 Office 中心和 OneNote 笔记移动版，在 Office Hub 中优化了文件检索机制，用户可以更加快速地找到相关 Office 文档，支持通过 OneNote 分享图片，提供语音笔记以及笔记搜索等功能，提供了广受好评的诺基亚地图服务，支持用于备份和推送的云计算及 OTA 技术，自定义 Live Tile 风格，对系统应用加锁的儿童乐园功能，提供了更便捷的人脉管理功能，整合 Xbox 用户体验，作为便携游戏主机解决方案，支持利用 SmartGlass 应用控制 Xbox 360 主机，提供钱包中心，支持用户可以通过手机绑定功能实现一键支付，用户可以在操作系统中快速地使用 Skype 收发信息，同时 Skype 联系人也被整合至人脉当中等。

WP8 围绕"以人为本"的主题对社交网络大力整合，给用户带来一种与"以应用为核心"的完全相反的用户体验，这对于消费者而言可以说是一个极具吸引力的卖点。

4. Android

Android 是谷歌与开放手机联盟合作开发的基于 Linux 平台的开源手机操作系统。谷歌通过与运营商、设备制造商、开发商和其他有关各方结成深层的合作伙伴关系，希望借助建立标准化、开放式的移动电话软件平台，在移动产业内形成一个开放式的生态系统。

Android 作为谷歌企业战略的重要组成部分，将进一步推进"随时随地为每个人提供信息"这一企业目标的实现。谷歌的目标是让移动通信不依赖于设备甚至平台，其奉行的移动发展战略是通过与全球各地的手机制造商和移动运营商结成合作伙伴，开发既有用又有吸引力的移动服务，并推广这些产品。

Android 平台包括操作系统、用户界面和应用程序。Andriod 系统框架中支持组件的重用与替换，提供的 Dalvik 虚拟机专门为移动设备做了优化。内部集成的浏览器基于开源的 WebKit 引擎，并提供了优化的图形库，包括 2D 和 3D 图形库，其中 3D 图形库基于 OpenGL ES 1.0（硬件加速可选），以及支持 SQLite 数据库。在多媒体支持方面，支持包括常见的音频、视频文件格式（如 MPEG4、H.264、MP3、AAC、AMR、JPG、PNG、GIF）等。不仅对蓝牙、EDGE、3G 和 WiFi 提供了良好的支持，而且对硬件如照相机、GPS、指南针和加速度计等均提供了支持。此外，在开发环境方面提供了包括设备模拟器、调试工具、内存及性能分析图表和 Eclipse 集成开发环境插件等工具，方便用户开发高效的应用程序。

最重要的是 Android 系统继承了 Linux 系统兼容性强的特点，支持包括 ARM、Intel、AMD 等在内的多种处理器。而且不同于 iOS 的封闭生态，Android 的另一优势在于其开放性和免费的服务。Android 是一个对第三方软件完全开放的平台，开发者在为其开发程序时拥有更大的自由度，所以获得了更多产商的支持。

5. BlackBerry

BlackBerry 即黑莓，是指由加拿大 Reserach In Motion（简称 RIM）公司推出的一种无线手持邮件解决终端设备，其特色是支持推动式电子邮件、手提电话、文字短信、互联网传真、网页浏览及其他无线资讯服务。因其使用了标准的 QWERTY 英文键盘，看起来像是草莓表面的一粒粒种子，所以得名“黑莓”。

RIM 推出的黑莓企业解决方案是针对高级白领和企业人士提供的企业移动办公的一体化解决方案。企业有大量的信息需要即时处理，出差在外时，也需要一个无线的可移动的办公设备。企业只要装一个移动网关，一个软件系统，用手机的平台实现无缝链接，无论何时何地，员工都可以用手机进行办公。它最大方便之处是提供了邮件的推送功能：即由邮件服务器主动将收到的邮件推送到用户的手持设备上，而不需要用户频繁地连接网络查看是否有新邮件。黑莓手机往往在对移动办公和安全性要求较高的政府部门和大型企业应用最为广泛。

其他手机操作系统还包括 Palm OS、HP WebOS、MeeGo、BADA 等，由于用户量较少，这里不做过多的介绍。

1.2　Android 起源

Android 是一家成立于 2003 年的美国公司，主要业务是手机软件和手机操作系统，后来被谷歌公司收购。Android 开发者之一安迪 • 鲁宾（Andy Rubin）在一次会议上表示，他们最初的目标是利用 Android 创建一个能够与 PC 联网的“智能相机”世界，由于智能手机市场开始爆炸性增长，Android 被改造为一款面向手机的操作系统。该操作系统是基于 Linux 核心的开源手机软件平台，该平台由操作系统、中间件、用户界面和应用软件组成，号称是首个为移动终端打造的真正开放和完整的移动软件。

下面简单介绍一下 Android 的起源和发展过程。

谈到 Android 的起源，首先需要了解“开放手机联盟”这个组织。其英文名称为 Open Handset Alliance，是美国谷歌公司于 2007 年 11 月 5 日宣布组建的一个全球性的联盟组织，该联盟共同开发名为 Android 的开放源代码的移动系统，是第一个完整的、开放的、免费的智能移动开发平台。目的是加速智能移动设备的发展，提供用户更多更好更便宜的服务。开放手机联盟包括手机制造商、手机芯片厂商和移动运营商几类，创始成员有 Aplix、Ascender、Audience、Broadcom、中国移动、eBay、Esmertec、谷歌、宏达电、英特尔、KDDI、Living Image、LG、Marvell、摩托罗拉、

NMS、NTT DoCoMo、Nuance、Nvidia、PacketVideo、高通、三星、SiRF、SkyPop、Sonic Network、Sprint Nextel、Synaptics、TAT、意大利电信、西班牙电信、德州仪器、T-Mobile 和 Wind River 等，目前，联盟成员数量已经达到了 43 家。

开放手机联盟建立后，Android 的发展也加快了速度，下面介绍一下 Android 历程中一些非常重要的时期，如表 1.1 所示。

表 1.1　　Android SDK 发展过程

版　本	代　号	发布时间	API 级别	备　注
4.3	Jelly Bean	2013.07	18	通知栏更加开放、支持蓝牙 4.0 LE 模式、提升硬件加速等
4.2		2012.10	17	全景拍照、无线显示共享、语音输出和手势模式导航功能、恶意软件扫描功能等
4.1		2012.06	16	发扬 Holo 风格的设计理念
4.0.3	Ice Cream Sandwich	2011.12	15	全新的 UI、自带照片应用、人脸识别功能、语音功能、全新的 3D 驱动、更多的感应器支持
4.0		2011.10	14	
3.2	Honeycomb	2011.07	13	针对平板优化、全新设计的 UI 增强网页浏览功能、支持 Google TV 等
3.1		2011.05	12	
3.0		2011.02	11	
2.3.4	Gingerbread	2011.05	10	提升游戏体验、提升多媒体能力、增加官方进程管理、改善电源管理、支持 NFC 近场通信等
2.3.3		2011.02	10	
2.3		2010.12	9	
2.2.x	Froyo	2010.05	8	提供对 Flash 10.1 的完整支持等
2.1.x	Eclair	2010.01	7	提升硬件速度，更多屏幕以及分辨率选择，大幅度的用户界面改良等
2.0.1		2009.12	6	
2.0		2009.10	5	
1.6	Donut	2009.09	4	集成语音搜索应用，对非标准分辨率有了更好的支持等
1.5	Cupcake	2009.04	3	第一个主要版本，用户界面得到了极大的改良
1.1	Base	2009.02	2	
1.0		2008.09	1	
0.9	m5-0.9	2008.08		
	m5-rc15	2008.03		
	Android	2007.11		项目启动

1.3　Android 特征

相对于其他智能操作系统，Android 系统具有以下几点显著的优势。

1. 开放性

Android 平台最大的优势就是其开放性，该平台允许任何移动终端厂商加入到 Android 联盟中来，从底层操作系统到上层的用户界面和应用程序都不存在任何阻碍产业创新的专有权障碍。

开放性对于 Android 的发展而言，会拥有更多的开发者队伍。而对于消费者来讲，可以享受日益丰富的软件资源。开放的平台也会带来更大竞争，如此一来，消费者将会有更多的选择。

2. 不受任何限制的开发商

由于 Android 的开放性，所以不同的厂商可以根据自身的需要对 Android 平台进行定制与扩展。这样对于用户来说，选择一款 Android 的手机将会有更大的抉择空间，更能展现自身的个性化。

3. 应用程序间的无界限

Android 打破了应用程序间的界限，用户可以将开发的程序与本地的联系人、日历、位置信息等应用很好地进行整合，使应用更加便利。此外，Android 中也允许系统程序可以被其他应用程序所替代，更具有多样化。

4. 丰富的硬件选择

由于 Android 的开放性，允许众多的厂商推出功能特色多种多样的各色产品。功能上的差异和特色，却不会影响到数据同步，甚至软件的兼容。

5. 紧密结合 Google 应用

全球最大的在线搜索服务商谷歌在过去的 10 年中，已经逐渐渗透进人们的日常生活中。谷歌提供的服务如地图、邮件、搜索和在线翻译等已经成为连接用户和互联网的重要纽带，人们不再仅仅满足于使用电脑终端享受这些服务，PC 到移动终端的延伸成了一种必然的趋势。而 Android 与谷歌服务的无缝集成，则可以很便利地满足人们的需求。

1.4 Android 体系结构

Android 的系统架构和其操作系统一样，采用了分层的架构。从架构图来看，Android 分为四个层，从高层到底层分别是应用程序层、应用程序框架层、系统运行库层和 Linux 核心层，其系统架构图如图 1.1 所示。

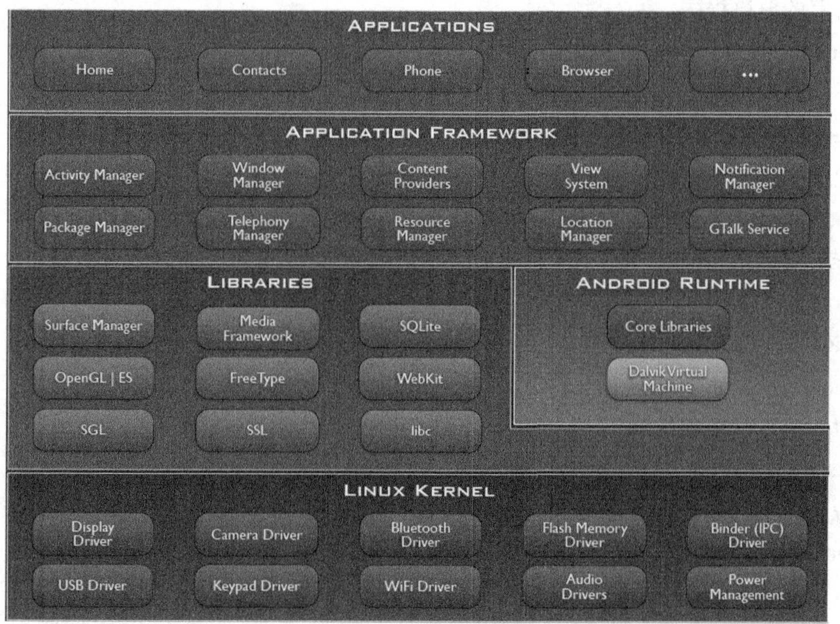

图 1.1 Android 系统架构图

下面对各层分别予以简单的介绍。

1.4.1 应用层

系统架构图的最上层是应用层（Applications），应用层包含的应用是用 Java 语言编写的运行在虚拟机上的程序。比如 SMS 短信客户端程序、电话拨号程序、图片浏览器、Web 浏览器等。Android 支持这些应用程序可以被用户开发的其他应用程序所替换，这点不同于其他手机操作系统固化在系统内部的系统软件，更加灵活和个性化。

系统支持的应用一般都是使用 Java 语言并基于 Android 的 SDK 进行开发的，但在很多应用中，尤其是游戏，需要进行大规模的运算和图形处理，以及使用 C、C++类库。这样通过 Java 来实现的话，会存在执行效率过低等问题，因此，Android 开发中，开发者可以使用 C、C++来实现底层模块，并使用 JNI 接口与上层 Java 实现进行交互，然后利用交叉编译工具生成类库添加到应用中。但整个应用例如界面绘制、进程调度等核心机制都是部署在框架层并通过 Java 来实现的，所以 C、C++程序仅能当作类库来使用。

1.4.2 应用框架层

应用框架层（Applications Framework）是用户进行 Android 开发的基础，是谷歌发布核心应用时所使用的 API 框架。应用框架层的架构设计简化了组件的重用，任何一个应用程序都可以发布它的功能块，并且其他的应用程序都可以在遵循框架的安全性限制下，使用其所发布的功能块。同样，应用框架层的重用机制也使用户可以方便的替换程序组件。

应用框架层包含了一系列的服务和系统，其中包括：

- Activity Manager（活动管理器）

管理各个应用程序生命周期以及通常的导航回退功能。

- Window Manager（窗口管理器）

管理所有的窗口程序。

- Content Provider（内容提供器）

使得不同应用程序之间可以存取或者分享数据。

- View System（视图系统）

构建应用程序的基本组件，如列表、网格、文本框、按钮，甚至可嵌入的 Web 浏览器等。

- Notification Manager（通知管理器）

应用程序可以在状态栏中显示自定义的提示信息。

- Package Manager（包管理器）

Android 系统内的程序管理。

- Telephony Manager（电话管理器）

管理所有的移动设备功能。

- Resource Manager（资源管理器）

提供应用程序使用的各种非代码资源，如本地化字符串、图片、布局文件、颜色文件等。

- Location Manager（位置管理器）

提供位置服务。

- XMPP Service（XMPP 服务）

提供 Google Talk 服务。

1.4.3 系统库层

系统库（Libraries）由一系列二进制动态库共同构成，通常使用 C、C++开发。与框架层的系统服务相比，系统库不能独立运行于线程中，需要被系统服务加载到其进程空间里，通过类库提供的 JNI 接口进行调用。系统库分为两部分，分别是函数库和 Android 运行时。

1. 系统库

系统库是应用程序框架的支撑，是连接应用程序框架层与 Linux 内核层的重要纽带。包含一些 C、C++库，这些库能被 Android 系统中不同的组件使用，组件通过 Android 应用程序框架为开发者提供服务。其主要分为以下几个部分：

- Surface Manager

执行多个应用程序时，负责管理显示与存取操作间的互动，另外也负责 2D 绘图与 3D 绘图进行显示合成。

- Media Framework

多媒体库，基于 PacketVideo OpenCore。支持多种常用的音频、视频格式录制和回放，编码格式包括 MPEG4、MP3、H.264、AAC、ARM。

- SQLite

小型的关系型数据库引擎。

- OpenGL|ES

根据 OpenGL ES 1.0 API 标准实现的 3D 绘图函数库，该库可以使用硬件 3D 加速或者使用高度优化的 3D 软加速。

- FreeType

提供位图和矢量字体的描绘与显示。

- LibWebCore

最新的 Web 浏览器引擎，支持 Android 浏览器和可嵌入的 Web 视图。

- SGL

底层的 2D 图形渲染引擎。

- SSL

在 Andorid 上的通信过程中实现握手。

- Libc

从 BSD 继承来的标准 C 系统函数库，专门为基于 embedded Linux 的设备定制。

2. Android 运行时

Android 应用程序是采用 Java 语言编写，程序在 Android 运行时中执行，其运行时分为核心库和 Dalvik 虚拟机两部分。

（1）核心库。

核心库提供了 Java 语言 API 中的大多数功能，同时也包含了 Android 的一些核心 API。常用库包含：

- android.app

提供高层的程序模型和基本的运行环境。

- android.content

包括各种设备上的数据进行访问和发布。

- android.database

通过内容提供者浏览和操作数据库。

- android.graphics

底层的图形库，包括画布、颜色过滤、点、矩阵，可以将其绘制到屏幕上。

- android.location

定位和相关服务的类。

- android.media

提供管理多种音频、视频的媒体接口。

- android.net

提供网络访问的类。

- android.os

提供了系统服务、消息传输和 IPC 机制。

- android.opengl

提供 OpenGL 的工具。

- android.provider

提供访问 Android 内容提供者的类。

- android.telephony

提供与拨打电话相关的 API 交互。

- android.view

提供基础的用户界面接口框架。

- android.util

设计工具性的方法，例如时间日期的操作。

- android.webkit

默认浏览器操作接口。

- android.widget

包含各种 UI 元素，在应用程序的布局中使用。

（2）Dalvik 虚拟机。

Android 程序不同于 J2ME 程序，每个 Android 应用程序都是一个独立的进程，并且并非多个程序运行在同一个虚拟机中，而是每个应用程序都运行在一个 Dalivik 虚拟机的实例中。Dalvik 虚拟机是一种基于寄存器的 Java 虚拟机，而不是传统的基于栈的虚拟机，并进行了内存资源使用的优化，以及支持多个虚拟机的特点。同时，Dalvik 虚拟机依赖于 Linux 内核的一些功能，例如线程机制和底层内存管理机制。需要注意的是，应用程序在虚拟机中执行的并非编译后的字节码，而是通过转换工具 dx 将 Java 字节码 class 文件转成 dex 格式的中间代码，其过程如图 1.2 所示。

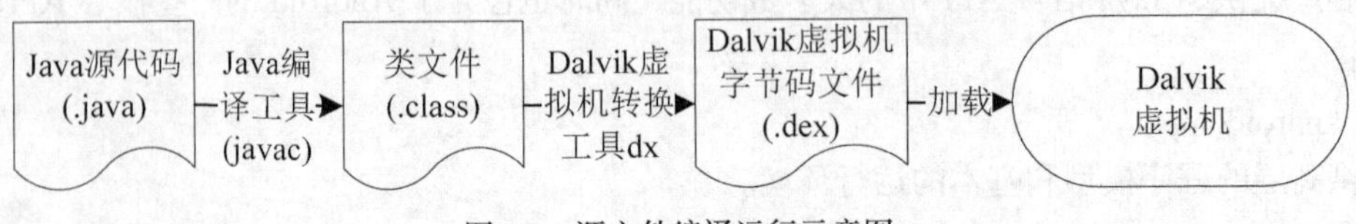

图 1.2　源文件编译运行示意图

1.4.4　内核层

Android 的核心系统服务基于 Linux 2.6 内核，如安全性、内存管理、进程管理、网络协议栈和驱动模型等，例如 Binder IPC 驱动。Linux 内核也同时作为硬件和软件栈之间的抽象层。

1.5　小　　结

本章首先对目前主流的手机系统做了简单的介绍，其中苹果推出的 iOS 系统、微软的 Windows Phone 和谷歌的 Android 是市场占有量最高的三款系统，虽然仍不断地有新系统的推出，但市场用户极少。

随后介绍了 Android 系统的起源，以及各版本间较大的改进和新功能的引入。可以看出从 2007 年开始到现在，短短的 6 年时间，Android 手机用户已经突破了 10 亿，发展确实极为迅猛。

然后简单地介绍了 Android 系统的特点，正是基于开放性的原则，Android 获得了运营商的大力支持，产业链条的热捧，使得手机厂商、开发人员、终端用户量实现了跨越式发展。

最后讲解了 Android 系统的体系结构。

练　　习

1. 简述 Android 平台的特征。
2. 描述 Android 平台体系结构的层次划分，并说明各个层次的作用。

第 2 章 Android 开发环境

本章主要介绍使用 Android SDK 进行开发所必需的硬件和软件需求，目标是掌握 Android 开发环境的安装配置方法，了解 Android SDK 的目录结构，了解各种 Android 开发工具的用途，并会使用向导建立一个 Android 工程，实现在模拟器上的输出。

前文已经介绍过 Android 开发语言是基于 Java 体系，所以开发环境包含 Java 开发包 JDK（Java Development Kit 5 或更新）、开发工具 Eclipse（推荐使用 3.5 版本或以上），以及 Android 开发包 Android Development Kit（简称 ADT），和其他组件如 Android 虚拟设备 Android Virtual Device（简称 AVD），也就是 Android 模拟器。

Android SDK 全部下载需要 3 GB 以上的空间，如果只下载常用版本的 SDK，大概需要 1.5 GB 左右空间，另外 JDK 需要大概 200 MB 左右空间，开发工具 Eclipse 及其插件大概需要 250 MB 左右空间，总共需要准备大于 2 GB 的空间，请用户预留足够的空间进行安装。

2.1 Java 开发环境安装

在搭建 Android 开发环境之前，首先要了解 Android 开发环境对操作系统的要求，如表 2.1 所示，本文是在 Windows 7 操作系统上进行 Android 开发环境的搭建。

表 2.1 Android 开发环境对操作系统的要求

操作系统	要求
Windows	Windows XP（32 位）; Windows Vista（32 位或 64 位）; Windows 7（32 位或 64 位）
Mac OS X	10.5.8 或更新（仅支持 x86）
Linux	在 Ubuntu 系统上，需要 8.04 版或更新; 64 位版本必须支持 32 位应用程序; 需要 GNU C 库（glibc）2.7 或更新

2.1.1 安装 JDK

JDK 是 Sun Microsystems 针对 Java 开发的产品。自从 Java 推出以来，JDK 已经成为使用最广泛的 Java SDK。JDK 是整个 Java 的核心，包括了 Java 运行环境、Java 工具和 Java 基础类库等，安装 Android 开发环境首先需要安装 JDK。JDK 版本 5.0 及以上均可，推荐使用最新版。

1. 下载 JDK

由于 Sun 公司已经被 Oracle 收购，因此可从 Oracle 官网下载，地址为：

http://www.oracle.com/technetwork/cn/java/javase/downloads/jdk7-downloads-1880260-zhs.html。

目前 JDK 最新版本是 JDK 7，请用户选择 JDK 7 的 32 位版下载即可，如图 2.1 所示。

Java SE Development Kit 7u15

您必须接受针对 Java SE 的 Oracle 二进制代码许可协议才能下载该软件。

感谢您接受针对 Java SE 的 Oracle 二进制代码许可协议；现在您可以下载该软件。

产品/文件说明	文件大小	下载
Linux x86	106.64 MB	jdk-7u15-linux-i586.rpm
Linux x86	92.97 MB	jdk-7u15-linux-i586.tar.gz
Linux x64	104.77 MB	jdk-7u15-linux-x64.rpm
Linux x64	91.68 MB	jdk-7u15-linux-x64.tar.gz
Mac OS X x64	143.75 MB	jdk-7u15-macosx-x64.dmg
Solaris x86（SVR4 软件包）	135.52 MB	jdk-7u15-solaris-i586.tar.Z
Solaris x86	91.94 MB	jdk-7u15-solaris-i586.tar.gz
Solaris SPARC（SVR4 软件包）	135.92 MB	jdk-7u15-solaris-sparc.tar.Z
Solaris SPARC	95.26 MB	jdk-7u15-solaris-sparc.tar.gz
Solaris SPARC 64 位（SVR4 软件包）	22.92 MB	jdk-7u15-solaris-sparcv9.tar.Z
Solaris SPARC 64 位	17.59 MB	jdk-7u15-solaris-sparcv9.tar.gz
Solaris x64（SVR4 软件包）	22.53 MB	jdk-7u15-solaris-x64.tar.Z
Solaris x64	14.96 MB	jdk-7u15-solaris-x64.tar.gz
Windows x86	88.75 MB	jdk-7u15-windows-i586.exe
Windows x64	90.4 MB	jdk-7u15-windows-x64.exe

图 2.1　JDK 7 下载页面

用户使用时版本可能已经发生变化，下载最新版即可。

2. 安装 JDK

（1）双击下载下来的 jdk-7u15-windows-i586.exe，打开如图 2.2 所示的安装向导，单击“下一步（N）”按钮。

（2）单击“更改(A)...”按钮后，修改 JDK 安装目录为“D:\android\Java\jdk1.7\”，如图 2.3 所示。

为了管理方便，Android 开发环境所涉及的软件均安装到 D:\android 目录下。

（3）鼠标单击“确定”按钮后，返回到功能选择窗体，如图 2.4 所示。

（4）继续单击“下一步（N）”按钮，开始 JDK 安装，如图 2.5 所示。

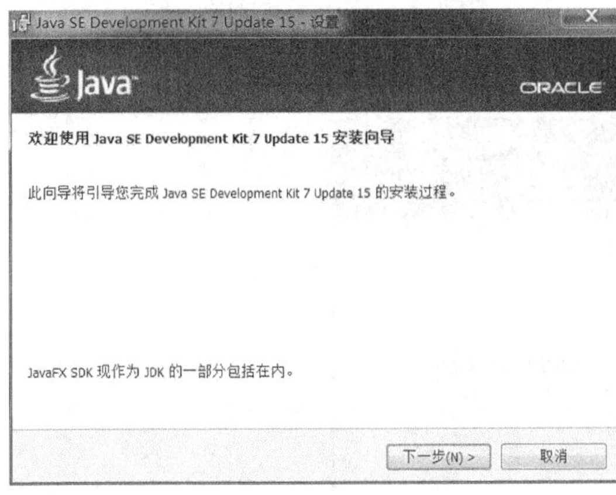

图 2.2　JDK 安装向导窗体

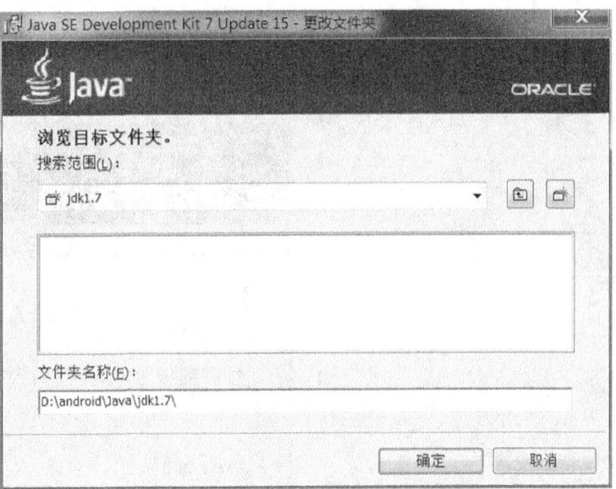

图 2.3　JDK 安装位置选择窗体

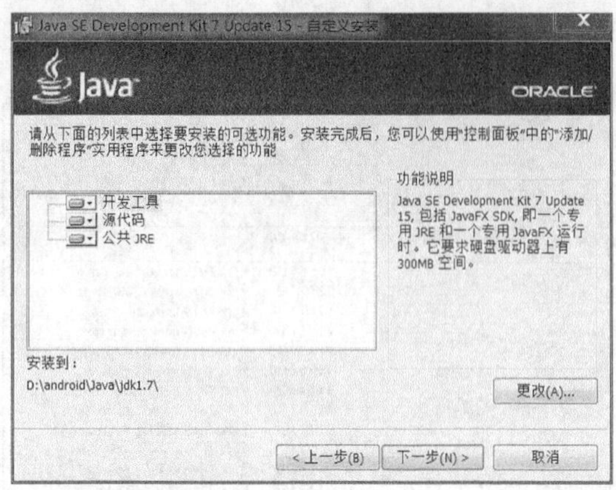

图 2.4　功能选择窗体

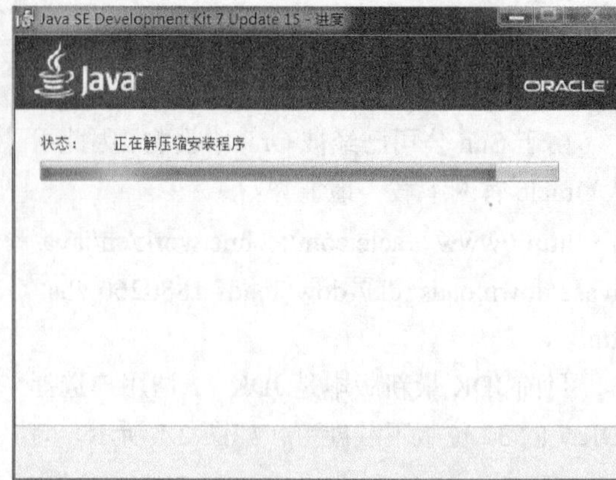

图 2.5　安装进度窗体

JDK 安装完毕后，紧接着进行 Java 运行环境 JRE（Java Run Environment）的安装。

3. 安装 JRE

（1）JRE 安装界面如图 2.6 所示，单击“更改(A)...”按钮，修改 JRE 安装目录为“D:\android\Java\jre1.7\”，单击“确定”按钮后返回 JRE 安装目录。

（2）继续单击“下一步（N）”按钮，如图 2.7 所示。

图 2.6　JRE 安装窗体

图 2.7　JRE 安装进度窗体

（3）JRE 安装完成，如图 2.8 所示，单击“关闭(C)”按钮，关闭该窗体。

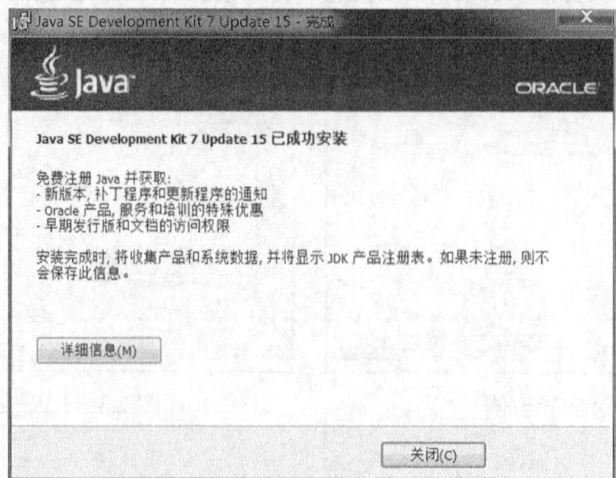

图 2.8　JRE 安装窗体

4. 设置环境变量

JDK 和 JRE 安装完成后，需要进行环境变量的设置。

（1）在资源管理器窗口下用鼠标右键单击“计算机”，如图 2.9 所示。

（2）单击“属性”项打开系统窗体，如图 2.10 所示。

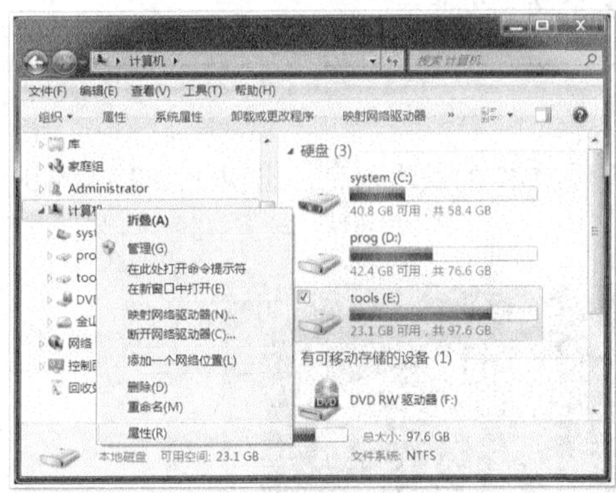

图 2.9 计算机属性项

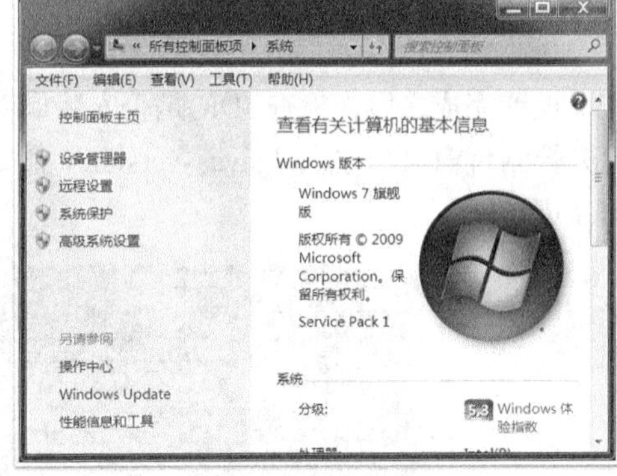

图 2.10 系统窗体

（3）单击系统窗体左侧的“高级系统设置”，打开系统属性窗体，如图 2.11 所示。

（4）依次单击系统属性窗体的“高级”选项卡，“环境变量(N)...”按钮，打开环境变量窗体，如图 2.12 所示。

（5）在“用户变量”区域单击“新建(N)...”按钮，添加变量名为“JAVA_HOME”，变量值为 JDK 的安装目录“D:\android\Java\jdk1.7”，如图 2.13 所示。

图 2.11 系统属性窗体

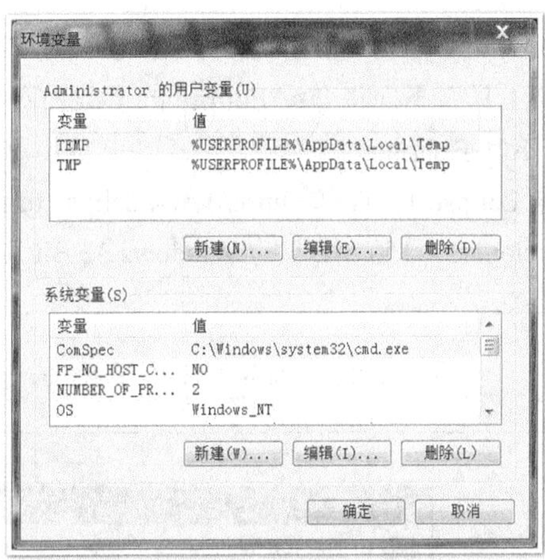

图 2.12 环境变量窗体

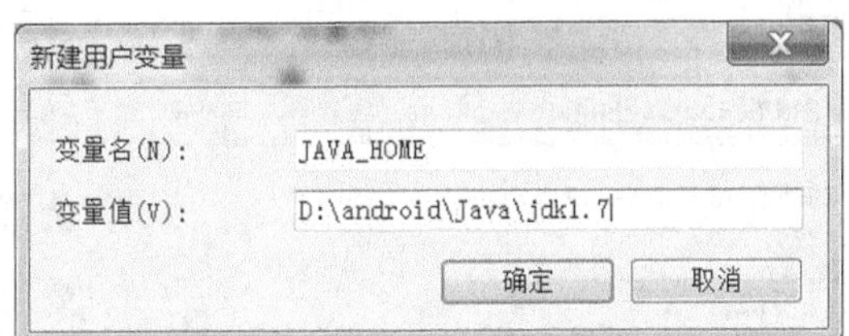

图 2.13 设置 Java 主目录窗口

（6）如上单击“新建(N)...”按钮，依次添加：

变量名为“Path”，变量值为“%JAVA_HOME%\bin;”。

变量名为“CLASSPATH”，变量值为“.;%JAVA_HOME%\lib\tools.jar; JAVA_HOME%\lib\dt.jar;”。

“CLASSPATH”的变量值中第一项是个英文句点。

（7）设置完成之后，检查 JDK 是否安装成功。

打开控制台窗口，输入“java -version”命令查看 JDK 的版本信息，出现如图 2.14 所示画面表示安装成功。

图 2.14　执行 Java 命令窗口

2.1.2　安装 Eclipse

Eclipse 是一个开源的集成开发环境，目前是开发 Java 项目主流的工具，同样也是 Andorid 开发的主流工具，下面将介绍 Eclipse 的下载和安装。

1. 下载 Eclipse

登录 Eclipse 官方网站 http://www.eclipse.org/downloads/，下载如图 2.15 所示的“Eclipse IDE for Java Developers，150 MB”的 Windows 32 Bit 版(即 eclipse-java-juno-SR2-win32.zip)。

用户下载时版本可能已经变化，下载最新版即可。

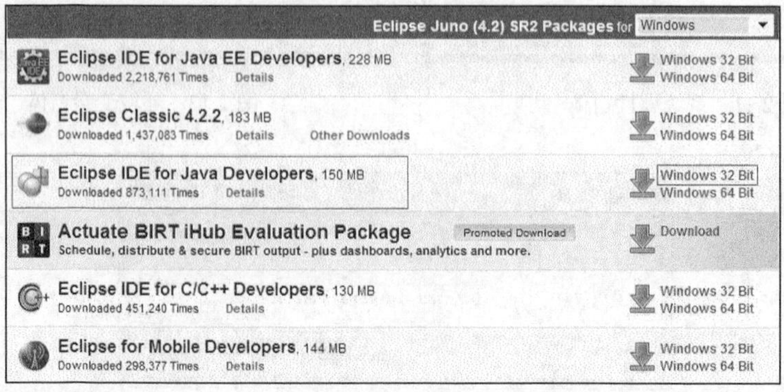

图 2.15　Eclipse 下载页面

2. 安装 Eclipse

（1）将压缩包内的 eclipse 文件夹解压到“D:\android”目录下。

（2）进入“D:\android\eclipse”目录，运行 eclipse.exe。

（3）选择自己的工作空间（Workspace）路径，工作空间即用户开发的工程所存放的位置。可以单击浏览“Browse...”按钮，选择自己的存放路径，如设置为“D:\android\eclipse\example”，若不希望下次打开 Eclipse 时再有该提示，可以单击“Use this as the default and do not ask again”前面的复选框，如图 2.16 所示。

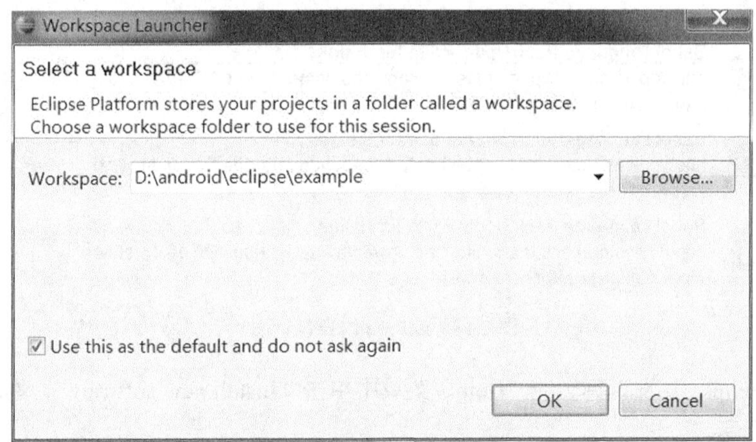

图 2.16　设置 Eclipse 工作空间窗口

（4）单击“OK”按钮，进入 Eclipse 主界面，如图 2.17 所示。

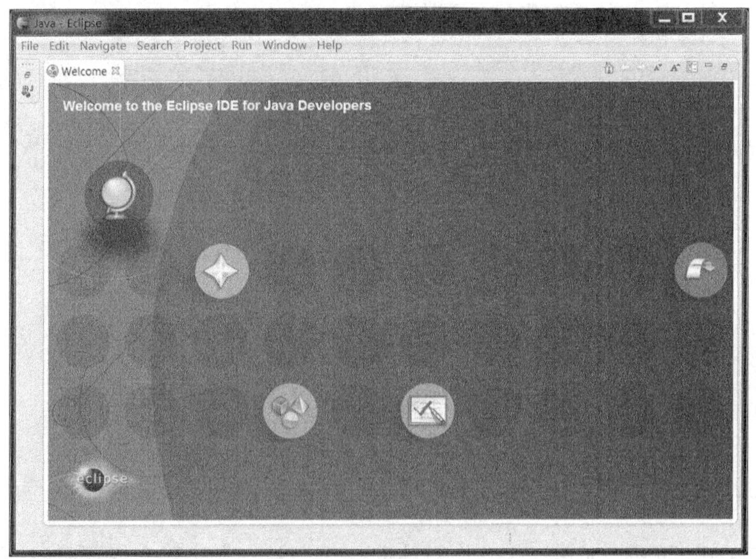

图 2.17　Eclipse 主界面

此时 Eclipse 安装完毕。

2.1.3　Eclipse 中文包的安装

Eclipse 界面提供对多种语言的支持，使用自己熟悉语言的界面能够提高开发效率，下面介绍如何对 Eclipse 进行汉化。

（1）访问 Eclipse 官方网站 http://www.eclipse.org/babel/downloads.php，复制 Babel 语言包的地址（注意选取 Eclipse 版本对应的语言包地址），即“http://download.eclipse.org/technology/babel/update-site/R0.11.0/juno”，如图 2.18 所示。

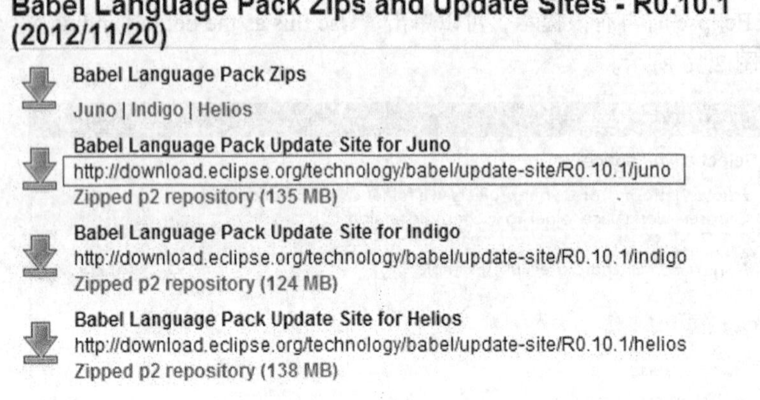

图 2.18　Eclipse 语言包页面

（2）打开 Eclipse 开发工具，在“Help”菜单中单击“install new software”，弹出“install”窗口，如图 2.19 所示。

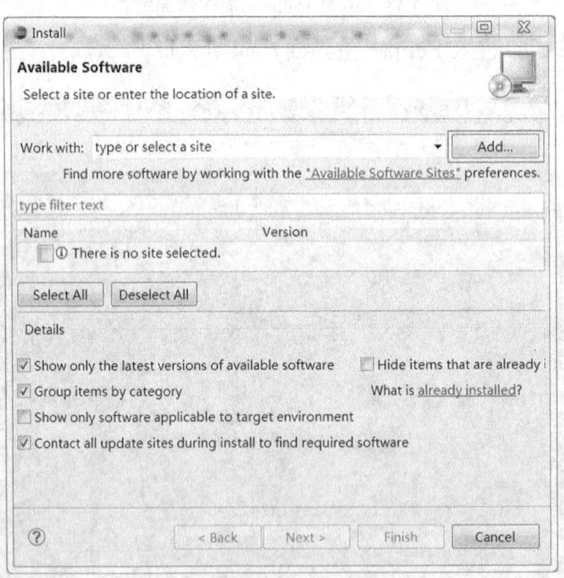

图 2.19　Install 窗体

（3）单击“Add”按钮，显示一个仓库对话框，在“Name”中输入语言包名称，如“中文包”（用户可自己设置一个合适的名称），在“Location”中粘贴第一步中复制的 Babel 语言包的地址，如图 2.20 所示。

（4）单击“OK”按钮，返回到上个界面，在“Work with”中选择刚刚配置的选项，并选择其中的中文包项“Babel Language Packs in Chinest(Simplified)”，如图 2.21 所示。

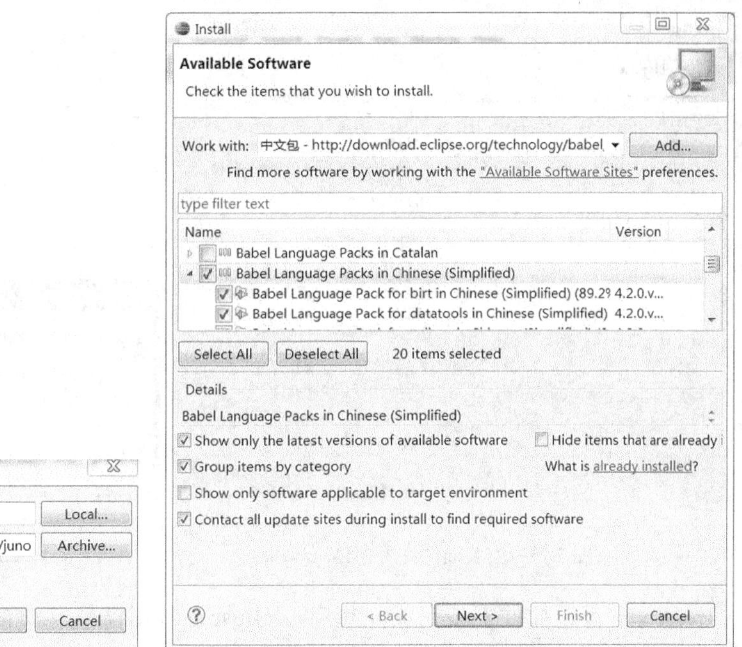

图 2.20　设置语言包仓库　　　　图 2.21　安装语言包窗体

（5）单击“Next”按钮，确定语言包的内容，如图 2.22 所示。

（6）单击“Next”按钮，打开选择安装插件许可窗体，选择接受协议“I accept the terms of the license agreement”，如图 2.23 所示。

（7）单击“Finish”开始安装，如图 2.24 所示。

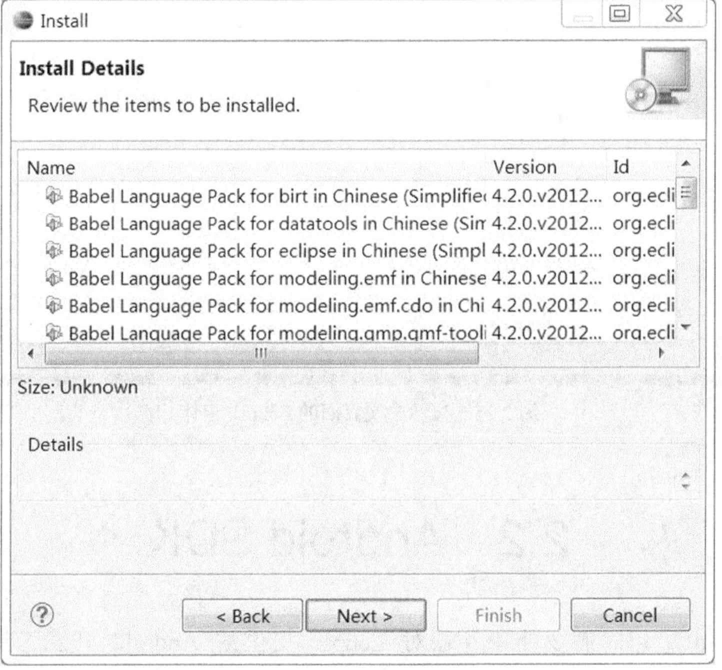

图 2.22　确定语言包窗体

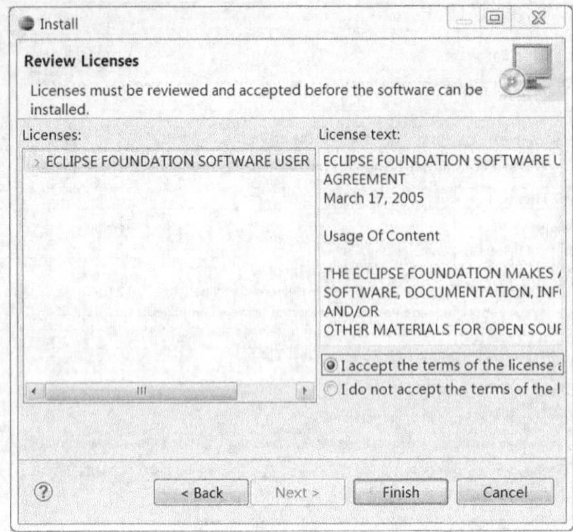

图 2.23 安装插件许可窗体

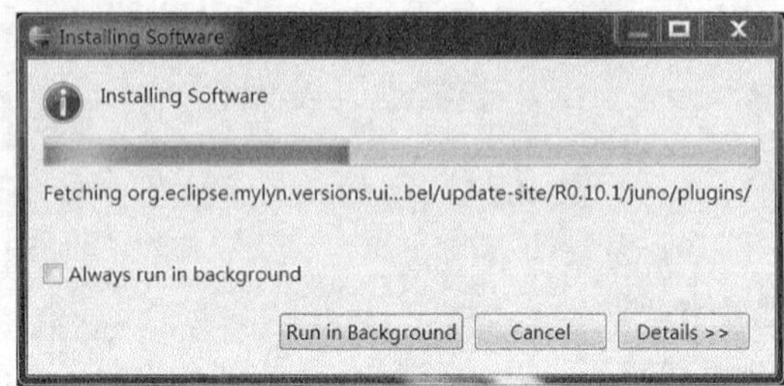

图 2.24 安装语言包插件窗体

（8）语言包插件安装完成，重启 Eclipse，常用菜单及设置项已经换成中文，如图 2.25 所示。

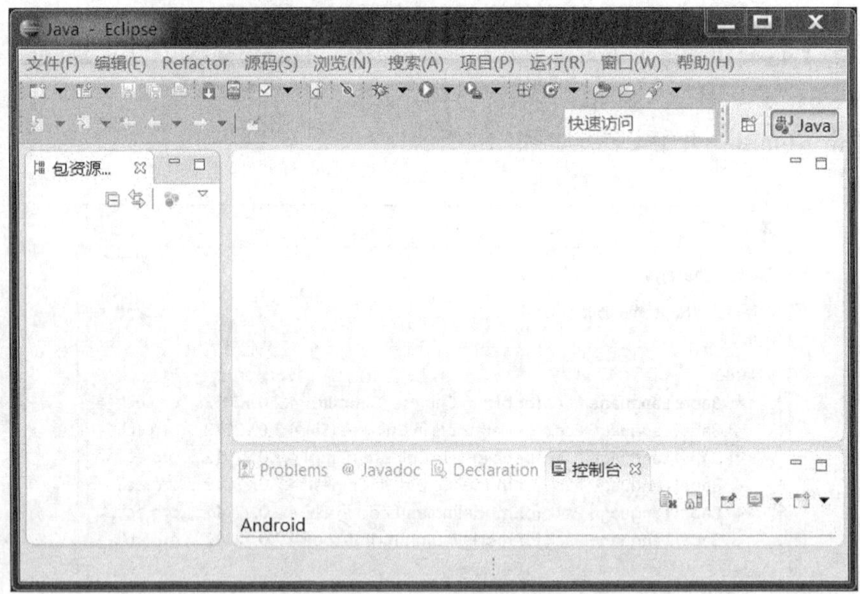

图 2.25 显示中文界面的 Eclipse 窗体

2.2 Android SDK

开发 Android 应用时需要使用谷歌对 Eclipse 提供的 Android 开发工具插件（Android Development Tools，ADT），以及 Android SDK，下面对安装分别予以介绍。

2.2.1　安装 ADT

打开 Eclipse 开发工具，安装 ADT 插件。

（1）在“Help”菜单中单击“install new software”，打开“Install”窗口，同前例。

（2）单击“Add...”按钮，在打开的对话框中 Name 处输入“ADT”，Location 栏输入“https://dl-ssl.google.com/android/eclipse”，如图 2.26 所示。

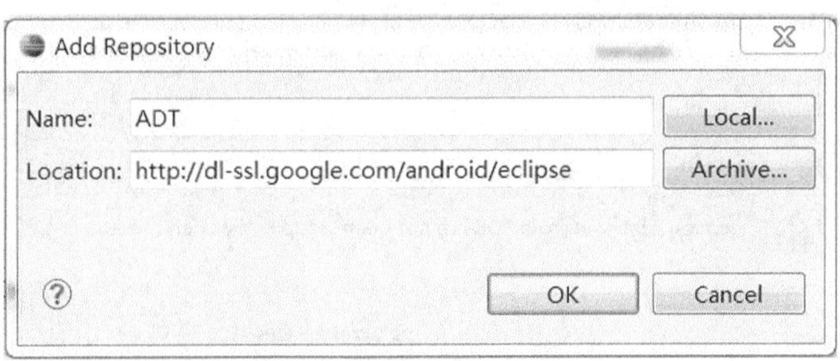

图 2.26　设置 ADT 仓库

（3）单击“OK”按钮返回后，在“Work with”中选择刚添加的 ADT 项，展开下方的“Developer Tools”，会出现“Android DDMS”、“Android Development Tool”、“Android Hierarchy Viewer”、“Android Traceview”选项，全部勾选，如图 2.27 所示。

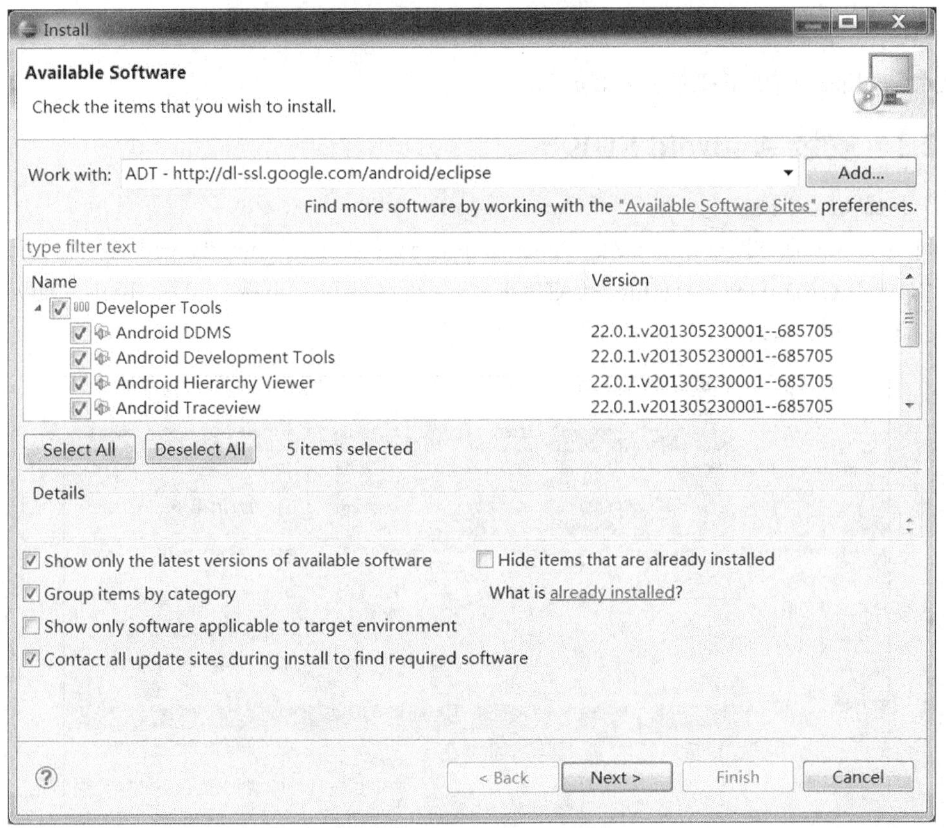

图 2.27　安装 ADT 插件窗体

（4）一路单击“Next>”按钮，并选择接受协议，安装完成会重启 Eclipse，这时会出现 Android 的 SDK 未设置提示，下一步将介绍 Android SDK 的安装，此时只需单击“Close”按钮关闭窗口即可，如图 2.28 所示。

图 2.28　未设置 Android SDK 提示窗口

至此，Eclipse 上的 ADT 插件安装成功。

2.2.2　安装 Android SDK

（1）下载 Android SDK。

访问 Android 官方网址“http://developer.android.com/sdk/index.html#download”，打开页面下部的“DOWNLOAD FOR OTHER PLATFORMS”折叠窗口，下载“SDK Tools Only”中的“android-sdk_r22.0.1-windows.zip”，如图 2.29 所示。

Developer Tools
Download
Setting Up the ADT Bundle
Setting Up an Existing IDE
Android Studio
Exploring the SDK
Download the NDK
Workflow
Tools Help
Revisions
Extras
Samples
ADK

DOWNLOAD FOR OTHER PLATFORMS

ADT Bundle

Platform	Package	Size	MD5 Checksum
Windows 32-bit	adt-bundle-windows-x86-20130522.zip	446736316 bytes	53345fa4121fa58cc048f66c3af3bae9
Windows 64-bit	adt-bundle-windows-x86_64-20130522.zip	446864400 bytes	b28817f62e7f54e3c683841b61b65564
Mac OS X 64-bit	adt-bundle-mac-x86_64-20130522.zip	409047751 bytes	3f4d05206d66e402e87b27a6b3dcf0f9
Linux 32-bit	adt-bundle-linux-x86-20130522.zip	439988972 bytes	1fdd8d7537ab9217d61d32ab912f0243
Linux 64-bit	adt-bundle-linux-x86_64-20130522.zip	440275051 bytes	e38751ff372a8d6208fcef5659133e98

SDK Tools Only

Platform	Package	Size	MD5 Checksum
Windows 32 & 64-bit	android-sdk_r22.0.1-windows.zip	113483496 bytes	cb7f7703450afa5914fb4b8b8332a9f3
	installer_r22.0.1-windows.exe (Recommended)	93479015 bytes	81621d3b164f81f91e066011b325f88f

图 2.29　Android SDK 下载地址页面

（2）将下载的 SDK 文件解压到 D:\android 目录下，并改名为 android-sdk。

（3）打开 Eclipse 开发工具，选择“Window”菜单中的“Preferences...”项，在打开的设置窗口中，选择左边面板上的“Android”项，设置“Android Perferences”中的“SDK Location”栏为“D:\android\android-sdk”，如图 2.30 所示。

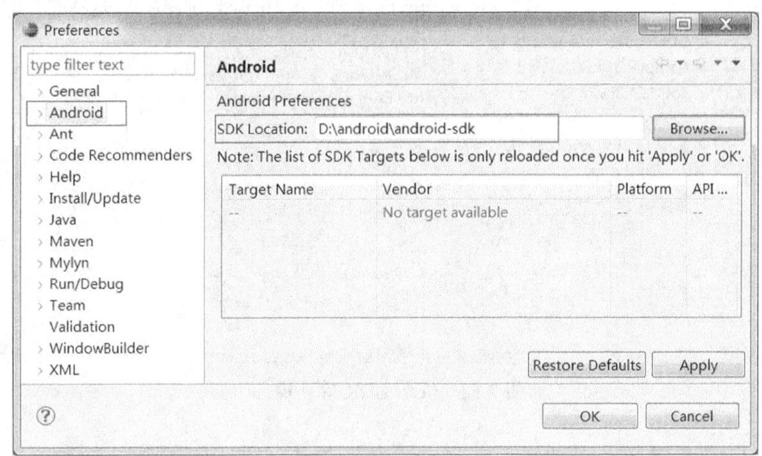

图 2.30　设置 SDK 路径窗口

（4）设置 Android SDK Manager。

选择“Window”菜单中的“Android SDK Manager”，在弹出的 Android SDK Manager 对话框中，选择 SDK 安装包。在 Android 4.2.2 项目中，选中 SDK 文档、SDK 平台、实例和基于不同硬件系统的模拟器映像插件等，选中 ARM 插件，而英特尔凌动 x86、MIPS 插件可根据用户实际需要选择。如只安装 Android 4.2.2 所有相关内容，单击“install 12 packages”按钮，如图 2.31 所示。也可根据需要选择 Android 其他版本的相关文档、实例和模拟器映像插件，如用户量仍然很大的 Android 2.3.3。

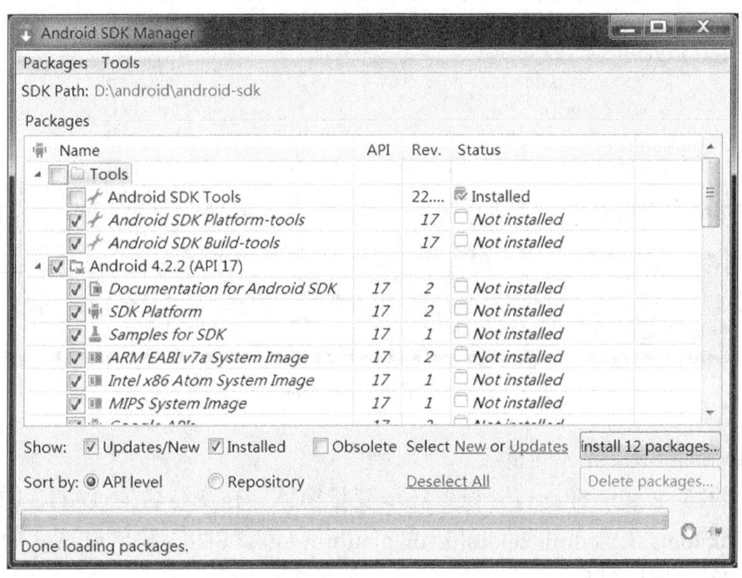

图 2.31　Android SDK Manager 窗口

（5）在弹出的确认框中，选中“Accept License”，单击“Install”按钮，如图 2.32 所示。

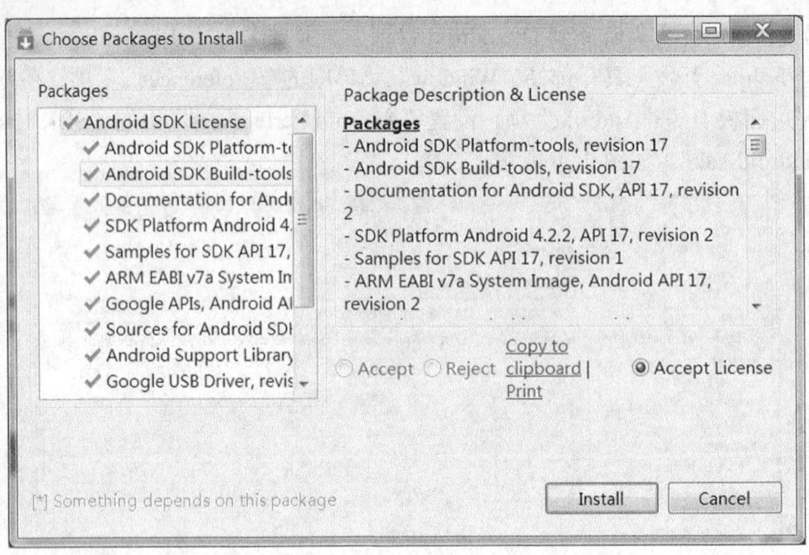

图 2.32　安装 SDK 确认窗体

（6）如组件已安装成功，则 Status 栏会显示 Installed 字样，如图 2.33 所示。如果还需要安装其他组件，以后可随时根据需要自行添减。

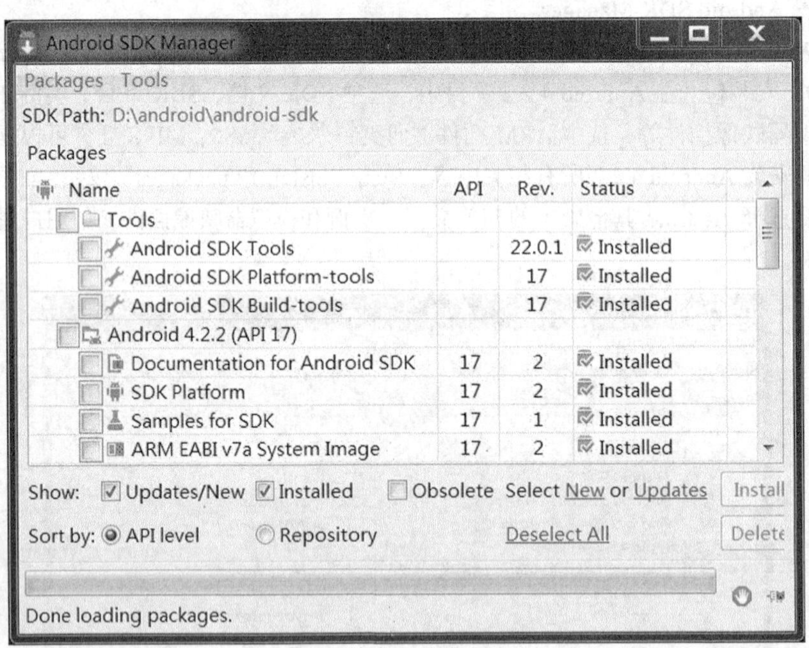

图 2.33　安装 SDK 后的 Manager 窗口

（7）设置 Android SDK 环境变量。

同 2.1.1 中的第 4 步一样，打开设置环境变量窗体，在变量名为 PATH 的值后添加“D:\android\android-sdk\tools;D:\android\android-sdk\platform-tools”SDK 路径，如图 2.34 所示，单击“确定”按钮，并重新启动计算机。

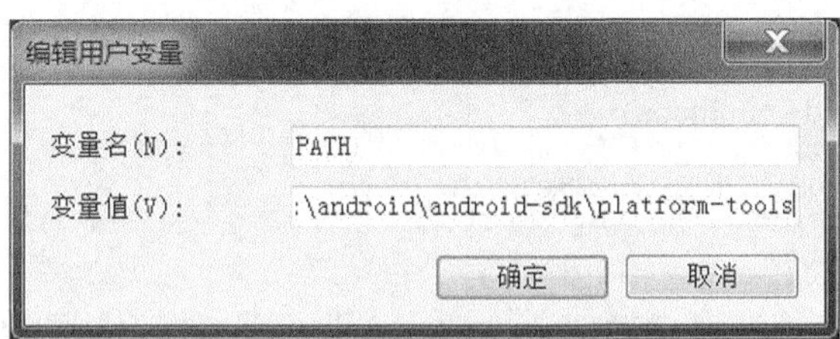

图 2.34 修改 PATH 环境变量

（8）重启后，打开控制台窗口，运行“android -h”，检查 SDK 是否安装成功。如果有图 2.35 所示的输出，表明安装成功。

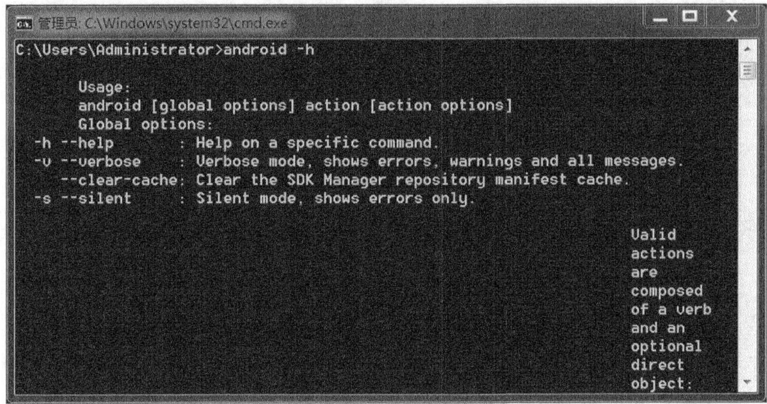

图 2.35 执行 Android 命令窗体

至此，Android SDK 安装完成。

除 2.1.2 到 2.2.2 的安装方法外，还可直接从 Android 官网下载集成工具来安装 Android 开发环境，该工具包含 Eclipse 开发工具、ADT 插件和 SDK 包，安装过程如下。

（1）打开“http://developer.android.com/sdk/index.html”网址，如图 2.36 所示。

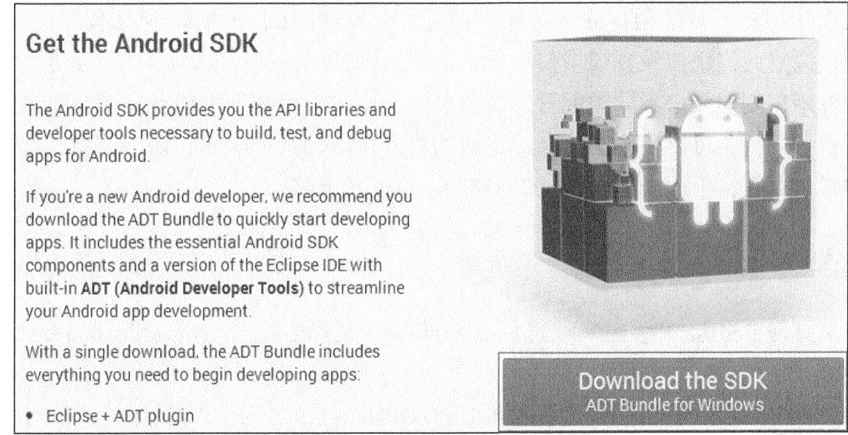

图 2.36 Android 集成工具页面

（2）单击“Download the SDK ADT Bundle for Windows”按钮，打开许可协议窗体，如图 2.37 所示。

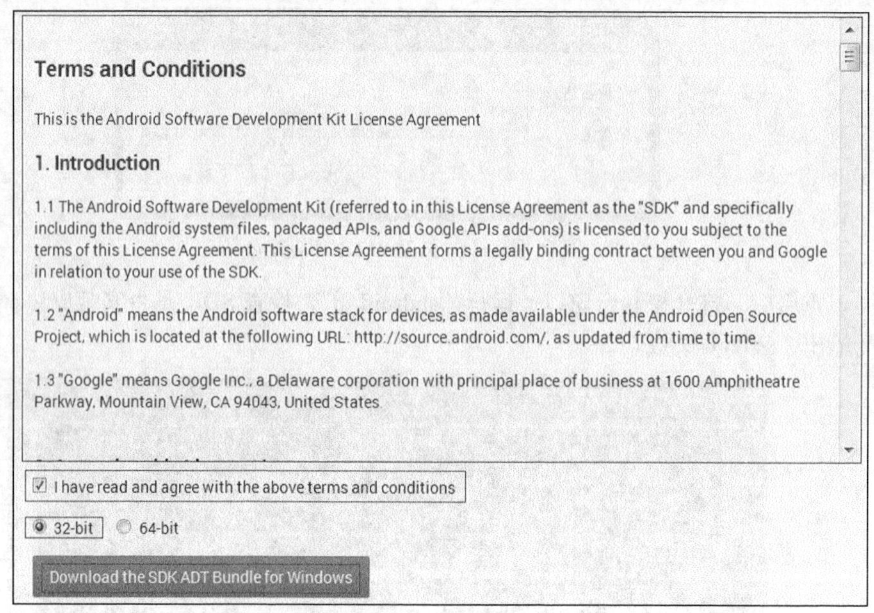
Terms and Conditions

This is the Android Software Development Kit License Agreement

1. Introduction

1.1 The Android Software Development Kit (referred to in this License Agreement as the "SDK" and specifically including the Android system files, packaged APIs, and Google APIs add-ons) is licensed to you subject to the terms of this License Agreement. This License Agreement forms a legally binding contract between you and Google in relation to your use of the SDK.

1.2 "Android" means the Android software stack for devices, as made available under the Android Open Source Project, which is located at the following URL: http://source.android.com/, as updated from time to time.

1.3 "Google" means Google Inc., a Delaware corporation with principal place of business at 1600 Amphitheatre Parkway, Mountain View, CA 94043, United States.

图 2.37　许可协议窗体

（3）选择 32 位版，单击“Download the SDK ADT Bundle for Windows”按钮下载。解压后打开 Eclipse，执行 2.2.2 第四步，安装相应的 SDK 文档、SDK 平台、实例和基于不同硬件系统的模拟器映像插件等即可。

2.3　Android 模拟器

Android 虚拟设备 AVD，也就是 Android 模拟器。它是一个可以运行在用户电脑上的模拟器，可以让用户不需使用物理设备即可预览、开发和测试 Android 应用程序。

该模拟器允许用户设置模拟移动设备的特定硬件属性(如 RAM 大小、分辨率、SD 卡大小等)，并且允许用户创建多个不同的配置以在各种 Android 平台和硬件组合下进行调试。应用程序一旦在模拟器上运行，可以使用 Android 平台的服务来启动其他应用、访问网络、播放音频视频、存储和检索数据等，下面介绍 AVD 的创建和基本使用方法。

2.3.1　创建 AVD

（1）打开 Eclipse 开发工具，选择“Window”菜单中的“Android Virtual Device Manager”项打开模拟器管理窗口，如图 2.38 所示。

（2）单击“New”按钮创建模拟器，AVD Name 命名为“AVD-4.2”，Device 为选择适配手机或分辨率，这里选择“Nexus S(4.0",480*800:hdpi)”，Target 为选择模拟器的版本，这里选择“Android 4.2.2-API Level 17”，SD Card 为设置外置内存卡的大小，这里设为 50 MB，如图 2.39 所示。

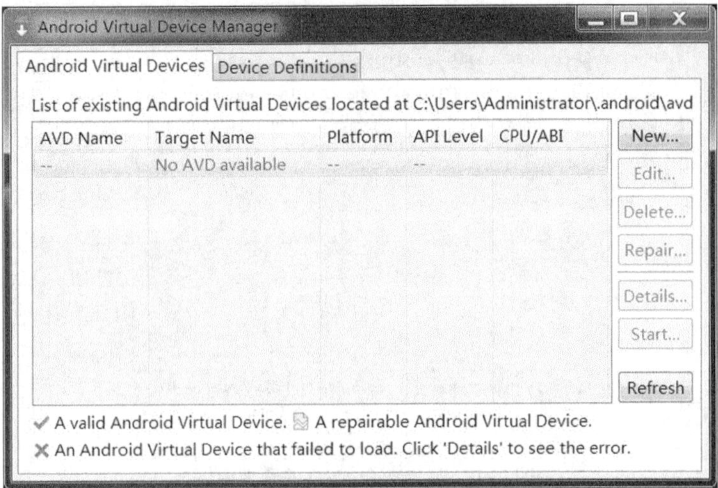

图 2.38 模拟器管理窗体

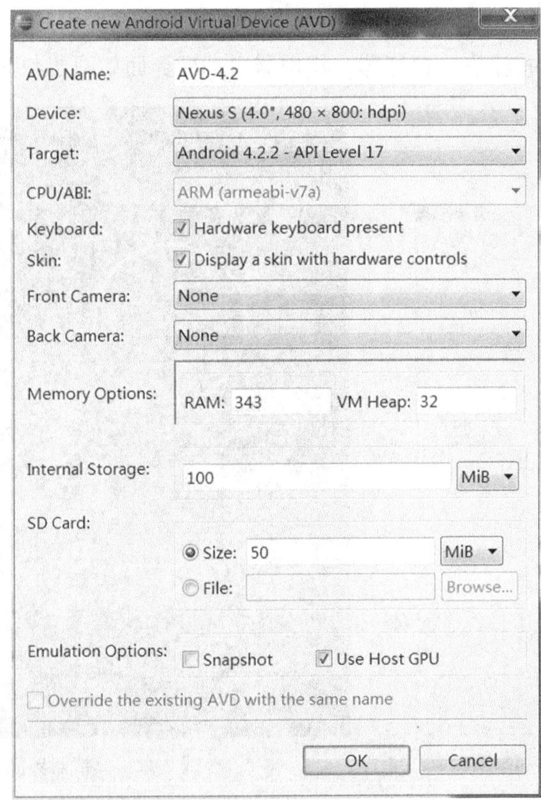

图 2.39 创建 AVD 窗体

同样方法再设置一个 Target 为 Android2.3.3 的模拟器“AVD2.3.3”，此时该模拟器管理器便设置好了两个模拟器，如图 2.40 所示，用户可在开发时选择使用。

建立模拟器 AVD2.3.3 之前，必须在 Android SDK Manager 中安装 2.3.3 的 SDK 平台相关资源。

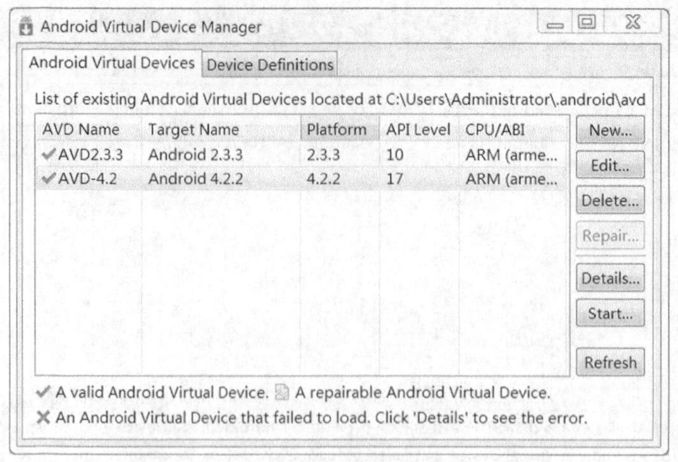

图 2.40　创建 2 个 AVD 的管理窗体

（3）模拟器的启动。

选中相应的 AVD 名称项，单击“Start”按钮启动模拟器，打开启动选项窗体，如图 2.41 所示。

（4）单击“Launch”按钮启动模拟器，模拟器启动后界面如图 2.42 所示。

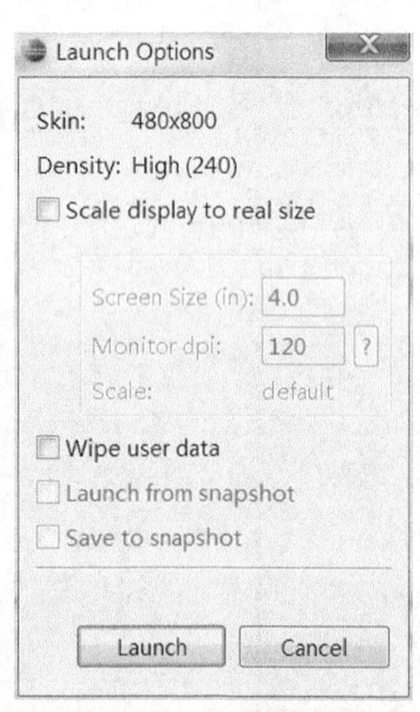

图 2.41　模拟器启动选项窗体

图 2.42　模拟器界面

（5）在模拟器中打开设置工具，在“语言和输入法”中将语言设置为简体中文，在“日期和时间”中将时区设为中国标准时间。

至此 Android 开发环境配置完成。

当网络不通畅时，上述安装过程可简化，从其他已安装好的计算机上将整个环境目录（如 D:\android）直接复制过来，该目录包含 Android SDK、JDK、JRE 和 Eclipse，然后设置环境变量即可（环境变量的设置可参考上文）。

2.3.2　开发环境测试

下面通过建立简单的“Hello World”例子来测试安装环境是否搭建成功。

（1）启动 Eclipse，单击“文件(F)”/“新建(N)”/“项目(R)”，打开如图 2.43 所示的窗口。

（2）选择 Android 文件夹中的“Android Application Project”选项，单击“下一步(N)>”按钮，打开如图 2.44 所示的窗口。

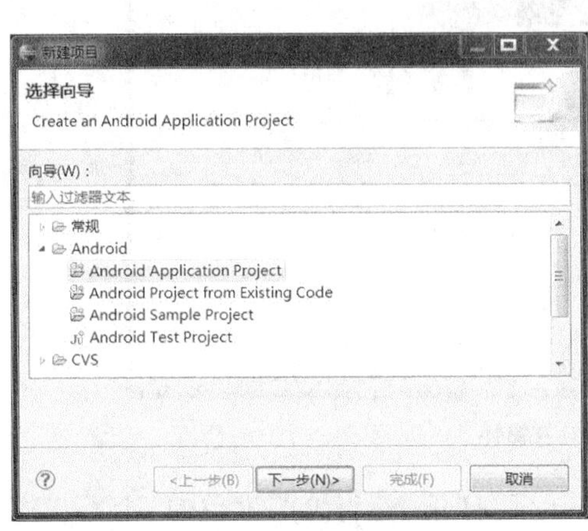

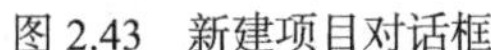
图 2.43　新建项目对话框

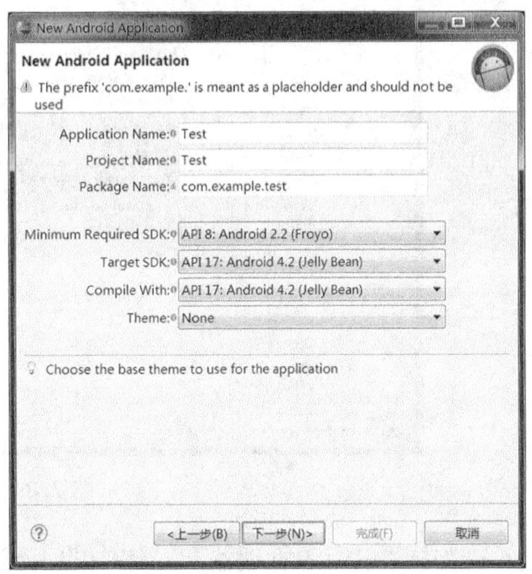

图 2.44　新建 Android 应用对话框

（3）在“Application Name:”栏中输入工程名称 Test，其余使用默认设置，在“Minimum Required SDK”栏中选择“API 8:Android 2.2(Froyo)”，在“Target SDK”和“Compile With”栏中选择“API 17:Android 4.2(Jelly Bean)”，在“Theme”栏中，选择“None”。单击“下一步(N)>”按钮，打开如图 2.45 所示的窗口。

（4）在后续窗口中，单击“下一步(N)>”按钮，最后单击“完成(F)”按钮，打开如图 2.46 所示的窗口。

图 2.45　配置 Android 工程对话框

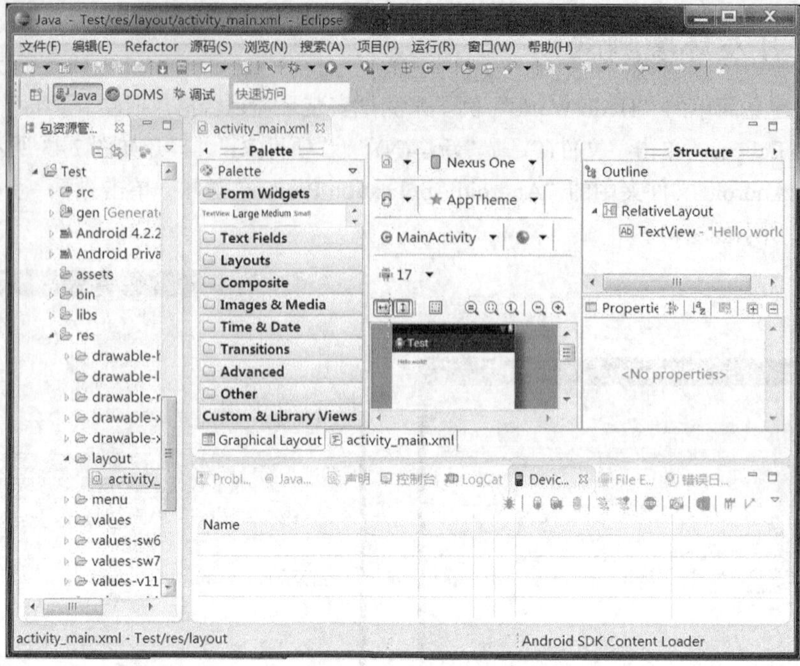

图 2.46　工程开发窗体

（5）选中“包资源管理器”窗体中的工程名 Test，再打开“运行(R)”菜单的“运行(R)”项，打开如图 2.47 所示的窗口。

（6）选择“Android Application”，单击“确定”按钮，打开模拟器，该 Android 工程运行结果如图 2.48 所示。

图 2.47　选择运行对话框

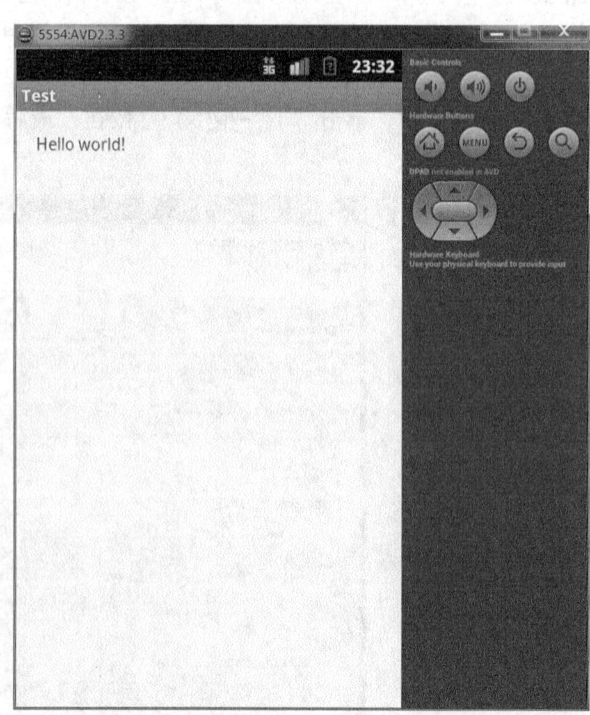

图 2.48　模拟器运行结果

2.3.3 模拟器的使用

模拟器除了不能拨打真实电话以外，可以模拟移动设备的所有软硬件特性。通过提供的多种导航和控制键，用户可以使用鼠标或键盘来为应用程序生成事件，模拟器还提供了屏幕来显示开发的应用程序，以及其他正在运行的 Android 应用。

1. 控制模拟器

用户可以使用启动选项和控制台命令来控制模拟器环境的行为和特征，当模拟器运行时，用户可以像使用真实移动设备那样使用模拟器，通过使用鼠标单击来模仿手指触摸设备屏幕，使用模拟器提供的键盘来模仿按下操作。

模拟器的按键与计算机键盘的对应关系如表 2.2 所示。

表 2.2 模拟器按键对应键盘的按键

模拟器按键	计算机键盘按键
Home	Home 键
Menu	F2 键或 PgUp 键
Start	Shift + F2 组合键或 PgDn 键
Back	Esc 键
Call/Dial	F3 键
Hangup/Light Off	F4 键
Search	F5 键
轨迹球模式	F6 键
Power	F7 键
手机网络开关	F8 键
Camera	Ctrl + F3 组合键或 Ctrl + 小数字键盘的“5”键
Volume Up（增大音量）	Ctrl + F5 组合键或小数字键盘的“+”键
Volume Down（减小音量）	Ctrl + F6 组合键或小数字键盘的“-”键
旋转屏幕方向	Ctrl + F11 组合键或小数字键盘的“7”键
全屏模式	Alt + Enter 组合键
方向键（上下左右）	小数字键盘的“8”“2”“4”“6”键
中心键	小数字键盘的“5”键

2. 模拟器可实现功能

当前版本中，模拟器可实现如下功能：

- 可模拟电话本、通话等功能。
- 内置的浏览器和 Google Maps 都可以联网。
- 可以使用键盘输入。
- 可单击模拟器按键输入。
- 可以使用鼠标单击、拖动屏幕进行操作。

3. **模拟器限制**

当前版本中，模拟器不能实现如下功能：

- 不支持呼叫和接听实际来电，但可以通过控制台模拟电话呼叫。
- 不支持 USB 连接。
- 不支持相机/视频捕捉。
- 不支持音频输入(捕捉)，但支持输出(重放)。
- 不支持扩展耳机。
- 不能确定连接状态。
- 不能确定电池电量水平和交流充电状态。
- 不能确定 SD 卡的插入/弹出。
- 不支持蓝牙。

4. **模拟器应用**

【例 2.1】 使用 DDMS 透视图管理模拟器上的 SD 卡

使用 Eclipse 的 ADT 插件管理模拟器，对于 SD 卡中的内容可完成导入、导出、建立目录、删除文件等操作，但非 SD 卡内容是只读的，仅仅能导出，不能实现其他操作。

（1）首先打开 Eclipse 和启动模拟器。

（2）单击“窗口(W)” / “打开透视图(O)” / “其他” / “DDMS”，打开 DDMS 窗体，如图 2.49 所示。

（3）双击“File Explorer”选项卡，并打开“mnt”文件夹中的“sdcard”文件夹，如图 2.50 所示。

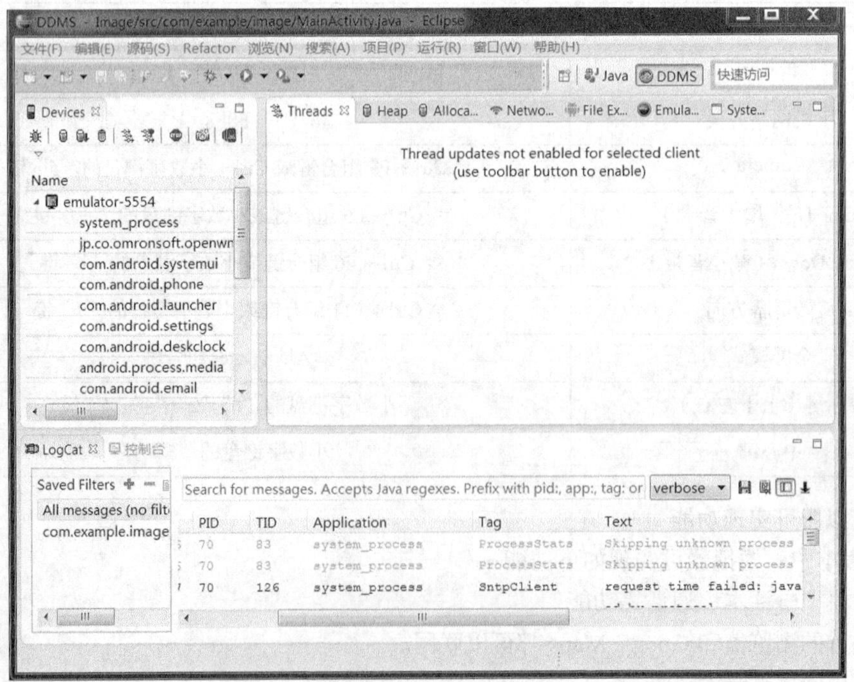

图 2.49 DDMS 透视图

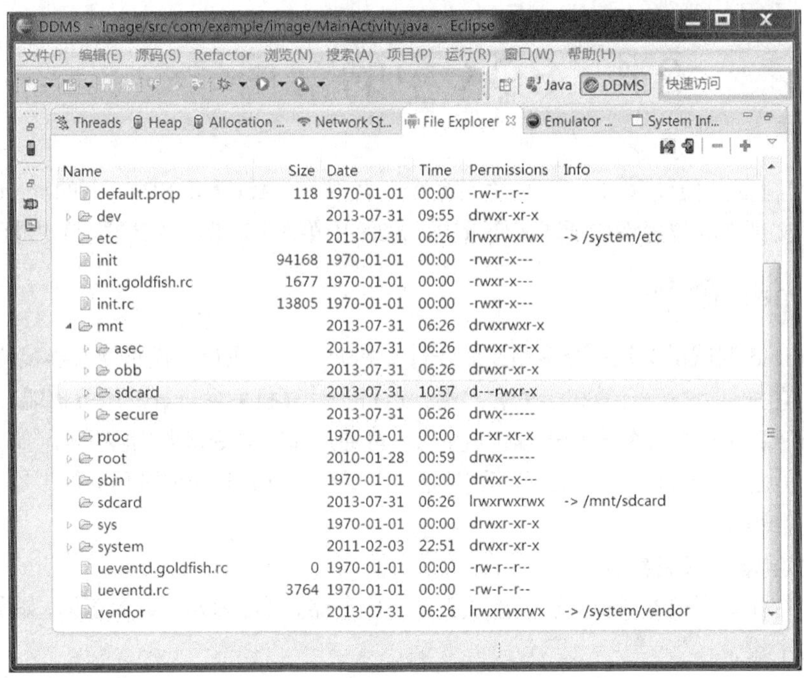

图 2.50　DDMS 透视图

（4）窗口右上角中图标用于从模拟器上导出文件，图标用于向模拟器上导入文件，图标用于删除模拟器上文件，图标用于在模拟器中建立文件夹。下面依次在“sdcard”文件夹中建立“MyApp”文件夹，并导入“360Helper.apk”文件，如图 2.51 所示。

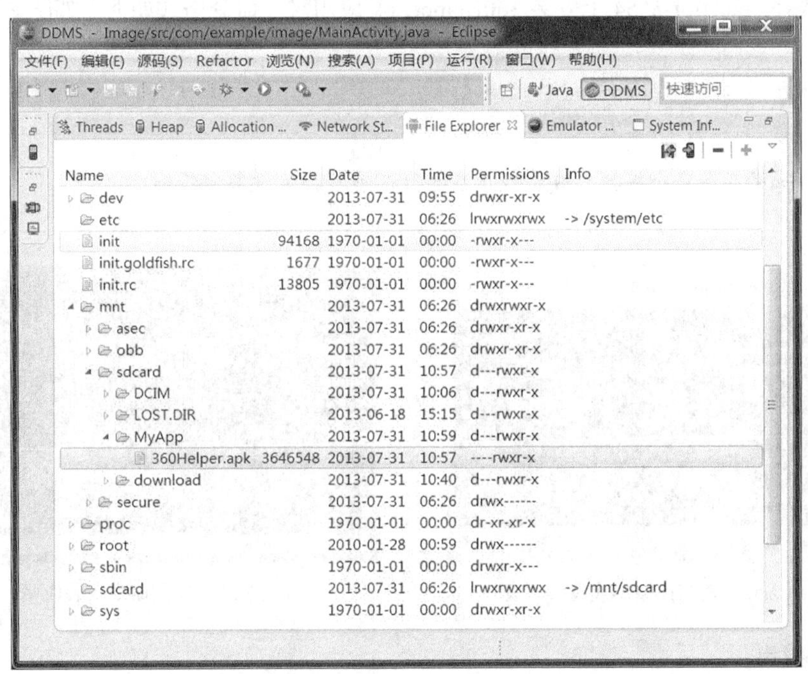

图 2.51　DDMS 管理模拟器文件

对于模拟器上的文件管理，可通过 DDMS 采用上述方法来完成。

2.4 SDK 中的常用命令

为了在控制台方便使用 SDK 的常用命令，需要将 Android SDK 中的 tools 文件夹和 platform-tools 文件夹的位置添加到环境变量中，2.2.2 中第（7）步已经讲过，不再赘述。

2.4.1 adb 命令

adb（Android 调试桥）是一个多用途命令行工具，它允许用户与模拟器或连接的 Android 设备进行通信。它是由运行于计算机上的客户端，运行于计算机后台进程的服务器端，和一个作为后台进程运行于各个模拟器的守护进程构成。用户可通过 adb 命令来调用客户端，如 ADT 插件和 DDMS 等工具也会创建 adb 客户端、服务器管理客户端和运行 adb 守护进程的模拟器之间的通信。下面介绍 adb 常用命令。

1. 查询模拟器或设备实例

获取目前有哪些模拟器或设备已经连接到 adb 服务器，可使用如下命令查询，如图 2.52 所示，表示目前有两个设备。

```
adb devices
```

2. 设定模拟器或设备实例

如果存在多个模拟器或实例正在运行，需要指定目标实例，命令格式如下，其中“<serialNumber>”表示序列号，“<command>”表示执行的命令。

```
adb -s <serialNumber> <command>
```

例如需要在 emulator-5554 上安装 360Helper.apk 应用时，命令格式如下。如图 2.53 所示，表示安装成功。

```
adb -s emulator-5554 install 360Helper.apk
```

注意

如果存在多个模拟器或实例正在运行，若不指定设备会报错。

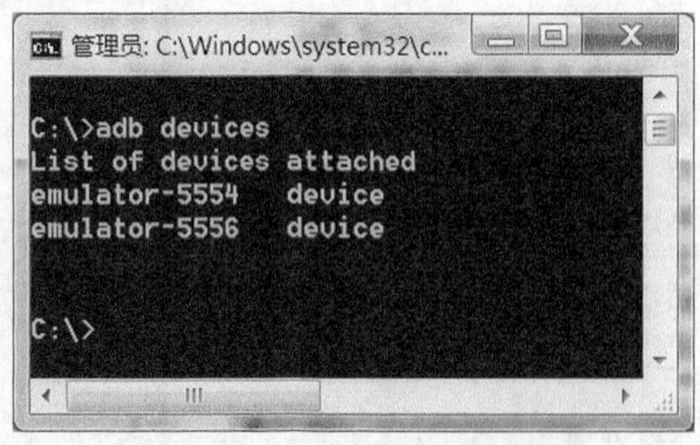

图 2.52 查询模拟器或设备列表

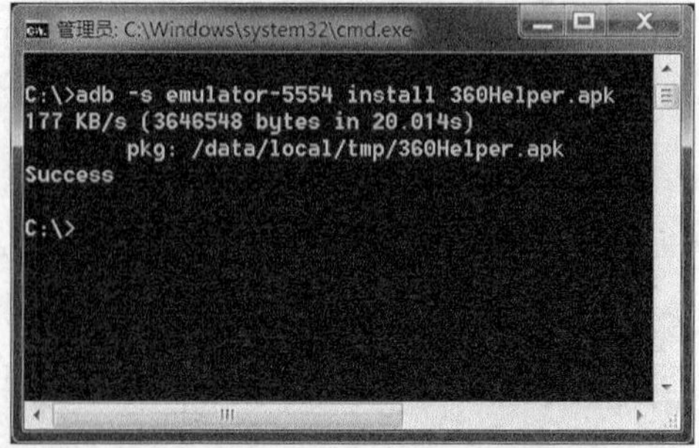

图 2.53 使用命令安装应用

3. 文件导出

将模拟器或设备上的文件导出到计算机，命令格式如下，如图 2.54 所示。

```
adb pull <remote> <local>
```

图 2.54　文件导出命令

4. 文件导入

将计算机本地文件导入到模拟器或设备，命令格式如下，其中“<local>”表示计算机上的文件位置，“<remote>”表示模拟器或设备实例上的文件位置。

```
adb push <local> <remote>
```

5. 打开 shell

Android 平台底层使用的是 Linux 内核，可以使用 shell 来进行操作，打开 shell 的命令格式如下。

```
adb shell
```

6. 安装应用程序

可以通过 adb 将计算机上的 apk 应用安装到模拟器或设备上，命令格式如下。

```
adb install <path_to_apk_file>
```

adb 提供了丰富的操作命令，这里不一一赘述，如有需要可通过输入 adb 查看帮助。

2.4.2　Android 命令

android 命令是一个非常重要的开发工具，功能包括创建、删除、查看 Android 虚拟设备（AVD），创建和更新 Android 项目，更新 Android SDK 等。也可以使用 Eclipse 中的 ADT 插件完成如上工作，会更加方便快捷。

1. 获取安装平台

安装 Android SDK 时下载了多个版本的平台，获取已安装的平台信息命令格式如下，执行结果部分如图 2.55 所示。

```
android list targets
```

2. 创建模拟器

可以使用 android 命令建立 AVD，命令格式如下，其中“<avd_name>”表示要创建的 AVD 名称，“<target_id>”表示 AVD 的 id 值，另外可指定模拟器 SD 卡的大小，用户数据文件位置等相关信息。

```
android create avd -n <avd_name> -t <target_id> [-<option> <value>] ...
```

3. 获取创建的模拟器

获取已创建的模拟器信息命令格式如下，执行结果如图 2.56 所示。

```
android list avd
```

```
选定 管理员: C:\Windows\system32\cmd.exe
C:\Users\Administrator>android list targets
Available Android targets:
----------
id: 1 or "android-10"
     Name: Android 2.3.3
     Type: Platform
     API level: 10
     Revision: 2
     Skins: HVGA, QVGA, WQVGA400, WQVGA432, WVGA800 (default), WVGA854
     ABIs : armeabi
----------
id: 2 or "Google Inc.:Google APIs:10"
     Name: Google APIs
     Type: Add-On
     Vendor: Google Inc.
     Revision: 2
     Description: Android + Google APIs
     Based on Android 2.3.3 (API level 10)
     Libraries:
      * com.android.future.usb.accessory (usb.jar)
          API for USB Accessories
      * com.google.android.maps (maps.jar)
          API for Google Maps
     Skins: WVGA854, WQVGA400, HVGA, WQVGA432, WVGA800 (default), QVGA
     ABIs : armeabi
```

图 2.55　列出模拟器类型

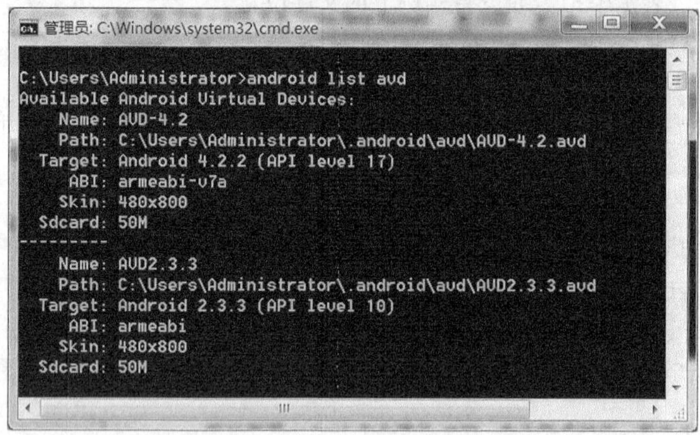

图 2.56　获取创建的模拟器

4．删除模拟器

删除指定的模拟器，命令格式如下，其中“<avd_name>”表示要删除的 AVD 名称。

```
android delete avd -n <avd_name>
```

Android SDK 中还提供了 emulator 等命令控制模拟器或设备，实现加载 AVD、加载磁盘镜像、调试、加载媒体文件、设置网络参数等功能，需要的用户可以在控制台下输入“android -help”查看帮助，这里不再说明。

2.5　小　结

本章重点介绍了如何搭建 Android 开发环境，并对于英语一般的用户，提供了 Eclipse 工具的汉化方法。目前，谷歌已经推出了多个版本的 Android 平台，文中讲解了 Android 各平台模拟器的建立和使用，在开发时可以建立多个模拟器进行调试，推荐使用 Eclipse 的 ADT 插件简化开发过程。开发过程中，一些功能的使用离不开命令的执行，因此，对于常用命令 adb 和 android 的基本用法予以介绍。

练　　习

1. 安装 Android 开发环境，创建针对不同分辨率的 Android 模拟器，并熟悉各选项的含义。
2. 在创建的模拟器中将语言改为中文，时区改为“GMT + 08:00，中国标准时间”。
3. 使用 adb 命令从模拟器上导入、导出文件。
4. 给模拟器安装应用程序，如 UC 浏览器、搜狗输入法等。
5. 使用 android 命令创建模拟器、删除模拟器。

第 3 章 Android 应用程序框架

本章将通过一个猜数字的小游戏程序来介绍 Android 应用程序项目的基本结构，分析并讲解 Android 应用程序项目的各组成元素。同时，也将介绍 Android 应用程序安装文件（APK 文件）的结构，Android 应用程序权限的设置，以及 Activity 的概念和 Activity 的生命周期。

3.1 第一个 Android 应用程序

在上一章学习搭建 Android 开发环境时，已经创建并运行了 Hello World 应用程序，该程序的目的主要是验证开发环境是否搭建成功。本章中将首先指导用户学习编写第一个 Android 应用程序——猜数字游戏。该游戏会随机生成一个 1 ~ 10 之间的整数，让用户来猜。如果用户输入的数字大于或小于该随机数，则提示用户重新猜。如果用户猜对了，则可以选择再重新玩一次。

该游戏界面如图 3.1 ~ 图 3.4 所示。

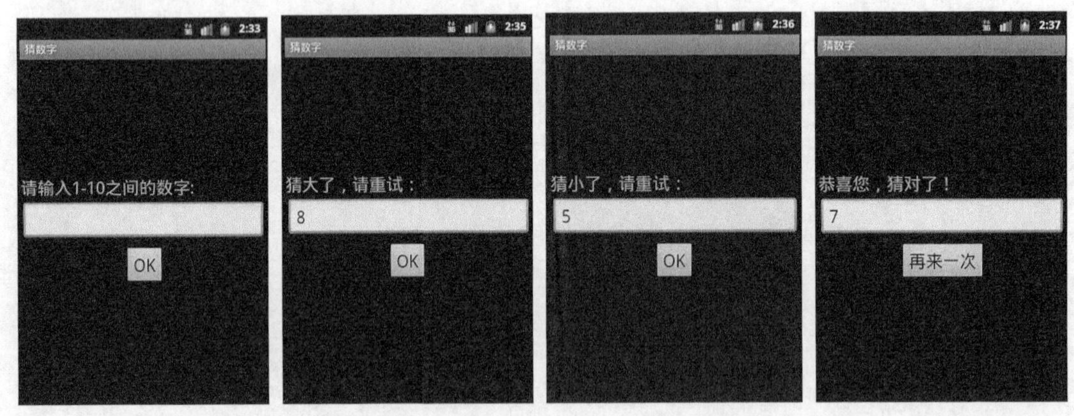

图 3.1 猜数字起始界面　　图 3.2 猜数字界面 1　　图 3.3 猜数字界面 2　　图 3.4 猜数字界面 3

下面将一步一步介绍如何编写这个猜数字游戏，这里不解释具体细节的内容，用户只要能按照说明将这个游戏编写出来即可。后面会逐步解释程序中的细节内容。

【例 3.1】 猜数字游戏。

（1）首先按照上一章中创建 Hello World 程序的步骤，在 Eclipse 中创建一个 Android 工程，工程名称为 GuessGame。工程创建后，在 Eclipse 的 Package Explorer 中可以看到如图 3.5 所示的工程结构。

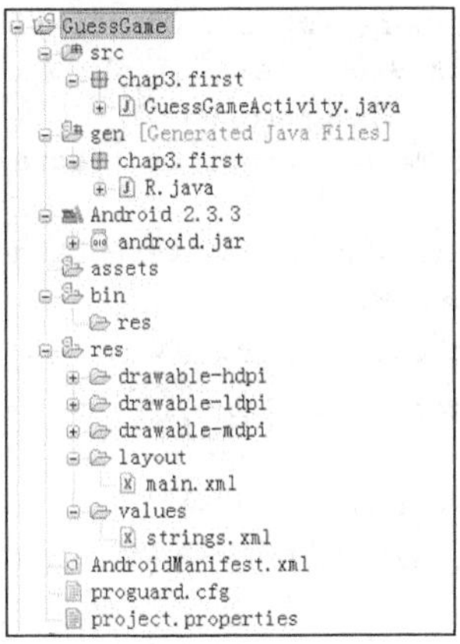

```
GuessGame
  src
    chap3.first
      GuessGameActivity.java
  gen [Generated Java Files]
    chap3.first
      R.java
  Android 2.3.3
    android.jar
  assets
  bin
    res
  res
    drawable-hdpi
    drawable-ldpi
    drawable-mdpi
    layout
      main.xml
    values
      strings.xml
  AndroidManifest.xml
  proguard.cfg
  project.properties
```

图 3.5 Android 工程结构

（2）修改布局文件，将默认的布局管理器改为 LinearLayout 线性布局，垂直排列。依次放置文本框、文本编辑框和按钮，代码如下所示。

```
<?xml version = "1.0" encoding = "utf-8"?>
<LinearLayout xmlns:android = "http://schemas.android.com/apk/res/android"
    android:layout_width = "match_parent"
    android:layout_height = "match_parent"
    android:gravity = "center"
    android:orientation = "vertical" >
    <TextView android:id = "@+id/info"
        android:layout_width = "match_parent"
        android:layout_height = "wrap_content"
        android:textSize = "10pt" />
    <EditText android:id = "@+id/input"
        android:layout_width = "match_parent"
        android:layout_height = "wrap_content"
        android:textSize = "10pt"
        android:layout_margin = "2pt"
        android:inputType = "number" />
    <Button android:id = "@+id/button"
        android:layout_width = "wrap_content"
        android:layout_height = "wrap_content"
        android:textSize = "10pt"
        android:layout_margin = "2pt" />
</LinearLayout>
```

（3）在主 Activity 文件 GuessGameActivity.java 中获取编辑框的数字与产生的随机数进行比较，单击提交按钮时打开提示信息，代码如下所示。

```
public class GuessGameActivity extends Activity {
    private TextView info;
    private EditText input;
    private Button button;
    private int target;
    @Override
    public void onCreate(Bundle savedInstanceState) {
        super.onCreate(savedInstanceState);
        setContentView(R.layout.main);
        info = (TextView)findViewById(R.id.info);
        input = (EditText)findViewById(R.id.input);
        button = (Button)findViewById(R.id.button);
```

```
        button.setOnClickListener(new View.OnClickListener() {
            public void onClick(View view) {
                if(button.getText().equals("OK")) {
                    if(input.getText().toString().equals(""))
                        return;
                    int tmp = Integer.parseInt(input.getText().toString());
                    if(tmp = =target) {
                        info.setText("恭喜您，猜对了！");
                        button.setText("再来一次");
                    }
                    else if(tmp<target)
                        info.setText("猜小了，请重试：");
                    else
                        info.setText("猜大了，请重试：");
                }
                else {
                    newGame();
                }
            }
        });
        newGame();
    }
}
```

（4）在主 Activity 文件中添加 newGame()方法，用于初始化各组件，并产生一个随机数，代码如下所示。

```
private void newGame() {
    info.setText("请输入 1-10 之间的数字:");
    input.setText("");
    button.setText("OK");
    target = (int)((Math.random())*9 + 1);
}
```

编译并运行本工程，结果如图 3.1 ~ 图 3.4 所示。如果编译或运行中有错误出现，请首先检查开发环境，包括 Android 模拟器设置等是否正确，然后检查代码编写是否正确。

3.2　Android 项目结构

通过编写和运行上一节的猜数字游戏，用户应该对开发一个 Android 应用程序有了初步的了解。观察上面的 Android 工程结构图可以看出，猜数字游戏工程中包括了五个文件夹：src、gen、assets、bin、res，以及 3 个文件：AndroidManifest.xml、proguard.cfg 和 project.properties。工程中所显示的 Android 2.3.3 子项目为本工程所引用的版本为 2.3.3 的 android.jar 文件。

下面将对这些文件夹和文件予以详细的介绍。

- src

文件夹 src 中存放 Java 源程序文件，可以有若干个包和类，如猜数字游戏中定义了一个包 chap3.first 和一个类文件 GuessGameActivity.java。

- bin

文件夹 bin 中存放源程序编译后得到的 Java 类文件，以及相关的资源文件，打包后得到的 apk

程序安装文件。

- assets

文件夹 assets 存放程序需要用到的辅助文件，如音频文件、视频文件、数据文件等。由于猜数字游戏较简单，并没有使用任何辅助文件。

- res

文件夹 res 存放程序中需要用到的资源文件，该文件夹通常包含若干子文件夹，如 drawable-XXX、layout、values、menu 等。

- res/drawable-XXX

drawable 子文件夹存放程序中用到的图片或者动画资源。由于不同手机的屏幕分辨率相差较大，要使程序在不同手机上都获得理想的显示效果，需要针对高分辨率、中等分辨率和低分辨率 3 种情况（注：Android API 版本不同，文件夹个数略有不同）准备 3 套不同的图片资源文件。所以对应的有 3 个 drawable 子文件夹：drawable-hdpi 中保存高分辨率图片，drawable-mdpi 中保存中等分辨率图片，drawable-ldpi 中保存低分辨率图片，猜数字游戏中没有使用图片，但是程序的图标是一个 png 类型的图片，打开 3 个 drawable 子文件夹就可以看到 3 个名称相同但是分辨率不同的图片。如果想用自己的图标替换这些系统图片，可以直接替换这些文件。

- res/layout

Android 应用程序开发的一个特点是将程序用户界面与对用户界面的控制分离开来。layout 子文件夹中存放的就是描述用户界面信息的若干 XML 文件，而对用户界面的控制则由 Java 程序来完成。一般来说，程序中的每个用户界面都对应一个 XML 文件，猜数字游戏只有一个界面，所以只有一个 main.xml 文件。该文件的内容见 3.1 节中（2）所示。

根标记<LinearLayout>说明本界面使用了线性布局。关于线性布局及其他更多的布局形式将在第 5 章中详细介绍。<LinearLayout>中包含了 3 个标记<TextView>、<EditText>和<Button>，分别对应游戏界面上的信息提示框、文本输入框和按钮。这 3 个标记都有一个 android:id 的属性值，用于从 Java 程序中获取到这 3 个界面控件，来对其进行操作。比如标记<TextView>的 android:id 属性值是"@+id/info"，在 Java 程序中可以通过如下的代码获得这个控件的对象。

```
TextView info = (TextView)findViewById(R.id.info);
```

除了 android:id 属性，这 3 个控件元素还包括了其他的一些属性，用来说明控件的宽度、高度、文字大小、控件边界大小等。详细的内容都将在第 4 章中进行介绍。

- res/values

子文件夹 values 用于存放一些程序中用到的数据，比如常量、字符串、尺寸、样式等，这些数据通常组织成 XML 格式的文件。猜数字游戏中有一个 strings.xml 文件，就是将程序的名字定义为一个可以从程序中获取的 XML 元素，如同一个常量的定义。这样用户可以从这个文件中去修改程序的名字，而不用修改其他的程序。

- gen

该文件夹名字后面有一个提示信息“Generated Java Files”。也就是说该文件夹里存放的是自动生成的 Java 文件，作为开发者不能自己在该文件夹里创建文件，或者去修改自动生成的文件。对于猜数字的游戏，自动生成的是一个包 chap3.first 和一个类 R.java。打开 R.java 文件，代码如下所示。

```
public final class R {
   public static final class attr {
   }
```

```
    public static final class drawable {
        public static final int ic_launcher = 0x7f020000;
    }
    public static final class id {
        public static final int button = 0x7f050002;
        public static final int info = 0x7f050000;
        public static final int input = 0x7f050001;
    }
    public static final class layout {
        public static final int main = 0x7f030000;
    }
    public static final class string {
        public static final int app_name = 0x7f040000;
    }
}
```

可以看到在类 R 中包含了若干个内部类，分别对应 drawable、layout 和 values 等资源文件。同时 main.xml 文件中定义的 3 个界面控件元素的 id 值也被包含进来。所有的这些信息都被赋了一个十六进制的整数值，用于唯一标示这些资源信息。文件夹 res 中的具体资源正是通过这些整数值与 Java 程序关联起来。比如对于 Java 程序中的如下代码。

```
TextView info = (TextView)findViewById(R.id.info);
```

可以看到方法 findViewById 需要的参数正是一个 R.id.info 这样的整数值。

介绍完 Android 工程中的所有文件夹之后，下面介绍另外 3 个独立的文件。

- proguard.cfg

创建工程时自动生成的用于混淆类文件的配置文件。混淆中保留了继承自 Activity、Service、Application、BroadcastReceiver、ContentProvider 等的基本组件，并保留了所有的 Native 变量名及类名，所有类中部分已设定了固定参数格式的构造函数、枚举等。

- project.properties

创建工程时自动生成的项目信息配置文件，记录了项目中所设定的环境信息，例如 Android 的版本等。

- AndroidManifest.xml

AndroidManifest.xml 是 Android 应用程序的最基本和最重要的全局配置文件。它保存应用程序中的 Activity、Intent、Service 等组件，以及应用程序权限声明等信息。打开猜数字游戏项目中的 AndroidManifest.xml 文件，代码如下所示。

```
<?xml version = "1.0" encoding = "utf-8"?>
<manifest xmlns:android = "http://schemas.android.com/apk/res/android"
    package = "chap3.first"
    android:versionCode = "1"
    android:versionName = "1.0" >
    <uses-sdk android:minSdkVersion = "10" />
    <application
        android:icon = "@drawable/ic_launcher"
        android:label = "@string/app_name" >
        <activity
            android:name = ".GuessGameActivity"
            android:label = "@string/app_name" >
            <intent-filter>
                <action android:name = "android.intent.action.MAIN" />
                <category android:name = "android.intent.category.LAUNCHER" />
            </intent-filter>
```

```
        </activity>
    </application>
</manifest>
```

由于本程序比较简单，AndroidManifest.xml 配置文件内容并不复杂。在根标记<manifest>中包含了两个子标记：<uses-sdk>和<application>。<uses-sdk>说明本应用程序所使用的 SDK 版本号。由于 Android 应用程序向后兼容，所以 Android 2.3.3 版本的程序可以在 Android 4.0 环境中运行，反之却不行。本程序是基于 Android 2.3.3 版本的 SDK 开发的，其 minSdkVersion 值为 10。<application>中的前两项数据 android:icon 和 android:label 分别说明该应用程序所使用的图标和程序标题。它们指向的正是文件夹 res 中的 drawable 图标和 strings.xml 中的程序标题字符串。<application>中最重要的一个子标记是<activity>，对于猜数字游戏创建的 Java 源程序 GuessGameActivity.java，就是一个继承了 android.app.Activity 类的子类。前面说过，layout 文件夹里定义了用户界面的样式，而对这些界面的操作，就是由相关的 Activity 类来完成的。一个 Android 程序可以包含多个 Activity 类，这些类都必须在 AndroidManifest.xml 文件中进行声明后才能使用，否则会出错。关于 Activity 的详细信息将在本章第 5 节介绍。

3.3　APK 文件结构

APK 是 Android Package 的缩写，即 Android 安装包，所有的 Android 程序都是以 APK 文件的形式发布的。可以在模拟器或手机上运行 APK 文件来安装程序。APK 文件的后缀为.apk，但是其格式是压缩文件 zip 的格式。可以通过 WinZip、WinRAR 等将其解压。在猜数字游戏的工程 bin 文件夹中可以看到 GuessGame.apk 文件，直接将该文件扩展名改为.zip 就可以解压该文件来了解 APK 文件的结构。

该 APK 文件包括了 2 个文件夹和 3 个文件。

- META-INF 文件夹

META-INF 文件夹下存放的 APK 文件的基本信息和签名信息，用来保证 APK 文件的完整性和系统的安全。类似于 jar 文件，META-INF 文件夹下有一个 MANIFEST.MF 文件，说明 Manifest 文件的版本和创建者，以及相关文件的校验信息。Android 系统在安装某个 APK 文件时会对 APK 中压缩的各个文件进行校验检查，如果校验结果与 META-INF 下的内容不一致，系统就不会安装这个 APK 文件，这就保证了程序安装包的安全性。例如解压一个 APK 文件，替换里面的一幅图片、一段代码，或一段版权信息，然后重新压缩打包，即使这个 APK 文件符合格式要求，但是也无法通过校验进行安装。所以这样就给病毒感染和恶意修改增加了难度，可以有效地保护系统安全。

- res 文件夹

res 文件夹存放资源文件，猜数字 APK 文件中的 res 文件夹里包括了 3 个子文件夹，分别存放前面提过的为了在不同屏幕分辨率下显示最佳效果的 3 个程序图标文件。由此也可以看到 APK 文件对资源文件的保护性较差，任何人都可以通过解压缩的方式从 APK 文件中提取相关的资源文件，如图片、音频等。

- AndroidManifest.xml

AndroidManifest.xml 是每个应用程序都必须定义和包含的全局配置文件，它描述了应用程序的名称、版本、权限、引用的库文件等信息。但是不同于系统开发时在项目文件夹中所看到的AndroidManifest.xml文件，在APK文件中，AndroidManifest.xml是经过压缩的，如果直接打开将看到乱码。可以通过相关工具将其解压。

- classes.dex

classes.dex 是 Java 源代码编译后生成的 Dalvik 虚拟机字节码文件，类似于 Java 虚拟机使用的.class 类文件。由于 Android 使用的 Dalvik 虚拟机与标准的 Java 虚拟机是不兼容的，因此 dex 文件与 class 文件相比，不论是文件结构还是操作码都不一样。如果编写的源程序有多个 Java 类，都会将其编译到同一个 classes.dex 文件中，也就是说每个 APK 文件只有一个 classes.dex 文件，而不论程序是简单还是复杂。目前常见的 Java 反编译工具都不能处理 dex 文件。Android 模拟器中提供了一个 dex 文件的反编译工具——dexdump，可以对其进行反编译。

- resources.arsc

resources.arsc 是编译后的二进制资源文件，内容包含了开发程序时项目文件夹中 res 子文件夹下 main.xml、strings.xml 等文件的信息。

除了以上文件夹和文件，在有些 APK 文件中，还有可能出现 lib 文件夹和 assets 文件夹。这两个文件夹中分别存放应用软件需要调用的库文件和其他资源配置文件。

3.4 Android 应用程序权限

在安装 Android 应用程序时经常会看到系统提示该程序需要获得一些访问权限，如访问 SD 卡、建立网络连接等，如果同意才能继续安装，否则该程序将不会被安装。这就是 Android 系统建立的通过权限设置的方式来保护系统文件、用户资料以及对系统进行各项操作的安全性控制。猜数字游戏并没有涉及安全的操作，所以没有进行权限设置。如果在开发应用程序时要进行权限设置，需要在项目的 Mainifest.xml 文件中声明所需权限。例如改进猜数字游戏，将用户猜数字所花费的时间保存到手机 SD 卡上，同时增加访问 Internet 的功能以上传最好成绩，则需要在 Mainifest.xml 文件中增加<uses-permission>标记，通过该标记的属性 android:name 来指定相应的权限，代码如下所示。

```
<?xml version = "1.0" encoding = "utf-8"?>
<manifest xmlns:android = "http://schemas.android.com/apk/res/android"
    package = "chap3.first"
    android:versionCode = "1"
    android:versionName = "1.0" >
    <uses-sdk android:minSdkVersion = "10" />
    <application
        android:icon = "@drawable/ic_launcher"
        android:label = "@string/app_name" >
        <activity
            android:name = ".GuessGameActivity"
            android:label = "@string/app_name" >
            <intent-filter>
                <action android:name = "android.intent.action.MAIN" />
                <category android:name = "android.intent.category.LAUNCHER" />
            </intent-filter>
```

```
        </activity>
    </application>
    <uses-permission android:name = "android.permission.WRITE_EXTERNAL_STORAGE" />
    <uses-permission android:name = "android.permission.INTERNET"/>
</manifest>
```

Android 中主要的权限设置名称及其描述如下。

- android.permission.ACCESS_COARSE_LOCATION

允许应用程序通过访问 CellID 和 WiFi 热点等方式获取粗略的本地位置。

- android.permission.ACCESS_FINE_LOCATION

允许应用程序通过访问 GPS 等方式获取较精确的本地位置。

- android.permission.ACCESS_NETWORK_STATE

允许应用程序访问（获取）网络信息。

- android.permission.ACCESS_WIFI_STATE

允许应用程序获取 Wi-Fi 网络的信息。

- android.permission.BATTERY_STATS

允许应用程序获取电池使用的统计信息，如剩余电量、各主要应用程序耗电占总耗电的百分比等。

- android.permission.BLUETOOTH

允许应用程序连接到已配对的蓝牙设备。

- android.permission.CALL_PHONE

允许应用程序不经过用户拨号界面而直接拨号。

- android.permission.CAMERA

允许应用程序访问摄像头设备。

- android.permission.CHANGE_NETWORK_STATE

允许应用程序更改网络连接状态。

- android.permission.CHANGE_WIFI_STATE

允许应用程序更改 Wi-Fi 连接状态。

- android.permission.DISABLE_KEYGUARD

允许应程序禁用键盘锁。

- android.permission.INTERNET

允许应用程序访问网络。

- android.permisson.MODIFY_AUDIO_SETTINGS

允许应用程序更改全局音频设置。

- android.permission.MOUNT_FORMAT_FILESYSTEMS

允许应用程序格式化移动存储设备。

- android.permission.MOUNT_UNMOUNT_FILESYSYTEMS

允许应用程序加载或卸载移动存储设备。

- android.permission.READ_CALENDAR

允许应用程序读取用户日历数据。

- android.permission.READ_CONTACTS

允许应用程序读取用户联系人列表。

- android.permission.READ_HISTORY_BOOKMARKS

允许应用程序读取浏览器的历史记录和书签。

- android.permission.READ_SMS

允许应用程序读取手机短消息。

- android.permission.RECEIVE_BOOT_COMPLETED

允许应用程序获取系统完全启动之后的广播。

- android.permission.RECEIVE_MMS

允许应用程序监控收到的彩信（MMS），能进行记录或处理。

- android.permission.RECEIVE_SMS

允许应用程序监控收到的短信（SMS），能进行记录或处理。

- android.permission.RECORD_AUDIO

允许应用程序录音。

- android.permission.SEND_SMS

允许应用程序发送短消息（SMS）。

- android.permission.SET_ORIENTATION

允许设置屏幕方向（旋转屏幕）。

- android.permission.SET_TIME

允许应用程序设置系统时间。

- android.permission.SET_TIME_ZONE

允许应用程序设置系统时区。

- android.permission.SET_WALLPAPER

允许应用程序设置桌面。

- android.permission.STATUS_BAR

允许应用程序打开、关闭、禁用状态栏和状态栏图标。

- android.permission.VIBRATE

允许应用程序访问振动器。

- android.permission.WAKE_LOCK

允许应用程序使用 PowerManager WakeLocks，避免处理器进入休眠，或屏幕变暗。

- android.permission.WRITE_CALENDAR

允许应用程序写用户日历数据。

- android.permission.WRITE_CONTACTS

允许应用程序写用户联系人数据。

- android.permission.WRITE_EXTERNAL_STORAGE

允许应用程序写数据到外部存储设备（主要是 SD 卡）。

- android.permission.WRITE_HISTORY_BOOKMARKS

允许应用程序写数据到用户浏览器历史记录和书签。

- android.permission.WRITE_SMS

允许应用程编写短消息。

更多的权限设置信息可查阅 Android 开发文档。需要注意的是，SDK 版本不同，其所支持的权限设置也不同。

3.5　Activity 及其生命周期

Activity 是编写 Android 应用程序需要掌握的最基本的概念，Activity 的生命周期是编写 Android 应用程序的基础。

3.5.1　什么是 Activity

Activity 是 Android 系统 API 的一个类，用于提供一个系统与用户交互的机制。在上一章的 Hello World 程序和本章的猜数字游戏中都用到了 Activity，如图 3.6 所示。

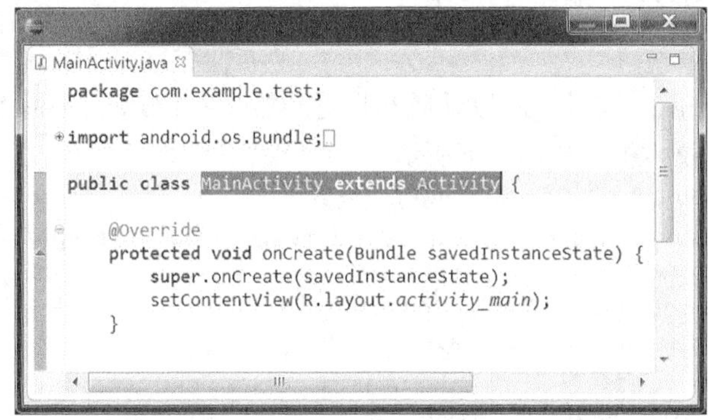

图 3.6　Activity 类

在 Android 程序中，凡是涉及需要与用户交互的，都应创建一个类，继承自 android.app.Activity。通过调用 Activity 类的 setContentView()方法，可以设置用户界面。代码如下所示，将项目 layout 文件夹中由 activity_main 文件声明的用户界面显示出来。

```
setContentView(R.layout.activity_main);
```

如果一个程序有多个用户界面，则需要在项目 layout 文件夹中将这些界面一一声明出来。通常对于每一个用户界面，需要编写与之对应的 Activity 子类程序。要从一个界面进入另外一个界面，可以调用 Activity 类的 startActivity()方法，该方法需要一个 Intent 类型的参数。

Android 系统采用栈的数据结构来管理一个程序的多个用户界面。当从界面 1 进入界面 2 时，界面 2 显示在界面 1 上面。此时再从界面 2 进入界面 3，则界面 3 在最上面，界面 1 在最下面，界面 2 在中间。如果单击手机上的返回按键，则可以从界面 3 回到界面 2，如果再次单击返回按键，则可以从界面 2 回到界面 1。下面通过例 3.2 介绍如何创建多用户界面程序，以及如何从一个界面进入另外一个界面。

【例 3.2】　创建多用户界面程序。

在 Eclipse 中创建 Android 项目 ActivityDemo，包名为 chap3.demo。系统将自动生成一个 Activity 类 ActivityDemoActivity，将默认生成的描述用户界面的 XML 文件改名为 main.xml。再手动添加一个 Activity 类 ActivityDemoActivity2，以及另外一个用户界面 main2.xml。这四个程序的内容如下。

（1）在主布局文件 main.xml 中，将默认的布局管理器改为 LinearLayout 线性布局，垂直排列。并在该布局中添加一个文本框、一个按钮，代码如下所示。

```
<?xml version = "1.0" encoding = "utf-8"?>
<LinearLayout xmlns:android = "http://schemas.android.com/apk/res/android"
    android:layout_width = "match_parent"
    android:layout_height = "match_parent"
    android:orientation = "vertical" >
    <TextView
        android:layout_width = "match_parent"
        android:layout_height = "wrap_content"
        android:textSize = "15pt"
        android:text = "DemoActivity1 " />
    <Button android:id = "@+id/btn1"
        android:layout_width = "wrap_content"
        android:layout_height = "wrap_content"
        android:textSize = "15pt"
        android:layout_margin = "12pt"
        android:text = "Start Activity2" />
</LinearLayout>
```

（2）在布局文件 main2.xml 中，将默认的布局管理器改为 LinearLayout 线性布局，垂直排列。并在该布局中添加一个文本框，代码如下所示。

```
<?xml version = "1.0" encoding = "utf-8"?>
<LinearLayout xmlns:android = "http://schemas.android.com/apk/res/android"
    android:layout_width = "wrap_content"
    android:layout_height = "wrap_content"
    android:orientation = "vertical" >
    <TextView
        android:layout_width = "246dp"
        android:layout_height = "wrap_content"
        android:text = "DemoActivity2"
        android:textSize = "15pt" />
</LinearLayout>
```

（3）在主 Activity 文件 ActivityDemoActivity.java 中，调用 setContentView()方法设置 main.xml 布局文件为当前界面，对提交按钮增加 OnClickListener 监听器（监听器内容下章会详细介绍），当单击提交按钮时会调用 onClick()方法。在该 onClick()方法中建立 Intent 的实例，在 Intent 构造方法中指定需打开的 Activity 类文件名，并通过 startActivity(intent)方法打开新的页面，代码如下所示。

```
public class ActivityDemoActivity extends Activity {
    @Override
    public void onCreate(Bundle savedInstanceState) {
        super.onCreate(savedInstanceState);
        setContentView(R.layout.main);
        final Button button = (Button) findViewById(R.id.btn1);
        button.setOnClickListener(new View.OnClickListener() {
            public void onClick(View view) {
                Intent intent = new Intent(ActivityDemoActivity.this, ActivityDemoActivity2.
class);
                startActivity(intent);
            }
        });
    }
}
```

（4）在 Activity 文件 ActivityDemoActivity2.java 中，调用 setContentView()方法设置 main2.xml 布局文件为当前界面，代码如下所示。

```
public class ActivityDemoActivity2 extends Activity {
   public void onCreate(Bundle savedInstanceState) {
      super.onCreate(savedInstanceState);
      setContentView(R.layout.main2);
   }
}
```

（5）另外，还需要在 AndroidManifest.xml 文件中注册 ActivityDemoActivity2，代码如下所示。

```
<?xml version = "1.0" encoding = "utf-8"?>
<manifest xmlns:android = "http://schemas.android.com/apk/res/android"
   package = "chap3.demo"
   android:versionCode = "1"
   android:versionName = "1.0" >
   <uses-sdk android:minSdkVersion = "10" />
   <application
      android:icon = "@drawable/ic_launcher"
      android:label = "@string/app_name" >
      <activity
         android:name = ".ActivityDemoActivity"
         android:label = "@string/app_name" >
         <intent-filter>
            <action android:name = "android.intent.action.MAIN" />
            <category android:name = "android.intent.category.LAUNCHER" />
         </intent-filter>
      </activity>
      <activity
            android:name = ".ActivityDemoActivity2"
            android:label = "@string/app_name" />
   </application>
</manifest>
```

运行本实例，显示结果如图 3.7、图 3.8 所示。

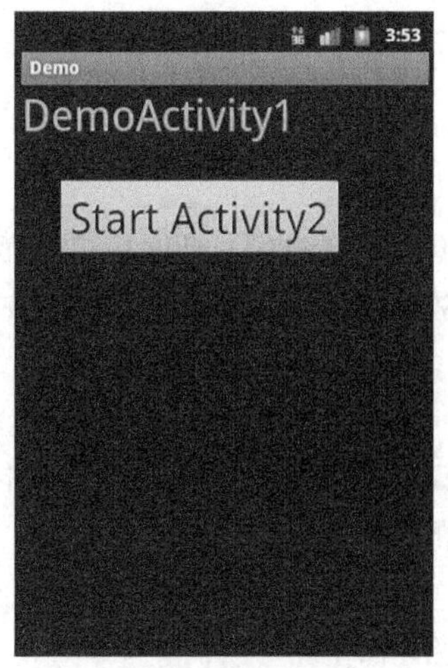

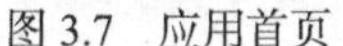
图 3.7　应用首页

图 3.8　应用第二页

3.5.2　Activity 生命周期

前面介绍过，Activity 是 Android 系统 API 的一个类，用于提供系统与用户交互的机制。

Android 程序中凡是涉及需要与用户交互的，都应创建一个类，继承自 Activity。如果一个程序有多个用户界面，则通常对于每一个用户界面，需要编写与之对应的 Activity 子类程序。Android 系统采用栈的数据结构来管理多个 Activity，那么这里有必要简单介绍一下 Activity 的各个状态和生命周期。

Activity 的状态有以下 5 种。

- started 状态

Activity 启动，并入栈，此时尚未在用户屏幕上将界面显示出来。

- running 状态

界面在屏幕上显示出来，且位于最上层，获得用户输入焦点。

- paused 状态

界面被遮挡，但仍然可见，失去用户输入焦点。

- stopped 状态

界面被另外一个 Activity 的界面遮挡，完全不可见，但仍然存活。

- exited 状态

Activity 出栈，停止运行，且被释放掉。

为了响应和处理 Activity 各个状态的变化，Activity 类提供了 7 个回调方法，供开发者覆盖。当 Activity 的状态发生变化时，这 7 个方法中的若干个会被调用，以进行相关处理。这些方法分别如下。

- onCreate()方法

Activity 启动后被调用，此时界面尚未显示出来，一般用于进行初始化的操作。前面各节编写的程序都使用了该方法，该方法需要一个 Bundle 类型的参数，当方法被调用时由系统传入该参数。

- onStart()方法

当 Activity 状态变为可见状态时被调用。

- onResume()方法

当 Activity 获得用户焦点时被调用。

- onPause()方法

当新的 Activity 启动，当前 Activity 失去用户焦点时被调用，此时用户界面仍然可见或部分可见。Activity 暂停后可能会被系统强制结束以释放内存空间，所以通常需要在此时保存程序的相关数据。

- onStop()方法

当前用户界面被新的界面遮挡，完全不可见时调用。此时该 Activity 有可能会被系统强制结束以释放内存空间，所以通常需要在此时保存程序的相关数据。

- onRestart()方法

当 Activity 从停止状态恢复时被调用，也就是说界面重新可见。

- onDestroy()方法

当 Activity 被释放时调用。

Activity 各个状态的变化和上述 7 个回调方法的关系，如图 3.9 所示。

下面将上一节的例子稍作修改，来学习上述 7 个回调方法的使用，并观察 Activity 各个状态的变化，如例 3.3 所示。

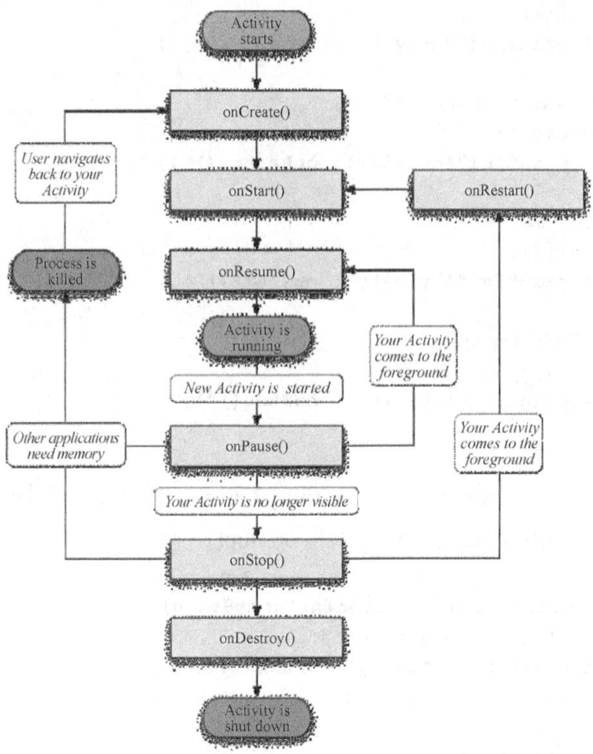

图 3.9　Activity 生命周期

【例 3.3】　Activity 生命周期实例。

（1）修改主 Activity 文件 ActivityDemoActivity.java，增加如下回调方法 onDestroy()、onPause()、onRestart()、onResume()、onStart()和 onStop()，代码如下所示。

```
public class ActivityDemoActivity extends Activity {
    @Override
    public void onCreate(Bundle savedInstanceState) {
        super.onCreate(savedInstanceState);
        System.out.println("Activity1: onCreate()");
        setContentView(R.layout.main);
        final Button button = (Button) findViewById(R.id.btn1);
        button.setOnClickListener(new View.OnClickListener() {
            public void onClick(View view) {
                Intent intent = new Intent(ActivityDemoActivity.this ,
                                     ActivityDemoActivity2.class);
                startActivity(intent);
            }
        });
    }
    protected void onDestroy() {
        super.onDestroy();
        System.out.println("Activity1: onDestroy()");
    }
    protected void onPause() {
        super.onPause();
        System.out.println("Activity1: onPause()");
    }
    protected void onRestart() {
```

```
        super.onRestart();
        System.out.println("Activity1: onRestart()");
    }
    protected void onResume() {
        super.onResume();
        System.out.println("Activity1: onResume()");
    }
    protected void onStart() {
        super.onStart();
        System.out.println("Activity1: onStart()");
    }
    protected void onStop() {
        super.onStop();
        System.out.println("Activity1: onStop()");
    }
}
```

（2）修改 Activity 文件 ActivityDemoActivity2.java，也增加如下回调方法 onDestroy()、onPause()、onRestart()、onResume()、onStart()和 onStop()，代码如下所示。

```
public class ActivityDemoActivity2 extends Activity {
    public void onCreate(Bundle savedInstanceState) {
        super.onCreate(savedInstanceState);
        System.out.println("Activity2: onCreate()");
        setContentView(R.layout.main2);
    }
    protected void onDestroy() {
        super.onDestroy();
        System.out.println("Activity2: onDestroy()");
    }
    protected void onPause() {
        super.onPause();
        System.out.println("Activity2: onPause()");
    }
    protected void onRestart() {
        super.onRestart();
        System.out.println("Activity2: onRestart()");
    }
    protected void onResume() {
        super.onResume();
        System.out.println("Activity2: onResume()");
    }
    protected void onStart() {
        super.onStart();
        System.out.println("Activity2: onStart()");
    }
    protected void onStop() {
        super.onStop();
        System.out.println("Activity2: onStop()");
    }
}
```

运行本实例，当界面 1 在模拟器上显示出来时，观察 LogCat，可以发现以下几行输出信息：

```
Activity1: onCreate()
Activity1: onStart()
Activity1: onResume()
```

这说明当一个 Activity 启动，并且直到其界面完全可见时，先调用创建方法 onCreate()，再调用启动方法 onStart()，最后调用恢复方法 onResume()。此时单击模拟器的返回键，即退出该程序，则能在 LogCat 里看到如下输出信息：

```
Activity1: onPause()
Activity1: onStop()
Activity1: onDestroy()
```

这说明该 Activity 从可见状态变到不可见并且被释放的过程中，上述 3 个方法被先后调用。以上的整个过程构成了该 Activity 的生命周期。下面做如下的操作：重新运行程序，从界面 1 中单击按钮打开界面 2，然后单击返回键返回界面 1，然后再单击返回键退出程序。此时 LogCat 中的输出信息如下，用户可结合本节前面所讲的内容自行加以分析，此处不再赘述。

```
Activity1: onCreate()
Activity1: onStart()
Activity1: onResume()
Activity1: onPause()
Activity2: onCreate()
Activity2: onStart()
Activity2: onResume()
Activity1: onStop()
Activity2: onPause()
Activity1: onRestart()
Activity1: onStart()
Activity1: onResume()
Activity2: onStop()
Activity2: onDestroy
Activity1: onPause()
Activity1: onStop()
Activity1: onDestroy()
```

上述例子中，界面 2 完全遮挡了界面 1。下面演示另外一种情况——界面被部分遮挡。

（3）创建布局文件 main3.xml，代码如下所示。

```
<?xml version = "1.0" encoding = "utf-8"?>
<LinearLayout xmlns:android = "http://schemas.android.com/apk/res/android"
    android:orientation = "vertical"
    android:layout_width = "match_parent"
    android:layout_height = "match_parent" >
    <TextView
        android:layout_width = "match_parent"
        android:layout_height = "wrap_content"
        android:textSize = "15pt"
        android:textColor = "#987654"
        android:text = "DemoActivity3" />
</LinearLayout>
```

（4）创建布局文件 main3.xml 对应的 Activity 文件 Activity Demo Activity 3.java，代码如下所示。

```
public class ActivityDemoActivity3 extends Activity {
    public void onCreate(Bundle savedInstanceState) {
        super.onCreate(savedInstanceState);
        System.out.println("Activity3: onCreate()");
        setContentView(R.layout.main3);
    }
    protected void onDestroy() {
        super.onDestroy();
        System.out.println("Activity3: onDestroy()");
```

```
    }
    protected void onPause() {
        super.onPause();
        System.out.println("Activity3: onPause()");
    }
    protected void onRestart() {
        super.onRestart();
        System.out.println("Activity3: onRestart()");
    }
    protected void onResume() {
        super.onResume();
        System.out.println("Activity3: onResume()");
    }
    protected void onStart() {
        super.onStart();
        System.out.println("Activity3: onStart()");
    }
    protected void onStop() {
        super.onStop();
        System.out.println("Activity3: onStop()");
    }
}
```

（5）在主布局 main.xml 中添加一个打开第 3 个界面的按钮，代码如下所示。

```
<Button android:id = "@+id/btn2"
    android:layout_width = "wrap_content"
    android:layout_height = "wrap_content"
    android:textSize = "15pt"
    android:layout_margin = "12pt"
    android:text = "Start Activity3" />
```

（6）在主 Activity 文件 ActivityDemoActivity.java 的 onCreate 方法里添加打开第 3 个界面的代码，代码如下所示。

```
final Button button2 = (Button) findViewById(R.id.btn2);
button2.setOnClickListener(new View.OnClickListener() {
    public void onClick(View view) {
    Intent  intent  =  new  Intent(ActivityDemoActivity.this,  ActivityDemoActivity3.
class);
    startActivity(intent);
    }
});
```

（7）修改 AndroidManifest.xml 文件，增加注册 ActivityDemoActivity3 类的<activity>标记，代码如下所示。

```
<activity android:name = ".ActivityDemoActivity3"
      android:theme = "@style/shadowed"
      android:label = "@string/app_name" />
```

注意，此处使用了 android:theme 属性用来说明该 Activity 的显示模式。通过值"@style/shadowed"来让其半透明显示，以部分遮挡前一个界面。

（8）在工程目录的 res/values 文件夹下新建样式文件 styles.xml，指定窗口背景，设定窗口无标题，设置宽度高度、动画样式等，代码如下所示。

```
<?xml version = "1.0" encoding = "utf-8"?>
<resources>
    <style name = "shadowed">
        <item name = "android:windowBackground" @drawable/background</item>
```

```
        <item name = "android:windowNoTitle">true</item>
        <item name = "android:layout_width">wrap_content</item>
        <item name = "android:layout_height">wrap_content</item>
        <item name = "android:windowIsTranslucent">true</item>
        <item name = "android:windowAnimationStyle">
@+android:style/Animation.Translucent</item>
    </style>
</resources>
```

（9）上面的 styles.xml 文件中 android:windowBackground 使用了 drawable 资源 background.xml，则在工程的 res/drawable-mdpi 目录下，新建 background.xml 文件，其代码如下所示。

```
<?xml version = "1.0" encoding = "utf-8"?>
<shape              xmlns:android = "http://schemas.android.com/apk/res/android"
android:shape = "rectangle">
    <padding   android:left = "30dp"   android:top = "60dp"   android:right = "40dp"
android:bottom = "80dp" />
    <stroke android:width = "5dp" color = "#000000" />
    <corners android:radius = "8dp" />
    <solid android:color = "#80aaaaaa" />
</shape>
```

至此，全部代码完成，运行结果如图 3.10 ~ 图 3.12 所示。可以看到启动 ActivityDemoActivity3 后，半透明的界面 3 出现，部分遮盖了界面 1，此时界面 1 仍然可见，但是失去了用户输入焦点。

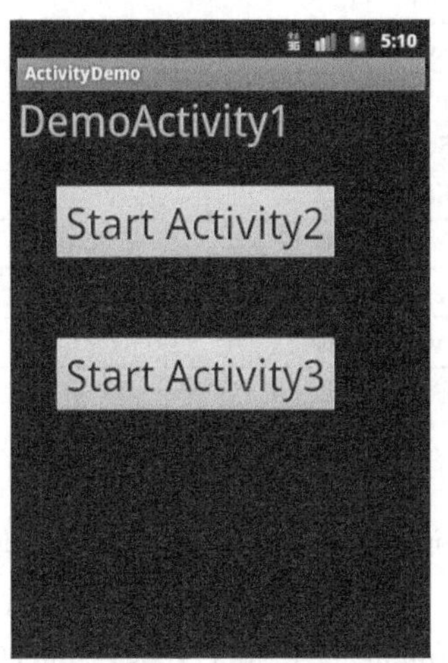

图 3.10　应用界面 1

图 3.11　应用界面 2

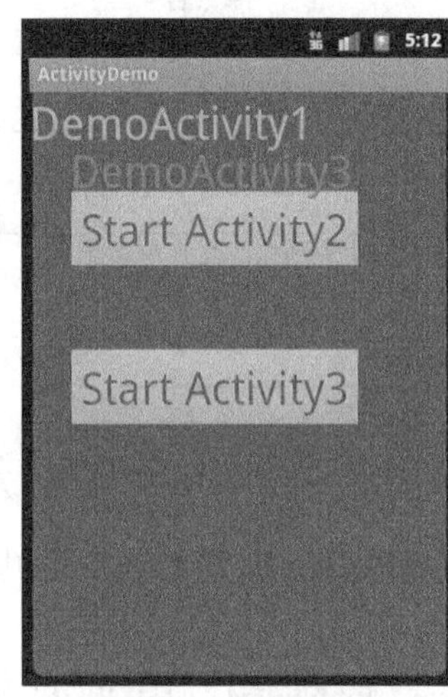

图 3.12　应用界面 3

当进行如下操作：运行程序，直接从界面 1 中启动 ActivityDemoActivity3，显示界面 3，并从界面 3 返回界面 1，再退出程序。从 LogCat 中可以观察到如下输入信息。用户可以将其与前面从界面 1 启动界面 2 的 LogCat 输出情形进行对比。从两者的不同之处可以看出 Activity 状态变化的不同会调用不同的回调方法，从而产生不同的生命周期。

```
Activity1: onCreate()
Activity1: onStart()
Activity1: onResume()
Activity1: onPause()
Activity3: onCreate()
Activity3: onStart()
Activity3: onResume()
Activity3: onPause()
```

```
Activity1: onResume()
Activity3: onStop()
Activity3: onDestroy
Activity1: onPause()
Activity1: onStop()
Activity1: onDestroy()
```

3.6 Intent 简介

Android 中利用 Intent 对象建立连接并实现组件通信的模式，称作意图机制。意图的名称来源于 Android 的 android.content.Intent 类名，本书简称为 Intent。Intent 类是 Android 组件连接的核心，一个 Intent 对象是对某个需要处理的操作所进行的封装和抽象的描述。它可以用来打开 Activity，在多个 Activity 之间传递数据，以及启动后台服务，与广播组件和后台服务交互等，集中体现了整个组件连接模型的设计思想。

Intent 启动不同组件的方法如表 3.1 所示。

表 3.1 Intent 组件的方法

组 件 名	方 法 名
Activity	startActivity()
	startActivityForResult()
Service	startService()
	bindService()
Broadcasts	sendBroadcast()
	sendOrderedBroadcast()
	sendStickyBroadcast()

3.6.1 Intent 属性与过滤器

一个 Intent 包含 6 个方面的属性：action、data、category、type、component 和 extras，下面逐一进行说明。

（1）action 属性描述 Intent 对象所要实施的动作，可以调用 Intent.setAction()方法为 Intent 对象来指定，常见的 action 如下。

- ACTION_MAIN：表示整个程序的入口。
- ACTION_VIEW：表示用于将一些数据显示给用户。
- ACTION_EDIT：表示允许用户对一些数据进行编辑。
- ACTION_DIAL：表示打电话面板
- ACTION_CALL：表示直接拨打电话。
- ACTION_SEND：表示发送短信。
- ACTION_SENDTO：表示选择发送短信。
- ACTION_BATTERY_LOW：表示电量低广播。

（2）data 属性描述 Intent 对象中用于进行操作的数据，例如向用户显示哪些信息，对哪个电话号码进行拨号等。可以通过 Intent.setData()或 Intent.Set Data AndType()来进行设置。

（3）category 属性描述 Intent 对象中的 action 属性属于哪个类别，也就是设置 Intent 对象进行某项操作时的约束，可以通过 Intent.addCategory()方法设置类别（即约束）。

（4）type 属性用来描述组件能够处理的请求类型（即数据的 MIME 类型），可以通过 Intent.setType()或 Intent.setDataAndType()来进行设置。它可以通过通配符*来表示整个类别的信息，如：

```
# 用MIME表示图片类型的数据
image/*
```

也可以具体的指明特定类型，如：

```
# 用MIME表示jpg图片类型的数据
image/jpg
```

（5）component 属性描述 Intent 对象中所使用的组件类的名字，可以通过 Intent.setComponent()方法利用类名进行设定，也可以通过 Intent.setClass()方法利用类型对象信息进行设置。当调用组件明确指定了 component 信息，组件管理服务就不再需要根据 action、data 等信息去寻找满足其需求的组件，只需要按照 component 信息实例化对应的组件作为功能实现者即可。一旦指定了 component，Intent 对象就变成了单纯的信息载体，只负责传递消息和数据。这种方式，通常用于内部组件的互连互通中。

（6）extras 属性以 Bundle 类的形式存储其他额外需要的数据，是以键值对的形式存放，可以通过使用 Intent.setExtra()方法设定。

Intent 的使用有两种形式：显式和隐式。

显式的方式是指创建 Intent 对象时直接指定了其所对应的 Activity 组件，这样就可以用这个 Intent 对象作为参数，来调用 startActivity()启动这个 Activity。本章例 3.2 中就使用了此方法，如：

```
Intent intent = new Intent(ActivityDemoActivity.this, ActivityDemoActivity2.class);
startActivity(intent);
```

这两行代码位于 ActivityDemoActivity.java 文件中，其作用就是创建一个 Intent 对象，通过构造方法的第二个参数指定需打开的 Activity 界面组件为 ActivityDemoActivity2.class，然后从当前 Activity 中调用 startActivity(intent)方法来启动这个新的 Activity。

另一种形式就是创建 Intent 时不指定其所对应的 Activity 组件，而是设置 action、category、type 等属性信息，由 Android 系统根据这些属性信息来选择最合适的 Activity 去运行这个 Intent。这种方式以弱耦合的形式将 Intent 和 Activity 关联起来，提供了更大的灵活性。要使用这种方式来使程序正确工作，需要在所编写的 Activity 类中说明这个 Activity 能够响应哪些 Intent，这样系统就可以根据某个 Intent 的属性与程序进行匹配，看能否由程序来运行这个 Intent。

前面章节讲过，每个应用程序中都有一个 AndroidManifest.xml 配置文件，用来对整个程序中的信息，尤其是 Activity 信息进行说明。程序中的每个 Activity 都需要在这个配置文件中通过<activity>...</activity>的元素进行说明。如果用户仔细观察过<activity>...</activity>元素，会发现在其中还包含了<intent–filter >...</intent-filter>的子元素。这就是上面所说的用来描述该 Activity 能够处理哪些 Intent 的过滤器。一个 Activity 中可以有一组 Intent 过滤器，也可以有多组，根据需要进行添加。每组过滤器中通常会包含 action、category、type 等属性信息。如下代码摘自上一章用于开发环境测试的 Test 工程中的 AndroidManifest.xml 文件，<activity>标记用于描述一个 Activity，有三项信息，分别是 name、label 和一组 Intent 过滤器。Intent 过滤器中有两项信息，分别是 action 和 category。action 属性的值“android.intent.action.MAIN”说明了这个 Activity 是整个程序的入口。而 category 属性的值“android.intent.category.LAUNCHER”说明了这个 Activity 是属于顶层启动器类的。

```
<activity
    android:name = ".Test"
    android:label = "@string/app_name" >
    <intent-filter>
        <action android:name = "android.intent.action.MAIN" />
        <category android:name = "android.intent.category.LAUNCHER" />
    </intent-filter>
</activity>
```

3.6.2 Intent 启动系统 Activity

上一小节主要介绍了 Intent 的基本概念,以及如何通过 Intent 启动 Activity 来打开一个新的界面。同样，也可以通过 Intent 的启动系统 Activity、服务、广播等系统组件。

下面通过例 3.4 来介绍如何使用 Intent 启动系统 Activity。

【例 3.4】 调用系统 Intent 打开手机主界面、拨号、发短信。

（1）创建 Android 工程，工程名为 SysIntentDemo。修改主布局文件，增加两个按钮，代码如下所示。

```
<LinearLayout xmlns:android = "http://schemas.android.com/apk/res/android"
    android:layout_width = "match_parent"
    android:layout_height = "match_parent"
    android:background = "#000000"
    android:orientation = "vertical">
    <Button android:id = "@+id/btnMain"
        android:layout_width = "wrap_content"
        android:layout_height = "wrap_content"
        android:text = "打开主界面" />
    <Button android:id = "@+id/btnDial"
        android:layout_width = "wrap_content"
        android:layout_height = "wrap_content"
        android:text = "     拨 号     " />
    <Button android:id = "@+id/btnSMS"
        android:layout_width = "wrap_content"
        android:layout_height = "wrap_content"
        android:text = "   发短信    " />
</LinearLayout>
```

（2）修改主 Activity 文件 SysIntentDemoActivity.java，添加处理单击按钮的事件，代码如下所示。

```
public class SysIntentDemoActivity extends Activity {
    private Button btnMain, btnDial, btnSMS;
    @Override
    protected void onCreate(Bundle savedInstanceState) {
        super.onCreate(savedInstanceState);
        setContentView(R.layout.activity_main);
        btnMain = (Button) findViewById(R.id.btnMain);
        btnDial = (Button) findViewById(R.id.btnDial);
        btnSMS = (Button) findViewById(R.id.btnSMS);
        btnMain.setOnClickListener(new View.OnClickListener() {
            @Override
            public void onClick(View view) {
```

```
                Intent intent = new Intent();
                intent.setAction(Intent.ACTION_MAIN);
                intent.addCategory(Intent.CATEGORY_HOME);
                startActivity(intent);
            }
        });
        btnDial.setOnClickListener(new View.OnClickListener() {
            @Override
            public void onClick(View view) {
                Intent intent = new Intent();
                intent.setAction(Intent.ACTION_DIAL);
                intent.addCategory(Intent.CATEGORY_DEFAULT);
                startActivity(intent);
            }
        });
        btnSMS.setOnClickListener(new View.OnClickListener() {
            @Override
            public void onClick(View view) {
                Intent intent = new Intent();
                intent.setData(Uri.parse("smsto:123456"));
                intent.putExtra("sms_body", "通过 Android 应用发送的短消息");
                startActivity(intent);
            }
        });
    }
}
```

注意程序中黑体字部分的代码，通过设定 intent 对象的 action 属性和 category 属性，可以启动所对应的系统 Activity。action 属性为“Intent.ACTION_MAIN”，category 属性为“Intent.CATEGORY_HOME”时，将启动系统 Home Activity，即返回 Android 系统 Home 主界面。action 属性为“Intent.ACTION_DIAL”，category 属性为“Intent.CATEGORY_DEFAULT”时，将启动系统的拨号面板 Activity。打开短消息面板 Activity 只要设置 data 属性即可。运行本实例，结果如图 3.13 ~ 图 3.16 所示。

图 3.13　程序首页

图 3.14　打开手机主界面

图 3.15　打开拨号界面

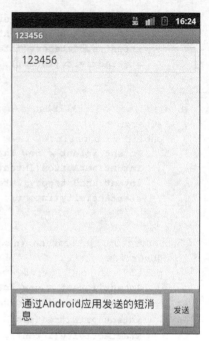

图 3.16　打开短信界面

3.7 小　结

本章首先介绍 Android 应用程序项目的基本结构，分析并讲解了 Android 应用程序项目的各组成元素。一个 Android 项目中通常会包括 5 个文件夹：src、gen、assets、bin、res 和 3 个文件：AndroidManifest.xml、proguard.cfg、project.properties。

接着介绍了 Android 应用程序安装文件 APK 文件的结构。一个 APK 文件包括 2 个文件夹和 3 个文件：META-INF 文件夹、res 文件夹、AndroidManifest.xml、classes.dex、resources.arsc。除了以上文件夹和文件，在有些 APK 文件中，还有可能出现 lib 文件夹和 assets 文件夹。这 2 个文件夹中分别存放应用软件需要调用的库文件和其他资源配置文件。

Activity 是编写 Android 应用程序需要掌握的最基本的概念，Activity 的生命周期是编写 Android 应用程序的基础。Activity 是 Android 系统 API 的一个类，用于提供一个系统与用户交互的机制。如果一个程序有多个用户界面，则通常对于每一个用户界面，需要编写与之对应的 Activity 子类程序。Android 系统采用栈的数据结构来管理多个 Activity。

Android 的意图机制是 Android 应用模型的核心，解决了组件间的连接问题。通过组件管理服务提供的 Intent 对象和 Intent Filter 对象的匹配策略，降低了组件间的耦合度，提高了平台的灵活性，增强了组件的复用性，从根本上减轻了应用开发的负担。

练　习

1. 一个新建的 Android 项目结构中包括哪些文件夹和文件？

2. R.java 是系统自动生成的文件，它里面包括哪些内容？

3. 如何声明 Android 应用程序权限？

4. 简述 Activity 的生命周期。

5. Activity 的生命周期有多少种状态？各种状态的变化关系如何？

6. 修改猜数字游戏，增加新的功能，如记录用户猜对数字所用的次数，记录用户猜对数字所用的时间等。

第 4 章 视图组件

视图组件是 Android 应用开发中最为基础、最为重要的一章。视图组件也常称为 widget 组件、UI 组件、View 组件，是构造 UI 界面的基本单元，为了更好地进行界面的设计，Android 中提供了大量的基础组件。用户需要了解这些组件的特性、使用场景和使用方法，才能更好更快地进行 UI 界面的开发。本章中将对视图组件的使用模式和常用的视图组件进行详细的介绍。

4.1 视图组件的使用模式

视图组件的使用模式，也就是 Android 应用程序的开发过程。一般过程是先通过 XML 布局文件或 Java 代码创建界面布局，设定组件显示样式，随后获取 UI 组件对象，并处理组件事件响应。下面对这一过程进行详细的介绍。

4.1.1 视图组件的定义

在设计 Android 应用时，用户首先需要通过各种视图组件来定义界面，包括设置组件的属性和布局方式等。Android 视图组件通常可通过下面两种方式进行定义。

1. 使用 XML 布局文件定义视图组件

Android 平台为大多数视图组件以及其子类提供了 XML 标记，可通过 XML 布局文件中的标记来定义视图组件。XML 中的每个元素代表了一个组件，即元素名称对应相应的 Java 类。如<Button>对应 Java 类就是 Button 类，<LinearLayout>对应的 Java 类是 LinearLayout 类。如下代码为一个使用 XML 文件定义界面布局的实例。

```
<LinearLayout xmlns:android = "http://schemas.android.com/apk/res/android"
    android:layout_width = "match_parent"
    android:layout_height = "match_parent"
    android:background = "#FFFAE4"
    android:orientation = "vertical">
    <Button android:id = "@+id/submit"
        android:layout_width = "match_parent"
        android:layout_height = "wrap_content"
        android:textColor = "@color/color1"
        android:textSize = "@dimen/dimen2"
        android:background = "@drawable/custom_button"
        android:text = "提交" />
    <Button style = "@style/title" />
```

```
</LinearLayout>
```

代码说明如下。

- LinearLayout 标记：表示当前界面采用线性布局管理器（布局管理器详细内容见下一章），即其内组件采用线性方向排列。
- xmlns:android = "http://schemas.android.com/apk/res/android"：表示命名空间，根节点必须有此项。
- android:layout_width 和 android:layout_height：表示布局的宽度和高度，可选值有fill_parent、match_parent 和 wrap_content。其中 fill_parent 表示该组件的宽度或高度与父容器相同；match_parent 和 fill_parent 功能完全相同，从 Android2.2 开始推荐使用；wrap_content 表示该组件的宽度或高度恰好能容纳它的内容。
- android:background：设置整个布局背景色。
- android:orientation = "vertical"：表示线性布局管理器内组件纵向排列。
- Button 标记：表示按钮组件。
- android:id：用于为当前组件指定一个 ID 属性，编译时由 aapt 工具自动生成到 R.java 文件中，在 Java 代码中可以通过 findViewById()方法来引用该组件。
- android:textColor：设置当前组件的字体颜色，本例从资源文件获取。
- android:textSize：设置当前组件的字体大小，本例从资源文件获取。
- android:background：设置当前组件的背景，本例从资源文件获取。
- style：设置当前组件使用样式，本例从资源文件获取。

为了方便界面设计，Android 开发工具也提供了界面设计工具，用户可以通过拖曳的方式向界面中添加各个组件，并在属性窗口中设置属性值，界面设计工具如图 4.1 所示。实际上对用户来说，用 XML 代码与 Java 代码设计界面具有同等效果。

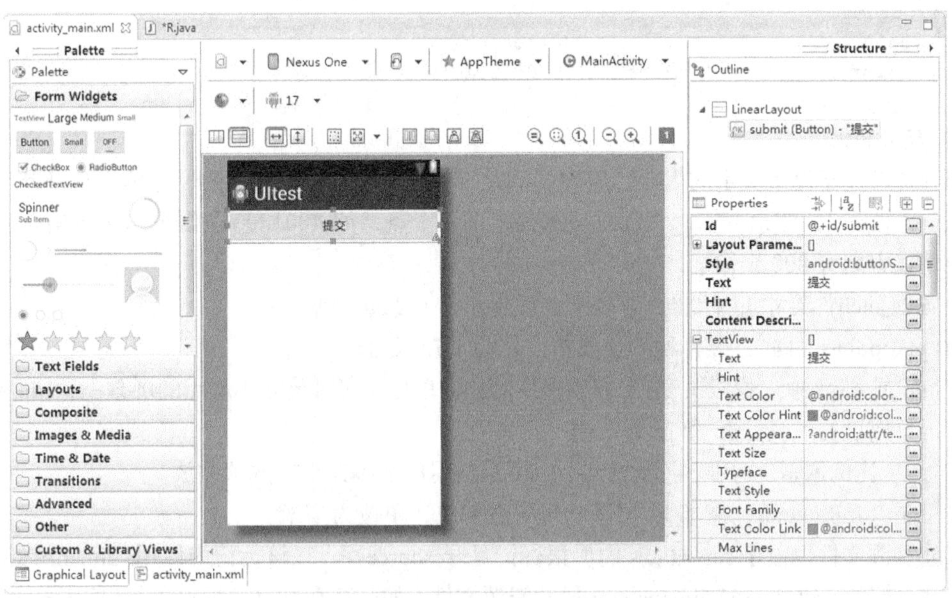

图 4.1　Android 界面设计工具

2. 使用 Java 代码定义视图组件

使用 Java 代码定义对象的方式来定义组件布局，通过使用 new 语句创建组件类的实例，再设置其属性，调用其方法。但该方式不够灵活，一般不推荐使用。实际上，通过使用 XML 定义组件的方式，同样会根据 XML 标记内容创建对应的类实例，只是实例化的过程是 Android 框架来实现的。当应用程序启动后，Android 运行库会解析布局文件的 XML 定义，自动实例化为对应的类对象，最后再通过应用程序框架来显示该组件。

4.1.2 资源的访问

Android 开发中针对视图组件显示风格，提供了多种资源来定义。如字符串资源、颜色资源、尺寸资源、Drawable 资源和样式资源等。下面对常用的几个资源予以简单介绍。

1. 颜色资源

颜色资源通常用于设置文字、背景的颜色等。在 Android 中，颜色值通过 RGB（红、绿、蓝）三原色和一个透明度（Alpha）值来表示。在设置颜色值时需要以“#”开头，其中透明度值可以省略，如果省略则表示完全不透明。颜色值一般使用以下 4 种方式定义。

- #RGB：使用红、绿、蓝三原色的值表示颜色，其中红、绿、蓝取值为 16 进制值 0 ~ F。例如，表示蓝色可以使用#00F。
- #ARGB：使用透明度和红、绿、蓝表示颜色，其中透明度和红、绿、蓝取值为 0 ~ F。例如，表示半透明蓝色可以使用#700F。
- #RRGGBB：同#RGB 方式，但红、绿、蓝取值为 00 ~ FF。例如，表示红色可以使用#FF0000。
- #AARRGGBB：同#ARGB 方式，但透明度和红、绿、蓝取值为 00 ~ FF。例如，表示半透明红色可以使用#77FF0000。

颜色资源文件位于 res/values 目录下，根标记是<resources>，在该标记内使用<color>标记定义各种颜色，颜色值可以使用上述四种方式之一。例如，建立一个 colors.xml 颜色资源文件，定义 color1 为红色，color2 为半透明红色，代码如下所示。

```
<resources>
    <color name = "color1">#FF0000</color>
    <color name = "color2">#77FF0000</color>
</resources>
```

2. 尺寸资源

尺寸资源用于设置文字的大小、组件的间距等。在 Android 中支持的尺寸单位有以下几种。

- px（pixel，像素）：每个 px 对应屏幕上的一个点，如 480*800 的屏幕水平方向有 480 个像素，垂直方向有 800 个像素。
- in（inch，英寸）：标准长度单位，每英寸等于 2.54 厘米。
- pt（points，磅）：屏幕物理长度单位，为 1/72 英寸。
- dp 或 dip（独立像素）：基于屏幕密度的抽象单位，在每英寸 160 点的屏幕上，1dp = 1px。但随着屏幕密度的改变，dp 和 px 的换算也会变化。
- sp（比例像素）：用于字体的大小，可根据用户字体大小首选项缩放。
- mm（millimeter，毫米）：屏幕物理长度单位，单位为毫米。

尺寸资源文件也位于 res/values 内，根标记是<resources>，在该标记内使用<dimen>标记定义各种尺寸。例如，建立一个 dimens.xml 尺寸资源文件（如已存在则修改即可），定义 dimen1 为 20 像素，dimen2 为 25dp，代码如下所示。

```
<resources>
    <dimen name = "dimen1">20px</dimen>
```

```
    <dimen name = "dimen2">25dp</dimen>
</resources>
```

3. Drawable 资源

Drawable 资源是 Android 应用中使用最为广泛、灵活的资源。不仅可以使用图片作为资源，也可以使用多种 XML 文件作为资源。Drawable 资源可针对屏幕不同分辨率，分别位于 res 目录下的 drawable-XXX 目录内。drawable-XXX 一般表示 drawable-hdpi、drawable-ldpi、drawable-mdpi、drawable-xhdpi 和 drawable-xxhdpi 这 5 个目录（部分 Android API 版本仅有前 3 个目录）。如不考虑分辨率的不同，而使用同一资源的话，资源文件可放入 drawable-mdpi 目录内（本书所有例子均放资源于此）。

Drawable 资源分为图片资源和 StateListDrawable 资源。Android 中图片资源支持 png、jpg、gif 和 9-Patch 等格式，这里不多做介绍。StateListDrawable 资源能根据状态来显示出不同的图像，例如，一个按钮会存在多种状态，如 pressed、enabled 或 focused 等。使用 StateListDrawable 资源可以为按钮的每个状态提供不同的显示效果、按钮图片等。StateListDrawable 资源文件根标记是<selector>，在该标记中使用<item>标记设置以下两种属性。

- android:drawable：用于指定 Drawable 资源。
- android:state_XXX：用于指定一种状态，常用状态如表 4.1 所示。

表 4.1　StateListDrawable 支持的常用状态属性

状 态 属 性	说　明
android:state_active	设置是否处于激活状态，取值 true、false(以下取值相同)
android:state_checked	设置是否处于选中状态
android:state_enabled	设置是否处于可用状态
android:state_first	设置是否处于开始状态
android:state_focused	设置是否处于获得焦点状态
android:state_last	设置是否处于结束状态
android:state_middle	设置是否处于中间状态
android:state_pressed	设置是否处于按下状态
android:state_selected	设置是否处于被选择状态

例如创建一个设置按钮按下前和按下后不同效果的 StateListDrawable 资源，资源文件名为 custom_button.xml，代码如下所示。

```
<selector xmlns:android = "http://schemas.android.com/apk/res/android">
    <!-- 单击未释放按钮时的颜色 -->
    <item android:state_pressed = "true">
        <shape>
            <gradient  android:startColor = "#F55030"  android:endColor = "#F55030"
                       android:angle = "270" />
            <stroke android:width = "1dp" android:color = "#FFFFFF" />
            <corners android:radius = "5dp" />
            <padding android:left = "10dp" android:top = "10dp" android:right = "10dp"
                     android:bottom = "10dp" />
        </shape>
    </item>
    <!-- 打开后的默认颜色 -->
```

```
        <item>
            <shape>
                <gradient android:startColor = "#378AE0" android:endColor = "#123456"
                      android:angle = "90" />
                <stroke android:width = "1dp" android:color = "#FFFFFF" />
                <corners android:radius = "5dp" />
                <padding android:left = "10dp" android:top = "10dp" android:right = "10dp"
                    android:bottom = "10dp" />
            </shape>
        </item>
    </selector>
```

上例中，通过<shape>标记设置按钮显示样式。其中<gradient>标记指定按钮背景的开始颜色、结束颜色（值不同的话即设置渐变色），<stroke>标记设置按钮的边框大小和颜色，<corners>标记设置按钮 4 个角的弧度，<padding>标记设置按钮四个方向的留白大小。

4. 样式资源

上面所讲的使用颜色资源和尺寸资源设置样式，组件可选取合适的颜色和尺寸组合搭配。而样式资源主要用于对组件显示样式的统一控制，包含文字的大小、颜色、宽高等。样式资源文件也位于 res/values 目录内，根标记是<resources>，在该标记内使用<style>、<item>标记定义样式。例如在工程中创建一个样式资源，文件名为 mystyles.xml，代码如下所示。

```
<resources>
    <style name = "title" parent = "AppBaseTheme">
        <item name = "android:layout_width">match_parent</item>
        <item name = "android:layout_height">wrap_content</item>
        <item name = "android:background">#000000</item>
        <item name = "android:textSize">30dp</item>
        <item name = "android:textColor">#0000FF</item>
        <item name = "android:text">注册</item>
    </style>
</resources>
```

上例中定义了组件的宽、高、背景色、字体大小、字体颜色、显示文字等。

4.1.3 生成视图组件资源标识

在 XML 资源实例化过程中，资源打包工具 aapt 将解析 XML 布局文件中的组件定义，随后会生成对应的资源标识，这样就建立起组件定义标识和组件实例的对应关系，该关系存放于工程 gen 目录下相应包的 R.java 文件中，可通过组件的定义标识获取对应的对象实例。如下代码就是根据上述 XML 布局文件和其他相关 XML 文件生成的部分资源标识信息。

```
public final class R {
    public static final class color {
        public static final int color1 = 0x7f040000;
        public static final int color2 = 0x7f040001;
    }
        publitic static final class dimen {
    public static final int dimen1 = 0x7f050000;
        publicc static final int dimen2 = 0x7f050001;
    }
    public static final class drawable {
        public static final int custom_button = 0x7f020000;
```

```
        public static final int ic_launcher = 0x7f020001;
    }
    public static final class id {
        public static final int action_settings = 0x7f090001;
        public static final int submit = 0x7f090000;
    }
    public static final class layout {
        public static final int activity_main = 0x7f030000;
    }
    public static final class style {
        public static final int AppBaseTheme = 0x7f060001;
        public static final int AppTheme = 0x7f060002;
        public static final int title = 0x7f060000;
    }
    ......
}
```

本例中 XML 布局文件名为 activity_main.xml，该布局文件表示的界面使用名为 activity_main 的标识进行定义，那么可使用 R.layout.activity_main 标识即可访问整个界面组件。界面中的按钮组件使用 submit 标识进行定义，同样可使用 R.id.submit 来访问该按钮。该 R.java 文件中还生成了关于菜单、样式等的标识，不再赘述。

4.1.4　视图组件的引用

在 XML 布局文件中，设置组件的 android:id 属性后，在 Java 代码中使用 Activity 组件或父组件的 findViewById()方法可以获取指定组件的对象。如下代码为使用 setContentView()方法通过布局 id 来显示整个界面，使用 findViewById()方法通过按钮的 id 来获取按钮对象。

```
protected void onCreate(Bundle savedInstanceState) {
    super.onCreate(savedInstanceState);
    setContentView(R.layout.activity_main);
    Button btn = (Button) findViewById(R.id.submit);
}
```

4.1.5　视图组件的事件响应

当用户通过按键、触摸、滑动等动作与应用程序交互时，需要组件对用户事件进行响应。同 JSE 平台一样，Android 系统为视图组件定义了用于捕获各种事件的监听器，视图组件可以使用这些监听器捕获用户事件并予以响应。

视图组件对用户事件的监听，一般可以通过以下 4 种方式实现。

1．定义事件监听器，再与组件绑定

在代码中先定义事件监听器，再通过组件的 setXXXListener()方法绑定监听器，代码示例如下所示。

```
public class MainActivity extends Activity {
    private View.OnClickListener myListener = new View.OnClickListener() {
        @Override
        public void onClick(View v) {
            if(v.getId() == R.id.submit)
                Toast.makeText(MainActivity.this, "数据提交中...",
                        Toast.LENGTH_LONG).show();
```

```
        }
    };
    @Override
    protected void onCreate(Bundle savedInstanceState) {
        super.onCreate(savedInstanceState);
        setContentView(R.layout.activity_main);
        Button btn = (Button) findViewById(R.id.submit);
        btn.setOnClickListener(myListener);  //对按钮组件设置监听器
    }
}
```

在上述代码中定义了单击事件监听器（对象为 myListener），按钮对象 btn 通过 setOnClickListener (myListener)方法将监听器与按钮组件绑定。在该监听器的定义中，实现了单击事件的回调方法（onClick），该方法判断是否是对应组件引发事件，若是则弹出一个提示信息。该实例运行后，单击提交按钮结果如图 4.2 所示。

图 4.2　按钮单击事件处理应用

2. 在与组件绑定时定义事件监听器

该方法是当通过调用组件的 setXXXListener()方法设置监听器时，定义只能当前组件使用的事件监听器，代码示例如下所示。

```
public class MainActivity extends Activity {
    @Override
    protected void onCreate(Bundle savedInstanceState) {
        super.onCreate(savedInstanceState);
        setContentView(R.layout.activity_main);
        Button btn = (Button) findViewById(R.id.submit);
        btn.setOnClickListener(new View.OnClickListener() {
            @Override
            public void onClick(View v) {
                Toast.makeText(MainActivity.this, "数据提交中...",
                        Toast.LENGTH_LONG).show();
            }
        });
    }
}
```

3. 当前 Activity 类实现监听器接口

该方法使当前 Activity 类实现相应的监听器接口，完成监听器中的抽象方法，组件绑定监听器时，监听器对象即为当前 Activity 对象，代码示例如下所示。

```
public class MainActivity extends Activity implements OnClickListener {
    @Override
    protected void onCreate(Bundle savedInstanceState) {
        super.onCreate(savedInstanceState);
        setContentView(R.layout.activity_main);
        Button btn = (Button) findViewById(R.id.submit);
        btn.setOnClickListener(this);
    }
    @Override
    public void onClick(View v) {
        if(v.getId() == R.id.submit)
            Toast.makeText(MainActivity.this, "数据提交中...",
                    Toast.LENGTH_LONG).show();
    }
}
```

4.XML 布局文件中设置回调方法名

该方法在 XML 布局文件中设置回调方法名，在 Java 代码中实现该方法即可。此方法不是一个通用的方法，但对按钮等常用组件的单击事件处理确实很方便，代码示例如下所示。

（1）activity_main.xml 布局文件。

```
<LinearLayout xmlns:android = "http://schemas.android.com/apk/res/android"
    android:layout_width = "match_parent"
    android:layout_height = "match_parent"
    android:orientation = "vertical">
    <Button android:id = "@+id/submit"
        android:layout_width = "match_parent"
        android:layout_height = "wrap_content"
        android:onClick = "myClick"
        android:text = "提交" />
</LinearLayout>
```

按钮的 android:onClick 属性设置回调方法名为 myClick。

（2）Java 代码。

```
public class MainActivity extends Activity {
    @Override
    protected void onCreate(Bundle savedInstanceState) {
        super.onCreate(savedInstanceState);
        setContentView(R.layout.activity_main);
    }
    public void myClick(View v) {
        Toast.makeText(MainActivity.this, "数据提交中...",
                Toast.LENGTH_LONG).show();
    }
}
```

Android 平台为组件提供了丰富的事件监听器，常用监听器如表 4.2 所示。

表 4.2　　组件常用监听器

监听器类型	对应事件	说明
OnClickListener	onClick	单击事件
OnLongClickListener	onLongClick	长点击事件
OnFocusChangeListener	onFocusChange	焦点改变事件
OnKey	onKey	按键单击事件
OnTouch	onTouch	触摸事件
OnCreateContextMenu	onCreateContextMenu	创建上下文菜单事件

4.1.6 组件的常用属性

Android 平台为视图组件提供了丰富的属性，部分属性大多组件都具备，这里简单地介绍下组件的常用属性，如表 4.3 所示。

表 4.3　　UI 组件的常用属性

属性名	说明
android:background	设置背景色或背景图片，以下两种值均可设置背景透明：@android:color/transparent 和@null。 注意 TextView 默认是透明的，不用此属性
android:id	组件唯一编号
android:layout_width	组件的宽度，取值为：match_parent、match_parent、wrap_content 和自定义大小
android:layout_height	组件的高度，取值同组件的宽度
android:layout_gravity	设置组件在布局中的位置，取值为：top、bottom、left、right、center_vertical、center_horizontal、center 等
android:drawingCacheQuality	设置绘图时半透明质量，取值为：auto（默认，由框架决定）、high（高质量，较高颜色深度，消耗内存多）、low（低质量，较低颜色深度，消耗内存少）
android:keepScreenOn	视图在可见的情况下是否保持唤醒状态
android:minHeight	设置视图最小高度
android:minWidth	设置视图最小宽度
android:onClick	单击时在对应的 Activity 中调用指定的方法
android:padding	设置边距。另外还提供属性 android:paddingTop、android:paddingBottom、android:paddingLeft、android: paddingRight，可分别设置距上下左右边距
android:saveEnabled	设置是否在窗口冻结时（如旋转屏幕）保存 View 的数据，默认为 true。注意需设置 id
android:visibility	设置是否显示组件，取值为：visible（默认值，显示）、invisible（不显示，但是仍然占用空间）、gone（不显示，不占用空间）

4.2 常用组件

Android 平台提供了一个 widget 包，其中包含了各种视图组件。图 4.3 所示为 Android 中一些常用组件间的继承关系，其中 View 类是最基本的视图类，几乎所有高级视图组件都是继承 View 类而实现的，例如文本框、按钮、编辑框、单选按钮、复选按钮等。下面对这些常用组件分别予以介绍。

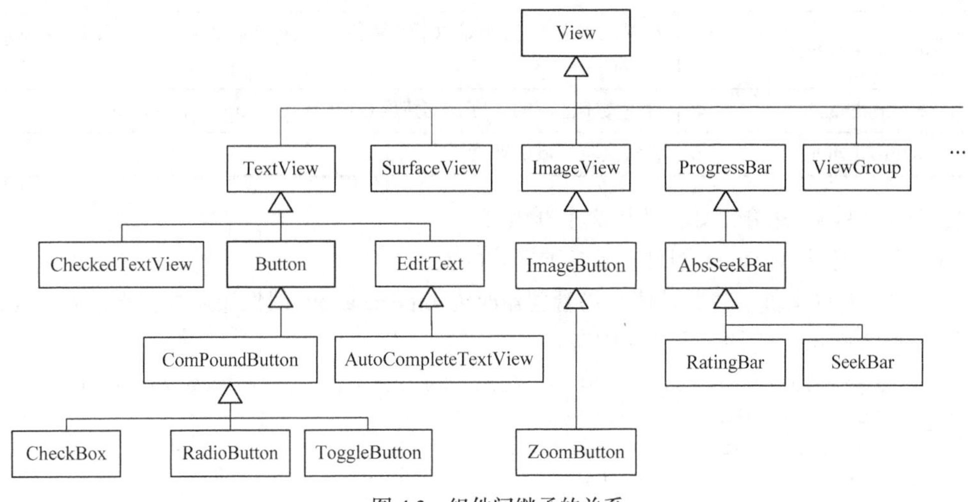

图 4.3　组件间继承的关系

4.2.1　文本框

文本框用 TextView 表示，用于在界面上显示文字信息。Android 中的 TextView 组件不仅能显示单行文本，也可以显示多行文本，甚至可以显示带图像的文本。向界面添加文本框，可以在 XML 布局文件中使用<TextView>标记添加，也可以在 Java 代码中通过 new 关键字创建，推荐采用第一种。后续介绍的其他组件均可通过上述两种方法创建。TextView 支持的常用 XML 属性如表 4.4 所示。

表 4.4　　TextView 的常用 XML 属性

属 性 名 称	描　　述
android:autoLink	设置当文本为 URL 链接、email、电话号码、map 时，文本是否显示为可单击的链接。可选值： none、web、email、phone、map、all
android:drawableBottom	在文字下方放置图片。另外提供属性 android:drawableLeft、android:drawableRight、android:drawableTop，分别在文字左方、右方、上方放置图片
android:gravity	设置文本位置，如设置成“center”，文本将居中显示
android:hint	Text 为空时显示的文字提示信息
android:maxLength	限制显示的文本长度，超出部分不显示
android:lines	设置文本的行数
android:singleLine	设置单行显示。当文本不能全部显示时，后面用“…”来表示
android:textColor	设置文本颜色
android:textSize	设置文字大小，推荐使用度量单位“sp”

续表

属 性 名 称	描 述
android:textStyle	设置字形，取值为：bold（粗体）0、italic（斜体）1、bolditalic（又粗又斜）2，可以设置一个或多个值，多值用“|”隔开
android:typeface	设置文本字体，必须是以下常量值之一：normal 0、sans 1、serif 2、monospace（等宽字体）3
android:height	设置文本区域的高度，支持度量单位：px、dp、sp、in、mm
android:width	设置文本区域的宽度

下面将通过例 4.1 来介绍如何使用文本框组件。

【例 4.1】 应用 TextView 组件显示多种样式的文本。

（1）修改主布局文件，将默认的布局管理器改为 LinearLayout 线性布局，垂直排列。在该布局中放置一个 TextView 组件，设置居中对齐。

```
<LinearLayout xmlns:android = "http://schemas.android.com/apk/res/android"
    android:layout_width = "match_parent"
    android:layout_height = "match_parent"
    android:background = "#FFFAE4"
    android:orientation = "vertical" >
    <TextView android:id = "@+id/tv1"
        android:layout_width = "wrap_content"
        android:layout_height = "wrap_content"
        android:layout_gravity = "center" />
    <!-- 待添加的组件 -- >
</LinearLayout>
```

在线性布局中再添加一个 TextView 组件，图片在文字上方，文字来源于资源“@string/notebook”，并且在该组件中居中。

```
<TextView
    android:layout_width = "wrap_content"
    android:layout_height = "wrap_content"
    android:drawableTop = "@drawable/note"
    android:gravity = "center"
    android:text = "@string/notebook" />
```

在线性布局中再添加两个 TextView 组件，一个设置带链接，另一个设置文字颜色和大小。

```
<TextView
    android:layout_width = "wrap_content"
    android:layout_height = "wrap_content"
    android:autoLink = "web"
    android:text = "点击访问百度:http://www.baidu.com" />
<TextView
    android:layout_width = "match_parent"
    android:layout_height = "wrap_content"
    android:gravity = "right"
    android:text = "居右自定义颜色和大小文本"
    android:textColor = "#FFFF0000"
    android:textSize = "20sp" />
```

在线性布局中再添加两个 TextView 组件，分别设置显示单行文本和多行文本。注意单行文本需要设置 android:singleLine 属性为 true，显示不下的文字以为“...”结尾。

```
<TextView
    android:layout_width = "wrap_content"
    android:layout_height = "wrap_content"
    android:singleLine = "true"
    android:text = "单行文本:Android 中 TextView 组件不仅能显示单行文本，也可以显示多行文本，甚至可以显示带图像的文本。"
    android:width = "300px" />
<TextView
    android:layout_width = "wrap_content"
    android:layout_height = "wrap_content"
    android:text = "多行文本:Android 中 TextView 组件不仅能显示单行文本，也可以显示多行文本，甚至可以显示带图像的文本。"
    android:width = "300px" />
```

（2）字符串资源文件 strings.xml 位于工程的 res/values 目录下，同目录下还有设置界面样式和外观的 styles.xml 文件，设置组件尺寸的 dimens.xml 文件和设置颜色的 colors.xml 文件等。字符串资源文件代码如下所示。

```
<resources>
    <string name = "app_name">文本框实例</string>
    <string name = "notebook">记事本</string>
    ......
</resources>
```

（3）主 Activity 文件中获取 id 为 tv1 的文本框，并通过该组件对象的 setText()方法设置该文本框显示文字。

```
public class MainActivity extends Activity {
    @Override
    protected void onCreate(Bundle savedInstanceState) {
        super.onCreate(savedInstanceState);
        setContentView(R.layout.activity_main);
        TextView tv = (TextView) findViewById(R.id.tv1);
        tv.setText("由 Java 代码设置文本内容");
    }
}
```

运行本实例，显示结果如图 4.4 所示。

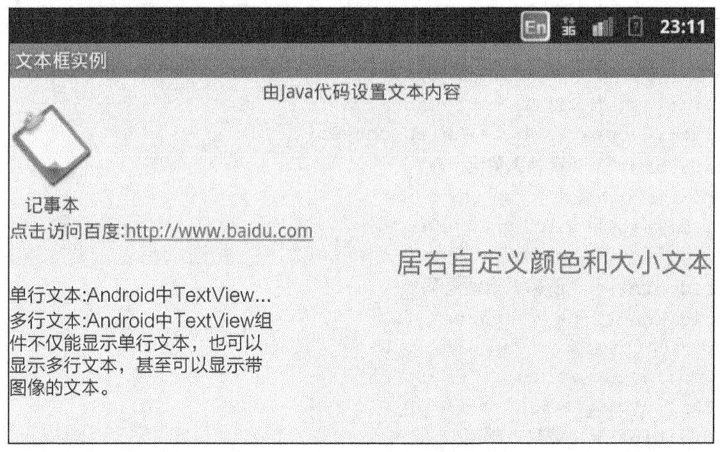

图 4.4　TextView 显示多种样式文本

4.2.2 编辑框

编辑框用 EditText 表示，用于在界面上输入文本。在 Android 中，编辑框组件不仅可以输入单行文本、多行文本，还可以输入指定格式的文本，如密码、电话号码、E-mail 地址等。从继承结构上看，EditText 是 TextView 的子类，所以表 4.4 中的属性对于 EditText 组件同样适用，另外还支持的 XML 属性如表 4.5 所示。

表 4.5　EditText 常用的其他 XML 属性

属性名称	描　述
android:digits	设置允许输入哪些字符，如“1234567890. + -*/%\n()”
android:inputType	设置文本的类型，取值为： text、textPassword、textEmailAddress、datetime、date 等
android:numeric	打开数字输入法

在界面上添加编辑框后，如需获取输入的文本时，可以通过编辑框组件提供的 getText()方法，也可以调用编辑框组件的 setText()方法设置文本，如下例所示。

```
EditText name = (EditText)findViewById(R.id.userName);
String msg = name.getText(); //获取文本
name.setText("angel"); //设置文本
```

下面将通过例 4.2 来介绍如何使用编辑框组件。

【例 4.2】　应用 EditText 组件填写个人信息。

（1）修改主布局文件，将默认的布局管理器改为 LinearLayout 线性布局，垂直排列。其中电话号码输入框设置 android:inputType 为 phone 仅允许输入电话号码，邮箱输入框设置 android:inputType 为 textEmailAddress 仅允许输入邮箱地址，代码如下所示。

```
<LinearLayout xmlns:android = "http://schemas.android.com/apk/res/android"
   android:layout_width = "match_parent"
   android:layout_height = "match_parent"
   android:orientation = "vertical" >
   <TextView
      android:layout_width = "wrap_content"
      android:layout_height = "wrap_content"
      android:layout_gravity = "center"
      android:text = "联系方式"
      android:textSize = "20sp"/>
   <EditText android:id = "@+id/name"
      android:layout_width = "200sp"
      android:layout_height = "wrap_content"
      android:hint = "请输入姓名" />
   <EditText android:id = "@+id/phone"
      android:layout_width = "200sp"
      android:layout_height = "wrap_content"
      android:hint = "请输入电话号码"
      android:inputType = "phone" />
   <EditText android:id = "@+id/email"
      android:layout_width = "200sp"
      android:layout_height = "wrap_content"
      android:hint = "请输入邮箱"
      android:inputType = "textEmailAddress" />
```

```
    <Button android:id = "@+id/submit"
        android:layout_width = "wrap_content"
        android:layout_height = "wrap_content"
        android:layout_gravity = "center"
        android:text = "提交" />
</LinearLayout>
```

（2）主 Activity 文件中获取各个编辑框的文本，单击提交按钮时弹出提示信息，代码如下所示。

```
public class MainActivity extends Activity {
    private EditText name, phone, email;
    private Button submit;
    @Override
    protected void onCreate(Bundle savedInstanceState) {
        super.onCreate(savedInstanceState);
        setContentView(R.layout.activity_main);
        name = (EditText) findViewById(R.id.name);
        phone = (EditText) findViewById(R.id.phone);
        email = (EditText) findViewById(R.id.email);
        submit = (Button) findViewById(R.id.submit);
        final StringBuilder sb = new StringBuilder();
        submit.setOnClickListener(new View.OnClickListener() {
            @Override
            public void onClick(View v) {
                if(sb.length() > 0)  sb.delete(0, sb.length()); //删除旧数据
                sb.append("用户名: ");
                sb.append(name.getText().toString());
                sb.append("\n电话号码: ");
                sb.append(phone.getText().toString());
                sb.append("\nEmail: ");
                sb.append(email.getText().toString());
                sb.append("\n数据提交中...");
                Toast.makeText(MainActivity.this, sb.toString(),
                            Toast.LENGTH_LONG).show();
            }
        });
    }
}
```

运行本实例，显示结果如图 4.5 所示，单击提交按钮后的结果如图 4.6 所示。

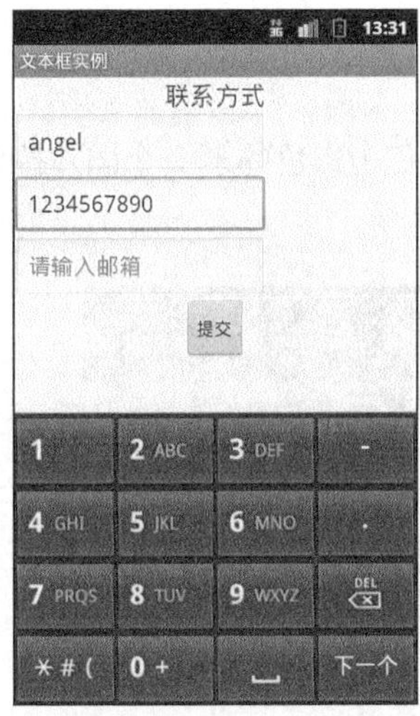

图 4.5　输入电话时为数字键盘

图 4.6　单击提交按钮后的结果

4.2.3 图片按钮

按钮用 Button 表示，用于在界面上触发一个指定的事件，是最为常用的组件之一。从继承结构上看，Button 也是 TextView 的子类，所以表 4.4 中的属性对于 Button 组件同样适用。图片按钮与普通按钮使用方法基本相同，使用 ImageButton 来定义，在布局文件中添加图片按钮的基本格式如下所示。

```
<ImageButton android:id = "@+id/imageBtn"
    android:layout_width = "wrap_content"
    android:layout_height = "wrap_content"
    android:src = "@drawable/play"
    android:background = "#FF000000"
    android:text = "提交" />
```

背景图片可以通过设置 android:src 属性或调用组件对象的 setImageResource()方法来指定。设置图片按钮的背景色，可使用 android:background 属性指定，但图片按钮中如果未设置背景色，则作为背景的图片会显示在一个灰色的按钮上，即图片按钮带有灰色边框，单击该按钮时按钮会变化；如果设置了背景色，则单击该图片按钮时该按钮不会变化。

下面将通过例 4.3 来介绍如何使用图片按钮组件。

【例 4.3】应用 ImageButton 组件显示播放、暂停按钮效果。

修改主布局文件，将默认的布局管理器改为 LinearLayout 线性布局，垂直排列。在线性布局中添加两个图片按钮，一个未设置背景色，另一个设置背景为黑色，代码如下所示。

```
<LinearLayout xmlns:android = "http://schemas.android.com/apk/res/android"
    android:layout_width = "match_parent"
    android:layout_height = "match_parent"
    android:gravity = "center_horizontal"
    android:orientation = "vertical" >
    <ImageButton
        android:layout_width = "240dp"
        android:layout_height = "wrap_content"
        android:src = "@drawable/play" />
    <ImageButton
        android:layout_width = "240dp"
        android:layout_height = "wrap_content"
        android:background = "#FF000000"
        android:src = "@drawable/pause" />
</LinearLayout>
```

运行该实例，显示结果如图 4.7 所示，其中设置背景色为黑色的第 2 个图片按钮单击时无动态效果。

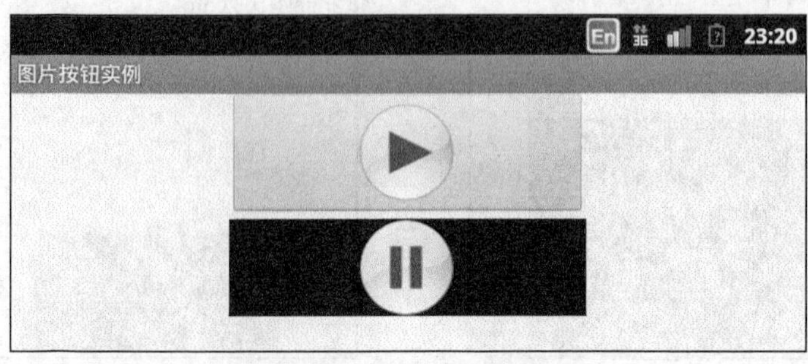

图 4.7 图片按钮实例

4.2.4　图片视图

图片视图用 ImageView 表示，用于在界面上显示图片信息，是最为常用的组件之一。在使用 ImageView 组件显示图片时，一般将待显示的图片放置在工程的 res 目录下的相应 drawable 目录内，通过 android:src 属性指定图片。

ImageView 组件支持的常用 XML 属性如表 4.6 所示。

表 4.6　ImageView 的常用 XML 属性

属 性 名 称	描　述
android:adjustViewBounds	设置 ImageView 是否调整边界来保持图片的长宽比（需要与 maxWidth、MaxHeight 一起使用，否则单独使用没有效果）
android:maxHeight	设置 ImageView 的最大高度，需要设置 android:adjustView-Bounds 为真
android:maxWidth	设置 ImageView 的最大宽度，需要设置 android:adjustView-Bounds 为真
android:scaleType	设置图片如何缩放或移动以适应 ImageView 的大小，属性值为常量：matrix（使用 matrix 方式缩放）、fitXY（不按比例缩放图片，使图片完全适应于此 ImageView 中）、fitStart（按比例缩放图片，使图片完全适应于此 ImageView，并位于 ImageView 左上角）、fitCenter（按比例缩放图片，使图片完全适应于此 ImageView，且显示在 ImageView 的中间）、fitEnd（按比例缩放图片，使图片完全适应于此 ImageView，且显示在 View 的右下角）、center（按原图大小显示图片，但图片宽高大于 ImageView 的宽高时，只显示图片中间部分）、centerCrop（按比例缩放图片，直至图片完全覆盖 ImageView）、centerInside（按比例缩放，直至 ImageView 能完全显示该图片）
android:src	设置图片来源
android:tint	设置图片着色

下面将通过例 4.4 来介绍如何使用图片视图组件。

【例 4.4】　应用 ImageView 组件显示四种位置和缩放效果。

修改主布局文件，将默认的布局管理器改为 LinearLayout 线性布局，水平排列。添加四个 ImageView 组件，代码如下所示。

```
<LinearLayout xmlns:android = "http://schemas.android.com/apk/res/android"
    android:layout_width = "match_parent"
    android:layout_height = "match_parent"
    android:gravity = "center_horizontal"
    android:orientation = "horizontal" >
    <ImageView
        android:maxWidth = "160px"
        android:maxHeight = "160px"
        android:layout_width = "wrap_content"
        android:layout_height = "wrap_content"
        android:adjustViewBounds = "true"
        android:background = "#FF000000"
        android:src = "@drawable/pic"
        android:layout_margin = "2dp" />
    <ImageView
        android:layout_width = "160px"
        android:layout_height = "160px"
```

```
        android:scaleType = "fitStart"
        android:src = "@drawable/pic"
        android:background = "#FF000000"
        android:layout_margin = "2dp" />
    <ImageView
        android:layout_width = "160px"
        android:layout_height = "160px"
        android:scaleType = "center"
        android:src = "@drawable/pic"
        android:background = "#FF000000"
        android:layout_margin = "2dp" />
    <ImageView
        android:layout_width = "160px"
        android:layout_height = "160px"
        android:tint = "#770000FF"
        android:src = "@drawable/pic"
        android:background = "#FF000000"
        android:layout_margin = "2dp" />
</LinearLayout>
```

第 1 个视图组件设置最大宽度和高度，允许自动缩放；第 2 个视图组件设置按比例缩放，缩放后位于组件的左上角；第 3 个视图组件设置按原大小显示图片，超出组件范围时，只显示图片中间部分；第 4 个视图组件为图片着色，设置的是半透明的蓝色。

运行本实例，显示结果如图 4.8 所示。

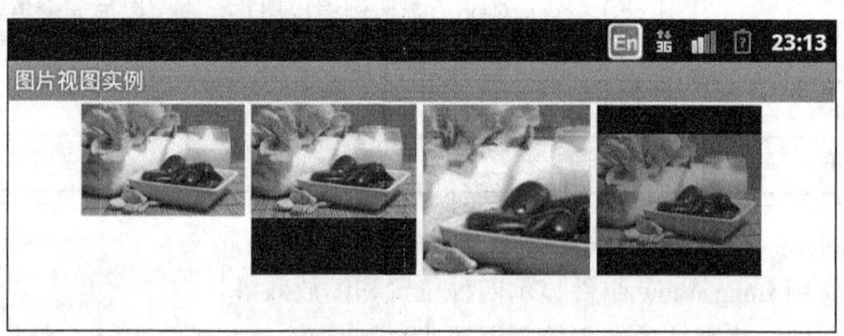

图 4.8 图片视图实例

4.2.5 单选按钮

单选按钮用 RadioButton 表示，是一种双状态的按钮，即选中或不选中。多个单选按钮通常与单选组（RadioGroup）同时使用。当一个 RadioGroup 包含几个单选按钮时，选中其中一个的同时将取消其他选中的单选按钮。由于 RadioGroup 是 Button 的子类，所以 RadioGroup 可以直接使用 Button 支持的属性。

下面将通过例 4.5 来介绍如何使用单选按钮组件。

【例 4.5】 应用 RadioButton 和 RadioGroup 组件选择用户所属的移动网络。

（1）修改主布局文件，将默认的布局管理器改为 LinearLayout 线性布局，垂直排列。在该布局中放置单选按钮组，其中包含 3 个移动网络名称的 3 个单选按钮，最后放置一个提交按钮，代码如下所示。

```
<LinearLayout xmlns:android = "http://schemas.android.com/apk/res/android"
    android:layout_width = "match_parent"
```

```
    android:layout_height = "match_parent"
    android:orientation = "vertical" >
    <TextView
        android:layout_width = "wrap_content"
        android:layout_height = "wrap_content"
        android:text = "请选择您使用的网络" />
    <RadioGroup android:id = "@+id/choosenet"
        android:layout_width = "wrap_content"
        android:layout_height = "wrap_content"
        android:orientation = "vertical" >
        <RadioButton android:id = "@+id/chinamobile"
            android:layout_width = "wrap_content"
            android:layout_height = "wrap_content"
            android:checked = "true"
            android:text = "中国移动" />
        <RadioButton android:id = "@+id/chinaunion"
            android:layout_width = "wrap_content"
            android:layout_height = "wrap_content"
            android:text = "中国联通" />
        <RadioButton android:id = "@+id/chinatelcom"
            android:layout_width = "wrap_content"
            android:layout_height = "wrap_content"
            android:text = "中国电信" />
    </RadioGroup>
    <Button android:id = "@+id/submit"
        android:layout_width = "wrap_content"
        android:layout_height = "wrap_content"
        android:onClick = "choosed"
        android:text = "确定" />
</LinearLayout>
```

（2）主 Activity 文件中提供两种方法获取选中项：一种是确定按钮单击后通过遍历单选按钮组找到被选中项；另一种是给单选按钮组添加选项改变监听器 OnCheckedChangeListener，当选择项改变时回调方法 onCheckedChanged()可获取到单选按钮的 ID，从而找到选中项。代码如下所示。

```
public class MainActivity extends Activity {
    private RadioGroup choosenet;
    private RadioButton chinamobile, chinaunion, chinatelcom;
    @Override
    protected void onCreate(Bundle savedInstanceState) {
        super.onCreate(savedInstanceState);
        setContentView(R.layout.activity_main);
        choosenet = (RadioGroup) findViewById(R.id.choosenet);
        chinamobile = (RadioButton) findViewById(R.id.chinamobile);
        chinaunion = (RadioButton) findViewById(R.id.chinaunion);
        chinatelcom = (RadioButton) findViewById(R.id.chinatelcom);
        //按钮组对象添加监听器
        choosenet.setOnCheckedChangeListener(new RadioGroup.OnCheckedChangeListener() {
            @Override
            public void onCheckedChanged(RadioGroup group, int checkedId) {
                RadioButton r = (RadioButton) findViewById(checkedId);
```

```
                Toast.makeText(MainActivity.this, "您使用的是:" + r.getText(),
                        Toast.LENGTH_LONG).show();
            }
        });
    }
    //按钮单击方法
    public void choosed(View view) {
        for(int i = 0; i < choosenet.getChildCount(); i + +) {
            RadioButton r = (RadioButton) choosenet.getChildAt(i);
            if(r.isChecked()) {
                Toast.makeText(MainActivity.this, "您使用的是:" + r.getText(),
                        Toast.LENGTH_LONG).show();
                break;
            }
        }
    }
}
```

运行本实例，显示结果如图 4.9 所示。

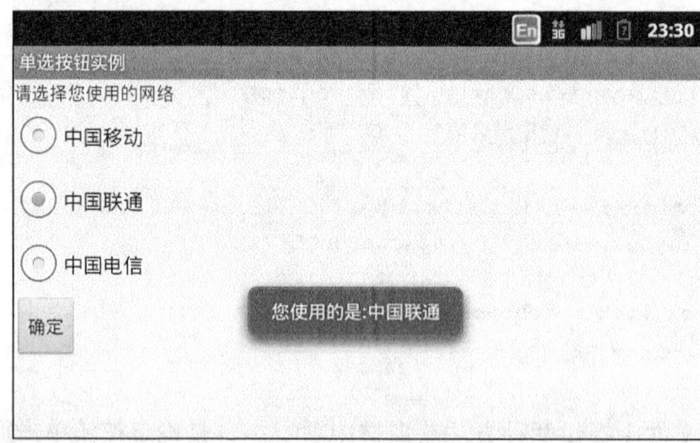

图 4.9　单选按钮实例

4.2.6　复选按钮

与单选按钮不同，复选按钮允许选择多项。每个复选按钮都提供选中或不选中两种状态。Android 中，复选按钮使用 CheckBox 表示，也是 Button 的子类，所以可以直接使用 Button 支持的属性。

下面将通过例 4.6 来介绍如何使用复选按钮组件。

【例 4.6】　应用 CheckBox 组件选择播放器支持的视频种类。

（1）修改主布局文件，将默认的布局管理器改为 LinearLayout 线性布局，垂直排列。在该布局添加 3 个视频格式的复选按钮，代码如下所示。

```
<LinearLayout xmlns:android = "http://schemas.android.com/apk/res/android"
    android:layout_width = "match_parent"
    android:layout_height = "match_parent"
    android:orientation = "vertical" >
    <TextView
        android:layout_width = "wrap_content"
```

```
        android:layout_height = "wrap_content"
        android:text = "请选择播放器支持的格式"
        android:textSize = "20sp" />
    <CheckBox android:id = "@+id/mp4"
        android:layout_width = "wrap_content"
        android:layout_height = "wrap_content"
        android:text = "MP4" />
    <CheckBox android:id = "@+id/mkv"
        android:layout_width = "wrap_content"
        android:layout_height = "wrap_content"
        android:text = "MKV" />
    <CheckBox android:id = "@+id/ts"
        android:layout_width = "wrap_content"
        android:layout_height = "wrap_content"
        android:text = "TS" />
    <Button
        android:layout_width = "wrap_content"
        android:layout_height = "wrap_content"
        android:onClick = "myClick"
        android:text = "确定" />
</LinearLayout>
```

（2）主 Activity 文件中定义 OnCheckedChangeListener 监听器对象，并在 3 个复选按钮中添加此监听器，当选中时显示提示信息。在单击确定按钮后，将所选中的所有格式予以输出。代码如下所示。

```
public class MainActivity extends Activity {
    private CheckBox mp4, mkv, ts;
    private CompoundButton.OnCheckedChangeListener listener =
                        new CompoundButton.OnCheckedChangeListener() {
        @Override
        public void onCheckedChanged(CompoundButton buttonView,
                                boolean isChecked) {
            if(isChecked)
                Toast.makeText(MainActivity.this, "您选中: " +
                            buttonView.getText().toString() + " 格式!",
                            Toast.LENGTH_SHORT).show();
        }
    };
    @Override
    protected void onCreate(Bundle savedInstanceState) {
        super.onCreate(savedInstanceState);
        setContentView(R.layout.activity_main);
        mp4 = (CheckBox) findViewById(R.id.mp4);
        mkv = (CheckBox) findViewById(R.id.mkv);
        ts = (CheckBox) findViewById(R.id.ts);
        //注册监听器
        mp4.setOnCheckedChangeListener(listener);
        mkv.setOnCheckedChangeListener(listener);
        ts.setOnCheckedChangeListener(listener);
    }
    public void myClick(View view) {
```

```
        String choosedItem = "";
        if(mp4.isChecked())
            choosedItem+ = mp4.getText().toString() + ",";
        if(mkv.isChecked())
            choosedItem+ = mkv.getText().toString() + ",";
        if(ts.isChecked())
            choosedItem+ = ts.getText().toString();
        Toast.makeText(MainActivity.this, "支持格式是：" + choosedItem,
                Toast.LENGTH_SHORT).show();
    }
}
```

运行本实例，结果如图 4.10 所示。

图 4.10　复选按钮实例

4.2.7　下拉列表

下拉列表组件用 Spinner 表示，使用该组件用户不需要输入数据，只需选择一个选项后即可完成数据输入工作。

下拉列表的 XML 属性 android:entries 是可选属性，用于指定列表项。如果列表项是已知不再变化的，可将其保存在数组资源文件中，然后通过数组资源来指定。这样，可在不编写 Java 程序的情况下为下拉列表指定列表项。

下面将通过例 4.7 来介绍如何使用下拉列表组件。

【例 4.7】　应用 Spinner 组件选择来电铃声。

（1）修改主布局文件，将默认的布局管理器改为 LinearLayout 线性布局，垂直排列。设置 Spinner 组件时，指定列表项来为数组资源，代码如下所示。

```
<LinearLayout xmlns:android = "http://schemas.android.com/apk/res/android"
    android:layout_width = "match_parent"
    android:layout_height = "match_parent"
    android:orientation = "vertical" >
    <TextView
        android:layout_width = "wrap_content"
        android:layout_height = "wrap_content"
        android:text = "请选择来电铃声"
        android:textSize = "20sp" />
    <Spinner
        android:id = "@+id/spinner"
```

```
        android:layout_width = "wrap_content"
        android:layout_height = "wrap_content"
        android:entries = "@array/ringList" />
    <Button
        android:layout_width = "wrap_content"
        android:layout_height = "wrap_content"
        android:text = "确定"
        android:onClick = "myClick" />
</LinearLayout>
```

（2）编写用于指定列表项的数组资源文件，名称为 arrays.xml，保存于工程的 res/values 目录下，其中添加的字符串数组名为 ringList，代码如下所示。

```
<resources>
    <string-array name = "ringList">
        <item>BirdWispher</item>
        <item>Cuckoo</item>
        <item>Ethereal</item>
        <item>Marimba</item>
    </string-array>
</resources>
```

（3）如果在用户选择不同的列表项后，需要立即执行相应的处理，则可以为该下拉列表添加 OnItemSelectedListener 监听器，并在 onItemSelected()方法中通过 spinner.getItemAtPosition(pos)方法获得选中项，也可以通过下拉列表对象的 getSelectedItem()方法获得选中项，在按钮单击处理中就是使用第 2 种方法。修改主 Activity 文件，代码如下所示。

```
public class MainActivity extends Activity {
    private Spinner spinner;
    protected void onCreate(Bundle savedInstanceState) {
        super.onCreate(savedInstanceState);
        setContentView(R.layout.activity_main);
        spinner = (Spinner) findViewById(R.id.spinner);
        spinner.setOnItemSelectedListener(new OnItemSelectedListener() {
            @Override
            public void onItemSelected(AdapterView<?> parent, View view, int pos,
                                   long id) {
                //通过位置获取下拉列表值
                String result = spinner.getItemAtPosition(pos).toString();
                Toast.makeText(MainActivity.this, result,
                            Toast.LENGTH_LONG).show();
            }
            @Override
            public void onNothingSelected(AdapterView<?> arg0) {}
        });
    }
    public void myClick(View view) {
        //通过选中项获得下拉列表值
        String result = spinner.getSelectedItem().toString();
        Toast.makeText(MainActivity.this, result, Toast.LENGTH_LONG).show();
    }
}
```

运行本实例，结果如图 4.11、图 4.12 所示。

图 4.11　下拉列表图起始界面

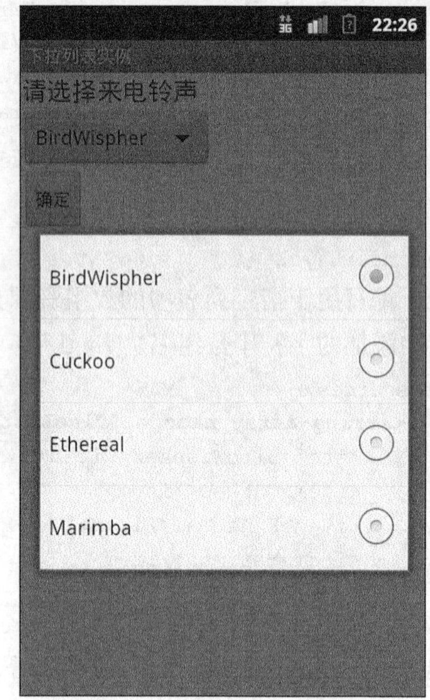

图 4.12　下拉列表单击后的界面

在使用下拉列表时，如果在布局文件中未指定 android:entries 属性，也可以在 Java 代码中通过适配器的方式来指定列表项。为下拉列表指定适配器，通常分为以下 3 个步骤。

（1）创建一个适配器对象，一般使用 ArrayAdapter 类。而在 Android 中，创建适配器对象又可以采用以下两种方法，一种是通过数组资源文件创建，另一种是通过字符串数组创建。

- 通过数组资源文件创建适配器对象

该方法需要使用 ArrayAdapter 类的 createFromResource()方法创建适配器，其中通过参数 android.R.layout.simple_dropdown_item_1line 指定显示样式，通过参数 R.array.ringList 指定资源文件，代码如下所示。

```
ArrayAdapter<CharSequence> adapter = ArrayAdapter.createFromResource(this,
                R.array.ringList, android.R.layout.simple_dropdown_item_1line);
```

- 通过字符串数组创建适配器对象

该方法需要先创建保存列表项的字符串数组，然后通过 ArrayAdapter 的构造方法 ArrayAdapter<String>(Context context,int textViewResourceId,String[] objects)指定数组来创建适配器对象，代码如下所示。

```
String[] ringArrays = { "BirdWispher", "Cuckoo", "Ethereal", "Marimba" };
ArrayAdapter<String> adapter = new ArrayAdapter<String>(this,
                android.R.layout.simple_spinner_item, ringArrays);
```

（2）为适配器设置下拉列表在下拉时的选项样式，代码如下所示。

```
adapter.setDropDownViewResource(android.R.layout.simple_spinner_dropdown_item);
```

（3）将适配器与选择列表关联，代码如下所示。

```
spinner.setAdapter(adapter);
```

可将例 4.7 布局文件中的 android:entries = "@array/ringList"这句删除，然后使用上述通过适配器的方式来指定列表项，再运行程序，功能相同但显示样式略有差异。

4.2.8　自动完成文本框

自动完成文本框用 AutoCompleteTextView 表示，用于当用户输入若干个字符后，能够根据输入字符显示一个下拉菜单，下拉菜单中的选项与用户输入的字符相关，供用户选择。HTML 中是以 AJAX 实现该功能，而 Android 平台则以自动完成文本框实现此功能。

下面将通过例 4.8 来介绍如何使用自动完成文本框组件。

【例 4.8】　应用 AutoCompleteTextView 组件显示搜索提示。

（1）修改主布局文件，将默认的布局管理器改为 LinearLayout 线性布局，水平排列。在该布局中放置一个自动完成文本框组件和一个搜索按钮，代码如下所示。其中语句 android:completion Threshold = "2"设置属性值为 2，表示当输入 2 个字符后开始提示。

```
<LinearLayout xmlns:android = "http://schemas.android.com/apk/res/android"
   android:layout_width = "match_parent"
   android:layout_height = "match_parent"
   android:orientation = "horizontal" >
   <AutoCompleteTextView android:id = "@+id/actv"
      android:layout_width = "wrap_content"
      android:layout_height = "wrap_content"
      android:layout_weight = "7"
      android:completionThreshold = "2"
      android:hint = "请输入搜索关键字"
      android:text = "" />
   <Button
      android:layout_width = "wrap_content"
      android:layout_height = "wrap_content"
      android:layout_weight = "1"
      android:onClick = "myClick"
      android:text = "搜索" />
</LinearLayout>
```

（2）修改主 Activity 文件，建立使用字符串数组的 ArrayAdapter 适配器，将适配器添加到自动完成文本框对象中。代码如下所示。

```
public class MainActivity extends Activity {
   private String[] keys= { "android", "android入门", "android实例",
          "android开发方法", "html", "html技术", "html元素", "html教程" };
   private AutoCompleteTextView actv;
   protected void onCreate(Bundle savedInstanceState) {
      super.onCreate(savedInstanceState);
      setContentView(R.layout.activity_main);
      actv = (AutoCompleteTextView) findViewById(R.id.actv);
      ArrayAdapter<String> adapter = new ArrayAdapter<String>(this,
                          android.R.layout.simple_list_item_1, keys);
      actv.setAdapter(adapter);
   }
   public void myClick(View view) {
      Toast.makeText(MainActivity.this, actv.getText().toString(),
                Toast.LENGTH_LONG).show();
   }
}
```

运行本实例，结果如图 4.13 所示。

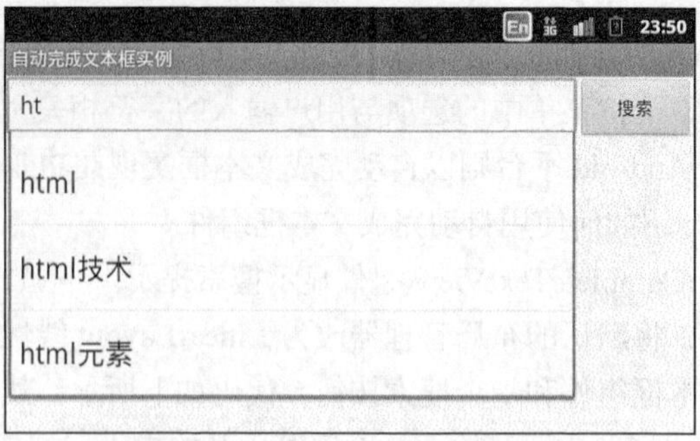

图 4.13　自动完成文本框

4.2.9　日期、时间选择器

Android 平台提供了日期选择器和时间选择器，便于用户对日期时间进行设置。其中日期选择器用 DatePicker 表示，时间选择器用 TimePicker 表示。为了在程序中获取用户选择的日期和时间，需要为 DatePicker 组件和 TimePicker 组件添加事件监听器。DatePicker 组件常用的事件监听器是 OnDateChangedListener，而 TimePicker 组件常用的事件监听器是 OnTimeChangedListener。

下面将通过例 4.9 来介绍如何使用日期、时间选择器组件。

【例 4.9】　应用 DatePicker 和 TimePicker 组件显示自定义时间。

本实例中，使用 DatePicker、TimePicker 和 TextView 3 种组件，其中 TextView 用来显示时间和日期，默认为当前系统的日期和时间，用 DatePicker 和 TimePicker 组件让用户动态调整日期和时间。当用户调整了日期和时间时，TextView 组件值会随着调整而改变。

（1）修改主布局文件，将默认的布局管理器改为 LinearLayout 线性布局，垂直排列。依次放置日期选择器、时间选择器和文本框，代码如下所示。

```
<LinearLayout xmlns:android = "http://schemas.android.com/apk/res/android"
    android:layout_width = "match_parent"
    android:layout_height = "match_parent"
    android:gravity = "center_horizontal"
    android:orientation = "vertical" >
    <DatePicker android:id = "@+id/datePicker"
        android:layout_width = "wrap_content"
        android:layout_height = "wrap_content" />
    <TimePicker android:id = "@+id/timePicker"
        android:layout_width = "wrap_content"
        android:layout_height = "wrap_content" />
    <TextView android:id = "@+id/showTime"
        android:layout_width = "wrap_content"
        android:layout_height = "wrap_content"
        android:textSize = "25sp" />
</LinearLayout>
```

（2）修改主 Activity 文件，首先获取当前时间作为日期和时间选择器的默认值，再对其分别添加 OnDateChangedListener 和 OnTimeChangedListener 事件监听器，当修改日期和时间时更新文本框数据。代码如下所示。

```
public class MainActivity extends Activity {
    private int year, month, date, hour, minute;
```

```
    private TextView showTime;
    @Override
    protected void onCreate(Bundle savedInstanceState) {
        super.onCreate(savedInstanceState);
        setContentView(R.layout.activity_main);
        showTime = (TextView) findViewById(R.id.showTime);
        DatePicker datePicker = (DatePicker) findViewById(R.id.datePicker);
        TimePicker timePicker = (TimePicker) findViewById(R.id.timePicker); //设置24小时制
        timePicker.setIs24HourView(true);
        Calendar cl = Calendar.getInstance();//获取当前的年、月、日、时、分
        year = cl.get(Calendar.YEAR);
        month = cl.get(Calendar.MONTH);
        date = cl.get(Calendar.DAY_OF_MONTH);
        hour = cl.get(Calendar.HOUR_OF_DAY);
        minute = cl.get(Calendar.MINUTE); //设置时间选择器默认为当前时间
        timePicker.setCurrentHour(hour);
        timePicker.setCurrentMinute(minute);
        //在init()方法中初始化日期选择器为当前日期，并添加监听器
        datePicker.init(year, month, date, new OnDateChangedListener() {
            public void onDateChanged(DatePicker dp, int year, int month, int day) {
                MainActivity.this.year = year;
                MainActivity.this.month = month;
                MainActivity.this.date = day;
                show(year, month, date, hour, minute);
            }
        });
        //对时间选择器添加监听器
        timePicker.setOnTimeChangedListener(new OnTimeChangedListener() {
            public void onTimeChanged(TimePicker tp, int hour, int minute) {
                MainActivity.this.hour = hour;
                MainActivity.this.minute = minute;
                show(year, month, date, hour, minute);
            }
        });
    }
    private void show(int year, int month, int day, int hour, int minute) {
        String str = year + "-" + (month + 1) + "-" + day + " " + hour + ":" + minute + " ";
                                                        //在TextView中显示修改过的时间
        showTime.setText(str);
    }
}
```

运行本实例，结果如图 4.14 所示。

图 4.14　设置日期时间

4.3 高级组件

上一节已经介绍了 Android 中基本视图组件的使用，本节将介绍 Android 中的部分高级组件。通过这些组件的应用，能极大地减轻用户开发 UI 界面的难度，提高开发效率。

4.3.1 进度条

进度条用 ProgressBar 表示。当在下载文件、更新数据、安装软件等费时操作时，通过进度条来提示用户执行的进度。进度条通常会和 Handler 联合实现后台进度显示效果。

ProgressBar 支持的常用 XML 属性如表 4.7 所示。

表 4.7 ProgressBar 常用 XML 属性

属 性 名 称	描 述
android:max	设置进度条最大值
android:progress	指定进度条完成的进度值
android:progressDrawable	设置进度条的轨道的绘制形式

除了 XML 属性之外，ProgressBar 还支持以下两个常用方法，用于操作进度。

- setProgress(int progress)：设置进度完成的百分比。
- incrementProgressBy(int diff)：设置进度条的进度增加或减少，当参数为正数时表示进度在递增，当参数为负数时表示进度在递减。

下面将通过例 4.10 来介绍如何使用进度条组件。

【例 4.10】 应用 ProgressBar 组件显示下载进度效果。

（1）修改主布局文件，将默认的布局管理器改为 LinearLayout 线性布局，垂直排列。依次放置显示进度值的文本框，一个水平进度条组件，一个圆形进度条组件，初始时这 2 个进度条通过属性赋值 android:visibility = "gone"均设置隐藏，最后放置一个按钮，代码如下所示。

```
<LinearLayout xmlns:android = "http://schemas.android.com/apk/res/android"
    android:layout_width = "match_parent"
    android:layout_height = "match_parent"
    android:gravity = "center_horizontal"
    android:orientation = "vertical" >
    <TextView android:id = "@+id/pbText"
        android:layout_width = "match_parent"
        android:layout_height = "wrap_content"
        android:textSize = "25sp"
        android:text = "准备下载中..." />
    <ProgressBar android:id = "@+id/pb1"
        style = "@android:style/Widget.ProgressBar.Horizontal"
        android:layout_width = "match_parent"
        android:layout_height = "wrap_content"
        android:layout_marginTop = "20dp"
```

```
        android:max = "100"
        android:visibility = "gone"/>
    <ProgressBar android:id = "@+id/pb2"
        style = "?android:attr/progressBarStyleLarge"
        android:layout_width = "wrap_content"
        android:layout_height = "wrap_content"
        android:layout_marginTop = "20dp"
        android:visibility = "gone"/>
    <Button
        android:layout_width = "wrap_content"
        android:layout_height = "wrap_content"
        android:background = "@drawable/custom_button"
        android:layout_marginTop = "20dp"
        android:onClick = "myClick"
        android:text = "开始下载"/>
</LinearLayout>
```

（2）修改主 Activity 文件，除了定义 XML 布局文件中的组件外，还定义了一个处理消息的 Handler 类对象。在定义 Handler 类对象中，重写了 handleMessage()方法，通过消息 Message 带来的值更新进度以及提示信息。当更新完成，提示“下载完成!”。代码如下所示。

```
public class MainActivity extends Activity {
    private TextView pbText;
    private ProgressBar pb1, pb2; //进度组件
    private Handler myHandler;
    private int pvalue;
    @Override
    protected void onCreate(Bundle savedInstanceState) {
        super.onCreate(savedInstanceState);
        setContentView(R.layout.activity_main);
        pbText = (TextView) findViewById(R.id.pbText);
        pb1 = (ProgressBar) findViewById(R.id.pb1);
        pb2 = (ProgressBar) findViewById(R.id.pb2);
        myHandler = new Handler() {
            @Override
            public void handleMessage(Message msg) {
             pb1.setProgress(msg.arg1); //更新进度值
             pbText.setText("正在下载中...(" + msg.arg1 + "%)"); //更新文本框提示
             if(msg.what == 0x222) {
                 Toast.makeText(MainActivity.this, "下载完成!",
                           Toast.LENGTH_SHORT).show();
             }
            }
        };
    }
}
```

在开始下载的单击事件处理中，首先设置两个进度条的可见性为显示，水平进度条初值设为 0。并建立一个线程，将随机数产生的进度值每 100ms 发送一条消息给主线程，用来更新当前进度值。将如下 myClick()方法放入 MainActivity 类中，代码如下所示。

```
    public void myClick(View view) {
        pb1.setVisibility(View.VISIBLE);
```

```
        pb2.setVisibility(View.VISIBLE);
        pb1.setProgress(0);// 设置初始进度值为 0
        new Thread(new Runnable() {
            @Override
            public void run() {
                pvalue = 0;
                while(true) {
                    pvalue+ = (int) (Math.random() * 10);
                    try {
                        Thread.sleep(100);
                    } catch(Exception e) {
                        e.printStackTrace();
                    }
                    Message mes = new Message();
                    if(pvalue < 100) {
                        mes.arg1 = pvalue;
                        mes.what = 0x111;
                        myHandler.sendMessage(mes);
                    } else {
                        mes.arg1 = 100;
                        mes.what = 0x222;
                        myHandler.sendMessage(mes);
                        break;
                    }
                }
            }
        }).start();
    }
```

运行本实例，单击开始下载之前的结果如图 4.15 所示，单击开始下载之后，进度值改变的结果如图 4.16 所示。

图 4.15　进度条开始之前

图 4.16　进度条开始

4.3.2　拖动条

拖动条用 SeekBar 表示，与进度条类似，只是增加了一个可拖动的滑块。用户可以通过拖动来改变值，通常用于对数值的调节，例如调整音量、亮度等。

拖动条中 android:thumb 属性可以改变拖动滑块的外观，该属性的属性值为一个 Drawable 对象，即用户可选择相应图片作为滑块。另外在用户通过移动滑块改变拖动条值的过程中，如果需要执行相应处理的话，可对拖动条添加监听器 OnSeekBarChangeListener，重写该监听器的开始拖动滑块时的 onStartTracking Touch 方法、拖动过程中进度值改变时的 onProgressChanged 方法和停止拖动滑块时的 onStopTrackingTouch 方法。

下面将通过例 4.11 来介绍如何使用拖动条组件。

【例 4.11】 应用 SeekBar 组件完成手动调整进度效果。

（1）修改主布局文件，将默认的布局管理器改为 LinearLayout 线性布局，垂直排列。依次放置显示进度值的文本框和一个拖动条组件，在拖动条中设置自定义滑块。其中 android:max = "100" 表示拖动条最大值，android:progress = "20"表示当前值，代码如下所示。

```
<LinearLayout xmlns:android = "http://schemas.android.com/apk/res/android"
    android:layout_width = "match_parent"
    android:layout_height = "match_parent"
    android:gravity = "center_horizontal"
    android:orientation = "vertical" >
    <TextView android:id = "@+id/seekBarValue"
        android:layout_width = "match_parent"
        android:layout_height = "wrap_content"
        android:textSize = "20sp"
        android:text = "拖动条状态" />
    <SeekBar android:id = "@+id/seekBar"
        android:layout_width = "match_parent"
        android:layout_height = "wrap_content"
        android:padding = "10dp"
        android:thumb = "@drawable/icon"
        android:max = "100"
        android:progress = "20" />
</LinearLayout>
```

（2）修改主 Activity 文件，对拖动条增加 OnSeekBarChangeListener 监听器，并在拖动开始、拖动过程中、拖动结束时修改文本框值，代码如下所示。

```
public class MainActivity extends Activity {
    private SeekBar seekBar;
    private TextView seekBarValue;
    @Override
    protected void onCreate(Bundle savedInstanceState) {
        super.onCreate(savedInstanceState);
        setContentView(R.layout.activity_main);
        seekBar = (SeekBar) findViewById(R.id.seekBar);
        seekBarValue = (TextView) findViewById(R.id.seekBarValue);
        seekBar.setOnSeekBarChangeListener(new
                        SeekBar.OnSeekBarChangeListener() {
            @Override
            public void onStopTrackingTouch(SeekBar seekBar) {
                seekBarValue.setText("停止拖动滑块");
            }
            @Override
            public void onStartTrackingTouch(SeekBar seekBar) {
```

```
                seekBarValue.setText("开始拖动滑块");
            }
            @Override
            public void onProgressChanged(SeekBar seekBar, int progress,
                    boolean fromUser) {
                seekBarValue.setText("当前拖动条的值是:" + progress);
            }
        });
    }
}
```

运行本实例，拖动中的结果如图 4.17 所示，拖动停止后的结果如图 4.18 所示。

图 4.17　拖动中的结果

图 4.18　拖动停止后的结果

4.3.3　评分条

评分条用 RatingBar 表示，与拖动条类似，都通过用户拖动来改变进度，区别在于评分条是以星表示进度。一般情况下，评分条用于评价对某一事物的支持度或对某种事物的满意程度等。

RatingBar 支持的常用 XML 属性如表 4.8 所示。

表 4.8　RatingBar 常用 XML 属性

属 性 名 称	描　述
android:isIndicator	设置该评分条是否允许用户修改，true(不允许)；false(允许)
android:numStars	设置该评分条共有多少个星
android:rating	设置该评分条默认星级
android:stepSize	设置每次最少晋级单位，默认 0.5 个

除了 XML 属性之外，RatingBar 还支持以下 3 个常用方法。

- getRating()：获取等级，表示选中了几颗星。
- getStepSize()：获取晋级单位。
- getProgress()：获取当前星级

下面将通过例 4.12 来介绍如何使用评分条组件。

【例 4.12】　应用 RatingBar 组件完成满意度调查。

（1）修改主布局文件，将默认的布局管理器改为 LinearLayout 线性布局，垂直排列。依次放置显示提示信息的文本框、一个评分条组件和一个提交按钮。代码如下所示。

```
<LinearLayout xmlns:android = "http://schemas.android.com/apk/res/android"
```

```
    android:layout_width = "match_parent"
    android:layout_height = "match_parent"
    android:gravity = "center_horizontal"
    android:orientation = "vertical" >
    <TextView android:id = "@+id/ratingBarValue"
        android:layout_width = "match_parent"
        android:layout_height = "wrap_content"
        android:textSize = "20sp"
        android:text = "请您对本应用打分:" />
    <RatingBar android:id = "@+id/ratingBar"
        android:layout_width = "wrap_content"
        android:layout_height = "wrap_content"
        android:isIndicator = "false"
        android:numStars = "5"
        android:rating = "3"
        android:stepSize = "0.5" />
    <Button android:id = "@+id/ratingBar"
        android:layout_width = "wrap_content"
        android:layout_height = "wrap_content"
        android:onClick = "myClick"
        android:text = "提交" />
</LinearLayout>
```

（2）修改主 Activity 文件，代码如下所示。

```
public class MainActivity extends Activity {
    private RatingBar ratingBar;
    @Override
    protected void onCreate(Bundle savedInstanceState) {
        super.onCreate(savedInstanceState);
        setContentView(R.layout.activity_main);
        ratingBar = (RatingBar) findViewById(R.id.ratingBar);
    }
    public void myClick(View view) {
        Toast.makeText(this, "您给了" + ratingBar.getRating() +
                "颗星,\n感谢您的参与!", Toast.LENGTH_LONG).show();
    }
}
```

运行本实例，结果如图 4.19 所示。

图 4.19　评分条

4.3.4 选项卡

选项卡用 TabWidget 表示，实现一个多标签页的用户界面。用户单击不同的选项卡，可以进行视图的切换，选项卡中包含的每个视图都是一个 Activity，因此开发过程中对每个选项卡都要实现相对应的 Activity。

Android 中使用选项卡功能，需要由 TabHost、TabWidget 和 FrameLayout 3 个组件共同完成。选项卡能实现对信息的分类显示和管理，有效地减少窗体的个数，开发过程一般需要以下几个步骤。

（1）在布局文件中添加 TabHost、TabWidget 和 FrameLayout 3 个组件。

（2）编写各选项卡页要显示内容的 XML 布局文件。

（3）在 Activity 中，获取并初始化 TabHost 组件。

（4）为 TabHost 对象添加选项卡项。

（5）将所有 Activity 注册到 AndroidManifest.xml 文件中。

下面将通过例 4.13 来介绍如何使用选项卡组件。

【例 4.13】 应用 TabHost、TabWidget 和 FrameLayout 组件完成用户登录、注册界面设计。

（1）修改主布局文件，将默认的布局内容删除。首先添加 TabHost 组件，然后在该组件内添加线性布局管理器，并在线性布局管理器中按纵向依次放入 TabWidget 组件和 FrameLayout 布局管理器，代码如下所示。

```
<?xml version = "1.0" encoding = "utf-8"?>
<TabHost xmlns:android = "http://schemas.android.com/apk/res/android"
    android:id = "@android:id/tabhost"
    android:layout_width = "match_parent"
    android:layout_height = "match_parent" >
    <LinearLayout
        android:layout_width = "match_parent"
        android:layout_height = "match_parent"
        android:orientation = "vertical">
        <TabWidget
            android:id = "@android:id/tabs"
            android:layout_width = "match_parent"
            android:layout_height = "wrap_content" />
        <FrameLayout
            android:id = "@android:id/tabcontent"
            android:layout_width = "match_parent"
            android:layout_height = "match_parent" />
    </LinearLayout>
</TabHost>
```

上述代码中 TabHost 组件的 ID 必须为“@android:id/tabhost”，TabWidget 组件的 ID 必须为“@android:id/tabs”，FrameLayout 布局管理器的 ID 必须为“@android:id/-tabcontent”，不能设为“@+id/tabhost”等。

（2）新建登录布局文件 logintab.xml，代码如下所示。

```
<?xml version = "1.0" encoding = "utf-8"?>
<LinearLayout xmlns:android = "http://schemas.android.com/apk/res/android"
```

```
    android:id = "@+id/loginTab"
    android:layout_width = "match_parent"
    android:layout_height = "match_parent"
    android:orientation = "vertical" >
    <EditText
        android:layout_width = "match_parent"
        android:layout_height = "wrap_content"
        android:hint = "请输入用户名" />
    <EditText
        android:layout_width = "match_parent"
        android:layout_height = "wrap_content"
        android:inputType = "textPassword"
        android:hint = "请输入密码" />
    <CheckBox
        android:layout_width = "wrap_content"
        android:layout_height = "wrap_content"
        android:checked = "true"
        android:text = "记住密码" />
    <Button
        android:layout_width = "match_parent"
        android:layout_height = "wrap_content"
        android:background = "@drawable/custom_button"
        android:text = "登录"
        android:onClick = "myclick" />
</LinearLayout>
```

（3）新建注册布局文件 registertab.xml，代码如下所示。

```
<?xml version = "1.0" encoding = "utf-8"?>
<LinearLayout xmlns:android = "http://schemas.android.com/apk/res/android"
    android:id = "@+id/registerTab"
    android:layout_width = "match_parent"
    android:layout_height = "match_parent"
    android:orientation = "vertical" >
    <EditText
        android:layout_width = "match_parent"
        android:layout_height = "wrap_content"
        android:hint = "请输入用户名" />
    <EditText
        android:layout_width = "match_parent"
        android:layout_height = "wrap_content"
        android:inputType = "textPassword"
        android:hint = "请输入密码" />
    <TextView
        android:layout_width = "match_parent"
        android:layout_height = "wrap_content"
        android:text = "请选择性别"
        android:textSize = "20sp" />
    <RadioGroup android:id = "@+id/sex"
        android:layout_width = "wrap_content"
        android:layout_height = "wrap_content"
        android:orientation = "horizontal" >
```

```
        <RadioButton android:id = "@+id/male"
            android:layout_width = "wrap_content"
            android:layout_height = "wrap_content"
            android:checked = "true"
            android:text = "男" />
        <RadioButton android:id = "@+id/female"
            android:layout_width = "wrap_content"
            android:layout_height = "wrap_content"
            android:text = "女" />
    </RadioGroup>
    <EditText
        android:layout_width = "match_parent"
        android:layout_height = "wrap_content"
        android:hint = "请输入联系电话"
        android:inputType = "phone" />
    <Button
        android:layout_width = "match_parent"
        android:layout_height = "wrap_content"
        android:background = "@drawable/custom_button"
        android:text = "注册"
        android:onClick = "myclick" />
</LinearLayout>
```

（4）修改主 Activity 文件，使 MainActivity 继承 TabActivity 类。通过 TabHost 的 addTab()方法增加选项卡名、显示选项卡名和设置内容 Intent，该内容 Intent 为登录 Activity 和注册 Activity。如果需要在选中选项卡时进行处理，则可以为 TabHost 添加 OnTabChangeListener 监听器。代码如下所示。

```
public class MainActivity extends TabActivity {
    private TabHost tabHost;
    @Override
    protected void onCreate(Bundle savedInstanceState) {
        super.onCreate(savedInstanceState);
        setContentView(R.layout.activity_main);
        tabHost = getTabHost();
        Intent loginIntent = new Intent(this, LoginTab.class);
        tabHost.addTab(tabHost.newTabSpec("tab1").setIndicator("登录").setContent (loginI
                ntent));
        Intent registerIntent = new Intent(this, RegisterTab.class);
        tabHost.addTab(tabHost.newTabSpec("tab2").setIndicator(" 注 册").setContent(register
                Intent));
        //当前显示第二个选项卡(通过编号)
        // mTabHost.setCurrentTab(1);
        //当前显示第二个选项卡(通过选项卡名)
        tabHost.setCurrentTabByTag("tab2");
        // 选项卡切换事件处理
        tabHost.setOnTabChangedListener(new OnTabChangeListener() {
            @Override
            public void onTabChanged(String tabId) {
                Toast.makeText(MainActivity.this, "当前选中: " + tabId + "标签",
                        Toast.LENGTH_LONG).show();
```

```
        }});
    }
}
```

（5）创建登录选项卡 Activity 文件，代码如下所示。

```
public class LoginTab extends Activity {
    @Override
    protected void onCreate(Bundle savedInstanceState) {
        super.onCreate(savedInstanceState);
        setContentView(R.layout.logintab);
    }
    public void myclick(View view) {
        // 登录处理
        Toast.makeText(this, "登录中...", Toast.LENGTH_LONG).show();
    }
}
```

（6）创建注册选项卡 Activity 文件，代码如下所示。

```
public class RegisterTab extends Activity {
    @Override
    protected void onCreate(Bundle savedInstanceState) {
        super.onCreate(savedInstanceState);
        setContentView(R.layout.registertab);
    }
    public void myclick(View view) {
        // 注册处理
        Toast.makeText(this, "注册中...", Toast.LENGTH_LONG).show();
    }
}
```

（7）在 AndroidManifest.xml 文件中，在 application 标签内添加登录选项卡 Activity 和注册选项卡 Activity 项，代码如下所示。

```
<?xml version = "1.0" encoding = "utf-8"?>
<manifest xmlns:android = "http://schemas.android.com/apk/res/android"
    package = "com.example.mytab"
    android:versionCode = "1"
    android:versionName = "1.0" >
    <uses-sdk
        android:minSdkVersion = "8"
        android:targetSdkVersion = "17" />
    <application
        android:allowBackup = "true"
        android:icon = "@drawable/ic_launcher"
        android:label = "@string/app_name"
        android:theme = "@style/AppTheme" >
        <activity
            android:name = "com.example.mytab.MainActivity"
            android:label = "@string/app_name" >
            <intent-filter>
                <action android:name = "android.intent.action.MAIN" />
                <category android:name = "android.intent.category.LAUNCHER" />
            </intent-filter>
        </activity>
```

```
        <!-- 增加两个 activity 项 -->
        <activity android:name = "com.example.mytab.LoginTab" />
        <activity android:name = "com.example.mytab.RegisterTab" />
    </application>
</manifest>
```

运行本实例，结果如图 4.20、图 4.21 所示。

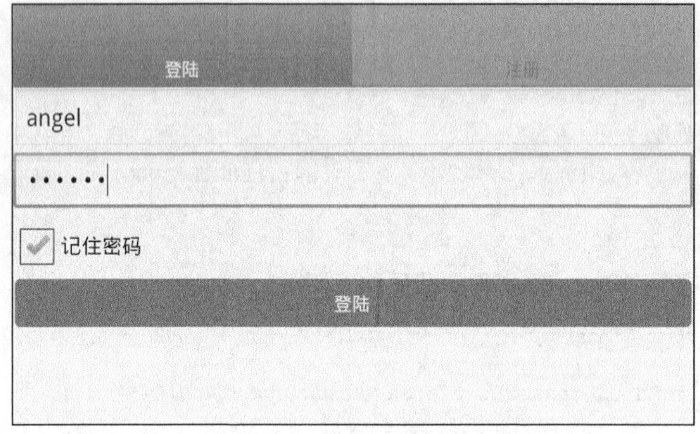

图 4.20　登录选项卡效果

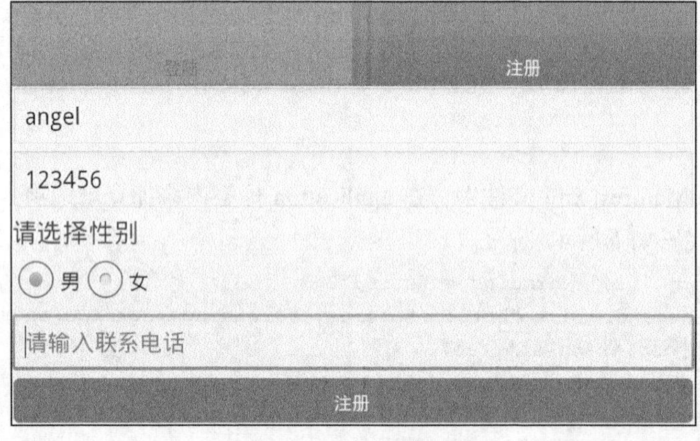

图 4.21　注册选项卡效果

4.4　提示框与警告对话框

前面各章节例子中，已经使用了 Toast 类显示一个简单的消息提示框。本节将对 Android 中用于消息提示的 Toast 类和 AlertDialog 类分别予以详细的介绍。

4.4.1　消息提示框

消息提示框用 Toast 类来完成，使用此方法实现的消息提示没有任何控制按钮和焦点，显示后间隔一段时间自动消失，常用于显示不需要用户确认操作的通知。

使用 Toast 显示消息提示，一般需要通过以下 3 个步骤。

（1）创建 Toast 对象，可以通过构造方法创建，也可以通过调用静态方法 makeText()来创建。

（2）调用 Toast 类的相应方法来设置消息提示框的对齐方式、页边距、显示内容等，常用方法如表 4.9 所示。

表 4.9　Toast 类常用方法

方法名称	描述
makeText(Context context, int resId, int duration)	创建一个 Toast 对象，提示文本从资源 id 获取，duration 为提示框显示时间常量
makeText(Context context, CharSequence text, int duration)	创建一个 Toast 对象，text 为提示文本
setDuration(int duration)	设置消息提示框的显示时间，取值为： Toast.LENGTH_SHORT 或 Toast.LENGTH_LONG
setGravity(int gravity, int xOffset, int yOffset)	设置消息提示框的位置，gravity 用于指定对齐方式，xOffset，和 yOffset 用于指定偏移值
setMargin(float horizontalMargin, float verticalMargin)	设置消息提示框的页边距
setText(CharSequence s)	设置消息提示框的文本信息
setView(View view)	设置消息提示框中显示的视图

（3）调用 Toast 类的 show()方法，显示消息提示框。

下面将通过例 4.14 来介绍如何使用 Toast 类来显示提示信息。

【例 4.14】　应用 Toast 组件显示 3 种提示框。

（1）修改主布局文件，将默认的布局管理器改为 LinearLayout 线性布局，垂直排列。依次添加 3 个按钮，代码如下所示。

```
<LinearLayout xmlns:android = "http://schemas.android.com/apk/res/android"
    android:layout_width = "match_parent"
    android:layout_height = "match_parent"
    android:background = "#FFFAE4"
    android:orientation = "vertical" >
    <Button
        android:layout_width = "match_parent"
        android:layout_height = "wrap_content"
        android:onClick = "click1"
        android:text = "默认提示框" />
    <Button
        android:layout_width = "match_parent"
        android:layout_height = "wrap_content"
        android:onClick = "click2"
        android:text = "带图标提示框" />
    <Button
        android:layout_width = "match_parent"
        android:layout_height = "wrap_content"
        android:onClick = "click3"
        android:text = "自定义提示框" />
</LinearLayout>
```

（2）建立布局文件 custom.xml 文件，该布局创建自定义提示框样式，代码如下所示。

```
<LinearLayout xmlns:android = "http://schemas.android.com/apk/res/android"
    android:id = "@+id/customLayout"
    android:layout_width = "wrap_content"
    android:layout_height = "wrap_content"
    android:background = "#FFFFFFFF"
    android:orientation = "vertical" >
    <TextView
        android:layout_width = "match_parent"
        android:layout_height = "wrap_content"
        android:layout_margin = "1dip"
        android:background = "#BB000000"
        android:gravity = "center"
        android:textColor = "#FFFFFFFF"
        android:text = "注意!"/>
    <LinearLayout
        android:layout_width = "wrap_content"
        android:layout_height = "wrap_content"
        android:layout_marginBottom = "1dip"
        android:layout_marginLeft = "1dip"
        android:layout_marginRight = "1dip"
        android:background = "#55000000"
        android:orientation = "vertical"
        android:padding = "15dip" >
        <ImageView
            android:layout_width = "wrap_content"
            android:layout_height = "wrap_content"
            android:src = "@drawable/send"
            android:layout_gravity = "center" />
        <TextView
            android:layout_width = "wrap_content"
            android:layout_height = "wrap_content"
            android:layout_gravity = "center_horizontal"
            android:padding = "5dip"
            android:textColor = "#FF000000"
            android:text = "自定义提示框"/>
    </LinearLayout>
</LinearLayout>
```

（3）修改主 Activity 文件，设置每个按钮触发后显示一种提示框样式，代码如下所示。

```
public class MainActivity extends Activity {
    @Override
    protected void onCreate(Bundle savedInstanceState) {
        super.onCreate(savedInstanceState);
        setContentView(R.layout.activity_main);
    }
    //建立默认样式的提示框
    public void click1(View view) {
        Toast.makeText(this, "默认提示框", Toast.LENGTH_LONG).show();
    }
```

```
    //建立带图标的提示框
    public void click2(View view) {
        Toast toast = new Toast(this);
        toast.setDuration(Toast.LENGTH_LONG);
        toast.setGravity(Gravity.CENTER, 0, 0);
        LinearLayout llayout = new LinearLayout(this);
        ImageView image = new ImageView(this);
        image.setImageResource(R.drawable.down);
        TextView text = new TextView(this);
        text.setText("带图标的提示框");
        llayout.addView(image);
        llayout.addView(text);
        toast.setView(llayout);
        toast.show();
    }
    //建立自定义样式的提示框
    public void click3(View view) {
        LayoutInflater inflater = getLayoutInflater();
        View layout = inflater.inflate(R.layout.custom,
                (ViewGroup) findViewById(R.id.customLayout ));
        Toast toast = new Toast(getApplicationContext());
        toast.setGravity(Gravity.RIGHT | Gravity.TOP, 50, 50);
        toast.setDuration(Toast.LENGTH_LONG);
        toast.setView(layout);
        toast.show();
    }
}
```

运行本实例，结果如图 4.22、图 4.23 和图 4.24 所示。

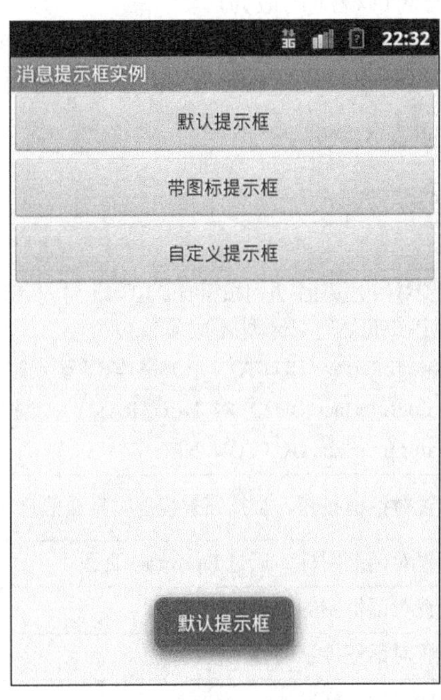

图 4.22　默认文本提示框

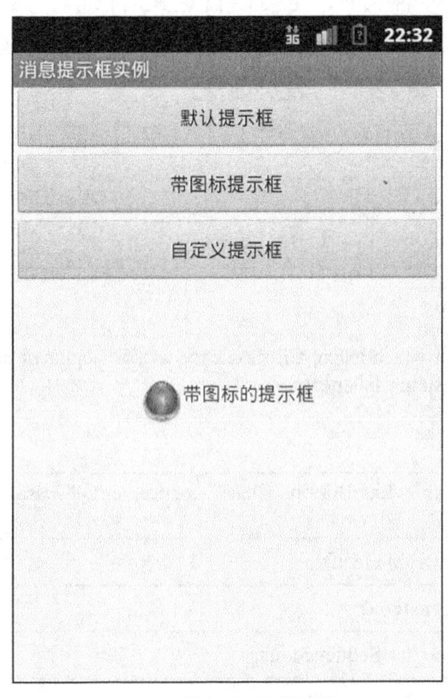

图 4.23　带图标的提示框

图 4.24　自定提示框

4.4.2　警告对话框

除了界面组件外，对话框（Dialog）也提供了丰富的交互功能。与界面组件不同，对话框必须依存于界面组件中，一般表现形式为出现在当前 Activity 之上的一个小窗口，提供用户选择或提示确认等与当前界面相关的功能。常用的对话框有警告对话框、进度对话框、日期选择对话框、时间选择对话框等，也可以自定义对话框，其中警告对话框应用最为广泛。

警告对话框使用 AlertDialog 类，可以生成带 1~3 个按钮的提示对话框，也可以生成带单选列表或多选列表的列表对话框。

使用 AlertDialog 类生成对话框时，常用的方法如表 4.10 所示。

表 4.10　　AlertDialog 类常用方法

方法名称	描述
setButton(int whichButton, CharSequence text, DialogInterface.OnClickListener listener)	设置对话框按钮，并对按钮增加监听器，其中 whichButton 指定按钮类型（int 型值），取值为： DialogInterface.BUTTON_POSITIVE（确定按钮） DialogInterface.BUTTON_NEUTRAL（中立按钮） DialogInterface.BUTTON_NEGATIVE（取消按钮）
setButton(int whichButton, CharSequence text, Message msg)	设置对话框按钮，当按钮被按下时发送消息
setIcon(Drawable icon)	设置对话框图标，通过 Drawable 资源
setIcon(int resIcon)	设置对话框图标，通过资源 ID
setMessage(CharSequence msg)	设置对话框提示内容
setTitle(CharSequence title)	设置对话框标题
show()	显示对话框

通常情况下，使用 AlertDialog 类只能生成带按钮的提示对话框，如果要生成带列表框的对话框，则需要使用 AlertDialog.Builder 类。AlertDialog.Builder 类的常用方法如下。

- create()：创建 AlertDialog 对话框类。
- setNegativeButton(CharSequence text, DialogInterface.OnClickListener listener)：在对话框中添加取消按钮，并添加监听器。
- setNeutralButton(CharSequence text, DialogInterface.OnClickListener listener)：在对话框中添加中立按钮，并添加监听器。
- setPositiveButton(CharSequence text, DialogInterface.OnClickListener listener)：在对话框中添加确定按钮，并添加监听器。
- setItems(CharSequence[] items, DialogInterface.OnClickListener listener)：在对话框中添加列表，列表项来源于字符串数组，并添加监听器。
- setMultiChoiceItems(CharSequence[] items, boolean[] checkedItems, DialogInterface.OnMultiChoiceClickListener listener)：在对话框中添加多选列表，列表项来源于字符串数组，并添加监听器。
- setSingleChoiceItems(CharSequence[] items, int checkedItem, DialogInterface.OnClickListener listener)：在对话框中添加单选列表，列表项来源于字符串数组，并添加监听器。
- setSingleChoiceItems(ListAdapter adapter, int checkedItem, DialogInterface.OnClickListener listener)：在对话框中添加单选列表，列表项来源于适配器，并添加监听器。

下面将通过例 4.15 来介绍如何使用警告对话框类。

【例 4.15】　应用 AlertDialog 类和 AlertDialog.Builder 类显示按钮对话框、列表对话框、单选对话框、多选对话框和带图标列表对话框。

（1）修改主布局文件，将默认的布局内容删除。添加线性布局管理器，在该布局管理器中依次添加 5 个按钮，并设置单击方法名为 myclick1 到 myclick5，分别打开不同类型的对话框，代码如下所示，界面如图 4.25 所示。

图 4.25　对话框应用界面

```
<LinearLayout xmlns:android = "http://schemas.android.com/apk/res/android"
   android:layout_width = "match_parent"
   android:layout_height = "match_parent"
   android:background = "#FFFAE4"
   android:orientation = "vertical" >
   <Button
      android:layout_width = "match_parent"
      android:layout_height = "wrap_content"
      android:layout_marginTop = "10dp"
      android:onClick = "myclick1"
      android:text = "按钮对话框" />
   <Button
      android:layout_width = "match_parent"
      android:layout_height = "wrap_content"
      android:onClick = "myclick2"
      android:text = "列表对话框" />
   <Button
```

```
        android:layout_width = "match_parent"
        android:layout_height = "wrap_content"
        android:onClick = "myclick3"
        android:text = "单选对话框" />
    <Button
        android:layout_width = "match_parent"
        android:layout_height = "wrap_content"
        android:onClick = "myclick4"
        android:text = "多选对话框" />
    <Button
        android:layout_width = "match_parent"
        android:layout_height = "wrap_content"
        android:onClick = "myclick5"
        android:text = "带图标列表对话框" />
</LinearLayout>
```

（2）修改主 Activity 文件，添加字符串数组成员变量，代码如下所示。

```
private CharSequence[] items= { "百度", "谷歌", "必应" };
```

（3）在主 Activity 文件中，添加"按钮对话框"单击方法 myclick1()，打开一个包含确定、中立、取消三个按钮的对话框。该方法中通过 AlertDialog.Builder 类的 create()方法创建 AlertDialog 对象，再通过 AlertDialog 的 setButton()方法添加相应按钮和按钮的监听器，代码如下所示。

```
public void myclick1(View view) { //按钮对话框单击方法
    AlertDialog alert = new AlertDialog.Builder(this).create();
    alert.setTitle("多选提示框应用");
    alert.setMessage("请选择一个按钮!");
    alert.setButton(DialogInterface.BUTTON_POSITIVE, "确定", new DialogInterface.
                OnClickListener() {
        @Override
        public void onClick(DialogInterface dialog, int which) {
            Toast.makeText(getApplicationContext(), "确定被点击",
                    Toast.LENGTH_LONG).show();
        }
    });
    alert.setButton(DialogInterface.BUTTON_NEUTRAL, "中立", new DialogInterface.
                OnClickListener() {
        @Override
        public void onClick(DialogInterface dialog, int which) {
            Toast.makeText(getApplicationContext(), "中立被点击",
                    Toast.LENGTH_LONG).show();
        }
    });
    alert.setButton(DialogInterface.BUTTON_NEGATIVE, "取消", new DialogInterface.
                OnClickListener() {
        @Override
        public void onClick(DialogInterface dialog, int which) {
            Toast.makeText(getApplicationContext(), "取消被点击",
                    Toast.LENGTH_LONG).show();
        }
    });
    alert.show();
}
```

单击“按钮对话框”，显示包含 3 个按钮的对话框，如图 4.26 所示。

（4）在主 Activity 文件中，再添加“列表对话框”单击方法 myclick2()，打开一个包含百度、谷歌、必应 3 个选项的列表对话框。在该方法中首先创建 AlertDialog.Builder 类对象，设置该对象的图标、标题和使用 items 字符串数组作为来源的列表项，并添加相应监听器，代码如下所示。

图 4.26　多按钮对话框

在设置完 AlertDialog.Builder 对象各项后，一定要调用 create()方法建立 AlertDialog 对象，再调用 show()方法显示该对话框。

```
public void myclick2(View view) {  //列表对话框单击方法
    Builder builder = new AlertDialog.Builder(MainActivity.this);
    builder.setIcon(R.drawable.search);
    builder.setTitle("列表对话框");
    builder.setItems(items, new DialogInterface.OnClickListener() {
        @Override
        public void onClick(DialogInterface dialog, int which) {
            Toast.makeText(getApplicationContext(), items[which],
                    Toast.LENGTH_LONG).show();
        }
    });
    builder.create().show();
}
```

单击“列表对话框”按钮后，显示的对话框如图 4.27 所示。

（5）在主 Activity 文件中，再添加“单选对话框”单击方法 myclick3()，打开一个包含百度、谷歌、必应 3 个单选选项的列表对话框。在该方法中首先创建 AlertDialog.Builder 类对象，通过调用 setSingleChoiceItems()方法，使用 items 字符串数组作为单选列表项来源，最后添加“确定”按钮，代码如下所示。

```
public void myclick3(View view) {  //单选对话框单击方法
    Builder builder = new AlertDialog.Builder(MainActivity.this);
    builder.setIcon(R.drawable.search);
    builder.setTitle("单选对话框");
    builder.setSingleChoiceItems(items, 0, new DialogInterface.OnClickListener() {
        @Override
        public void onClick(DialogInterface dialog, int which) {
            Toast.makeText(getApplicationContext(), items[which] + "被选中了!",
                    Toast.LENGTH_LONG).show();
        }
    });
    builder.setPositiveButton("确定", null);
    builder.create().show();
}
```

单击“单选对话框”按钮后，显示的对话框如图 4.28 所示。

图 4.27 列表对话框

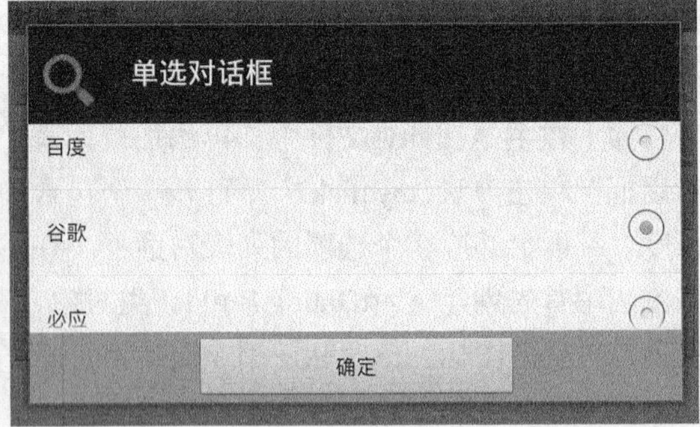

图 4.28 单选对话框

（6）在主 Activity 文件中，再添加“多选对话框”单击方法 myclick4()，打开一个包含百度、谷歌、必应 3 个复选选项的列表对话框。在该方法中首先创建 AlertDialog.Builder 类对象，通过调用 setMultiChoiceItems()方法，使用 items 字符串数组作为复选列表项来源，最后添加确定和取消按钮，代码如下所示。

```
//多选对话框单击方法
public void myclick4(View view) {
    boolean[] itemsChecked = new boolean[items.length];
    Builder builder = new AlertDialog.Builder(MainActivity.this);
    builder.setIcon(R.drawable.search);
    builder.setTitle("多选对话框");
    builder.setMultiChoiceItems(items, itemsChecked, new DialogInterface.OnMultiChoice
                              ClickListener() {
        @Override
        public void onClick(DialogInterface dialog, int which, boolean isChecked) {
            Toast.makeText(getApplicationContext(), items[which] + "被选中了!", Toast.
                      LENGTH_LONG).show();
        }
    });
    builder.setPositiveButton("确定", null);
    builder.setNegativeButton("取消", null);
    builder.create().show();
}
```

单击“多选对话框”按钮后，显示的对话框如图 4.29 所示。

（7）创建带图标列表对话框时，对话框中显示图标和文字布局的样式需要在布局文件中定义。创建布局文件 items.xml，该文件采用水平线性布局管理器，并在该布局管理器中添加一个 ImageView 组件和一个 TextView 组件，分别用于表示显示列表项的图标和文字，代码如下所示。

```
<LinearLayout xmlns:android = "http://schemas.android.com/apk/res/android"
  android:layout_width = "match_parent"
  android:layout_height = "match_parent"
  android:orientation = "horizontal">
  <ImageView android:id = "@+id/image"
     android:padding = "15dp"
     android:adjustViewBounds = "true"
     android:maxWidth = "216px"
     android:maxHeight = "202px"
     android:layout_gravity = "center"
     android:layout_width = "wrap_content"
     android:layout_height = "wrap_content" />
```

```
    <TextView android:id = "@+id/title"
        android:layout_width = "wrap_content"
        android:layout_height = "wrap_content"
        android:layout_gravity = "center"
        android:padding = "15dp"
        android:textSize = "20sp" />
</LinearLayout>
```

在主 Activity 文件中，再添加“带图标列表对话框”单击方法 myclick5()。在该方法中首先创建用于保存列表项中图片 ID 和文字的数组，并将这些图片 ID 和文字保存到 List 集合中，然后再建立一个 SimpleAdapter 适配器将列表项和 List 集合关联。代码如下所示。

```
public void myclick5(View view) { //带图标列表对话框的单击方法
    int[] imageId= { R.drawable.baidu, R.drawable.google, R.drawable.bing };
    List<Map<String, Object>> listItems = new ArrayList<Map<String, Object>>();
    Map<String, Object> map;
    listItems.clear();
    for(int i = 0; i < imageId.length; i + +) {
        map = new HashMap<String, Object>();
        map.put("image", imageId[i]);
        map.put("title", items[i]);
        listItems.add(map);
    }
    final SimpleAdapter adapter = new SimpleAdapter(this, listItems, R.layout.items,
new String[] { "image", "title" }, new int[] { R.id.image, R.id.title });
    Builder builder = new AlertDialog.Builder(MainActivity.this);
    builder.setIcon(R.drawable.search);
    builder.setTitle("带图标列表对话框");
    builder.setAdapter(adapter, new DialogInterface.OnClickListener() {
        @Override
        public void onClick(DialogInterface dialog, int which) {
            Toast.makeText(getApplicationContext(), items[which],
                    Toast.LENGTH_LONG).show();
        }
    });
    builder.create().show();
}
```

单击“带图标列表对话框”按钮后，显示的对话框如图 4.30 所示。

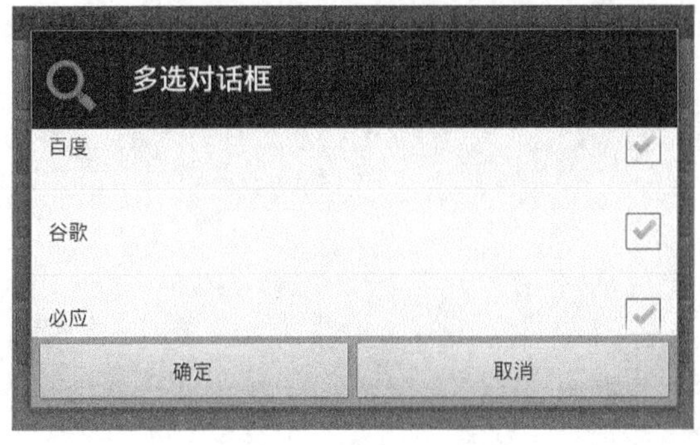

图 4.29　多选对话框

图 4.30　带图标列表对话框

4.5 小　结

本章介绍了 Android 平台提供的常用视图组件的使用，主要分为 3 个部分。首先介绍了通常情况下开发界面应用所遵从的开发步骤，即视图的使用模式。接着介绍了一些基本视图组件，主要包括文本框、编辑框、图片按钮、图片视图、单选按钮、复选按钮、下拉列表、自动完成文本框和日期、时间选择器，这些都是经常应用到的组件。随后又讲解了高级视图组件，包含进度条、拖动条、评分条和选项卡等，这些组件是进行特色开发时常用的组件。最后介绍了用于给用户进行提示的提示框和警告对话框，方便应用程序与用户的交互。

练　习

1. 使用基本组件完成用户调查问卷的界面设计。
2. 使用两个下拉列表组件，完成省市联动的应用，即在第 1 个下拉列表中选择省名，在第 2 个下拉列表会显示该省的市名。
3. 使用自动完成文本框实现课程名提示功能。
4. 使用进度条模拟图片加载过程，加载完成则显示图片。
5. 使用选项卡组件设计一个带有收件箱、发件箱、已删除 3 个标签页的选项卡应用。
6. 使用消息提示框，完成一个带图片的提示框的应用。
7. 使用警告对话框的多选框实现一个选择待播放音乐名的应用。

第 5 章 视图界面布局

为了管理 Android 应用中用户界面的各种组件，Android 提供了布局管理器。通过使用布局管理器，Android 可以很方便地控制各个组件的位置和大小，使之在不同的分辨率下和不同的版本之间都具有一致性。

5.1 界面布局设计

上一章已经介绍了所有可视化组件的基类 View 类，Android 平台提供了一个 ViewGroup 类，是 View 类的子类，可以充当其他组件的容器。同时布局管理器又作为 ViewGroup 的子类，在 Android 中提供了五种，包括线性布局管理器 LinearLayout、表格布局管理器 TableLayout、帧布局管理器 FrameLayout、相对布局管理器 RelativeLayout 和绝对布局管理器 AbsoluteLayout。对应于五种布局管理器，Android 提供了五种布局方式，布局可以相互嵌套。常用的布局管理器及其类结构如图 5.1 所示。

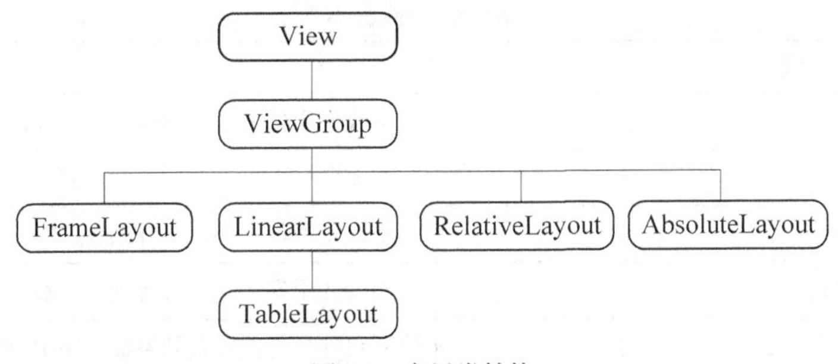

图 5.1 布局类结构

在 Android 中，可以在 XML 文件中定义布局管理器，也可以使用 Java 代码来创建，推荐使用在 XML 布局文件中定义布局管理器。在 XML 布局文件中定义，其基本语法如下。

```
<布局管理器名
    xmlns:android = "http://schemas.android.com/apk/res/android"
    android:id = "layoutId"
    android:layout_width = "match_parent"
    android:layout_height = "match_parent"
    其他属性......
>
```

布局管理器中的组件列表

</布局管理器名>

上述代码中布局管理器名即为 LinearLayout、TableLayout、FrameLayout、RelativeLayout 和 AbsoluteLayout 之一。布局管理器还支持嵌套，可以将一个布局放在另一个布局中。

另外代码中的“布局管理器中的组件列表”为各个先后有序的视图组件。

下面将分别对每个布局管理器进行详细的介绍。

5.1.1 线性布局

线性布局使用<LinearLayout>标签来表示，是最简单的布局之一，它可以设置组件在垂直或者水平方向排列。如果垂直排列，则每行仅包含一个视图组件，如果水平排列，则每列仅包含一个视图组件。同时使用该布局时可以通过设置组件的 weight 参数控制各个组件在容器中的相对大小。LinearLayout 布局的属性既可以在布局文件（XML）中设置，也可以通过成员方法进行设置。LinearLayout 常用的 XML 属性如表 5.1 所示。

表 5.1 线性布局常用 XML 属性

XML 属性	说　明
android:orientation	设置线性布局的方向，存在 horizontal（水平）和 vertical（垂直）两种排列方式
android:gravity	设置线性布局的内部组件的布局对齐方式

orientation 设置布局排列方式时，水平或垂直分布如果超过一行或一列，不会自动换行或换列，超出屏幕的组件将不会被显示，除非将其放到 ScrollView 中。

gravity 属性用来在线性布局中设置内部组件的对齐方式，当要为 gravity 设置多个值时，用“|”来分隔。gravity 取值及其说明如表 5.2 所示。

表 5.2 gravity 取值说明

属 性 值	说　明
top	不改变组件大小，对齐到容器顶部
bottom	不改变组件大小，对齐到容器底部
left	不改变组件大小，对齐到容器左侧
right	不改变组件大小，对齐到容器右侧
center_vertical	不改变组件大小，对齐到容器纵向中央位置
center_horizontal	不改变组件大小，对齐到容器横向中央位置
center	不改变组件大小，对齐到容器中央位置
fill_vertical	若有可能，纵向拉伸以填满容器
fill_horizontal	若有可能，横向拉伸以填满容器
fill	若有可能，纵向横向同时拉伸以填满容器

下面将通过例 5.1 来介绍如何使用线性布局管理器。

【例 5.1】 实现采用线性布局显示一组按钮。

（1）修改工程中的主布局文件，将默认的布局管理器改为 LinearLayout，线性布局的方向 android:orientation 设置为垂直 vertical，代码如下所示。

```
<LinearLayout xmlns:android = "http://schemas.android.com/apk/res/android"
    android:layout_width = "match_parent"
    android:layout_height = "match_parent"
    android:background = "#FFFAE4"
    android:orientation = "vertical">
    <!-- 待添加组件 -->
</LinearLayout>
```

（2）在该线性布局内添加四个按钮，将按钮的 android:layout_width 属性设置为 match_parent，并设置背景样式为蓝底圆角，其中 custom_button 是位于工程 res/drawable 下自定义的样式文件 custom_button.xml。

```
    <Button android:id = "@+id/b1"
        android:layout_width = "match_parent"
        android:layout_height = "wrap_content"
        android:background = "@drawable/custom_button"
        android:text = "按钮 1" />
    <Button android:id = "@+id/b2"
        android:layout_width = "match_parent"
        android:layout_height = "wrap_content"
        android:background = "@drawable/custom_button"
        android:text = "按钮 2" />
    <Button android:id = "@+id/b3"
        android:layout_width = "match_parent"
        android:layout_height = "wrap_content"
        android:background = "@drawable/custom_button"
        android:text = "按钮 3" />
    <Button android:id = "@+id/b4"
        android:layout_width = "match_parent"
        android:layout_height = "wrap_content"
        android:background = "@drawable/custom_button"
        android:text = "按钮 4" />
```

运行本项目，将屏幕设成横屏，显示结果如图 5.2 所示。

（3）修改主布局文件，线性布局的方向 android:orientation 设置为水平 horizontal，并增加布局管理器内部组件的对齐方式 android:gravity 值为 right|center_vertical（注意竖线“|”），同时将按钮宽度 android:layout_width 改成根据内容大小自动设定 wrap_content，表示组件居右对齐，垂直居中，代码如下所示。

```
<LinearLayout
    xmlns:android = "http://schemas.android.com/apk/res/android"
    android:layout_width = "match_parent"
    android:layout_height = "match_parent"
    android:orientation = "horizontal"
    android:gravity = "right|center_vertical">
    <Button
        android:layout_width = "wrap_content"
        android:layout_height = "wrap_content"
        android:background = "@drawable/custom_button"
```

```
        android:text = "按钮 1" />
    <Button
        android:layout_width = "wrap_content"
        android:layout_height = "wrap_content"
        android:background = "@drawable/custom_button"
        android:text = "按钮 2" />
    <Button
        android:layout_width = "wrap_content"
        android:layout_height = "wrap_content"
        android:background = "@drawable/custom_button"
        android:text = "按钮 3" />
    <Button
        android:layout_width = "wrap_content"
        android:layout_height = "wrap_content"
        android:background = "@drawable/custom_button"
        android:text = "按钮 4" />
</LinearLayout>
```

重新运行，显示结果如图 5.3 所示。

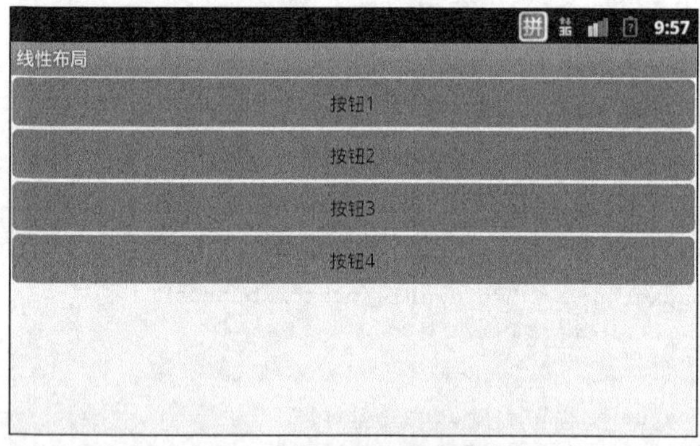

图 5.2　线性布局垂直效果

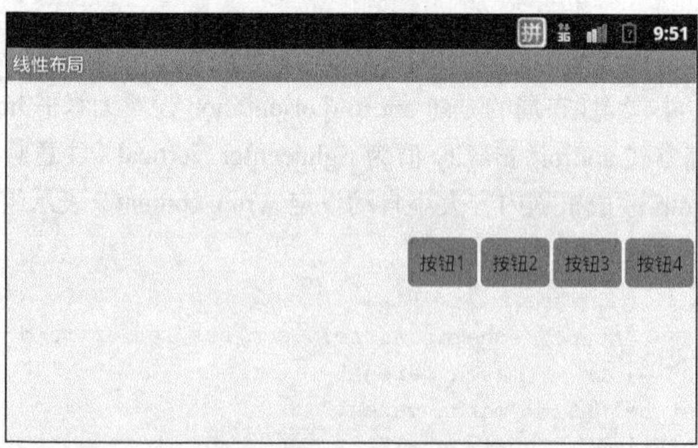

图 5.3　线性布局水平效果

5.1.2　表格布局

表格布局使用<TableLayout>标签来表示，类似常见的表格，通过指定行和列的形式来管理视图组件。在表格布局中可以添加多个<TableRow>标签，每个<TableRow>标签就是一个表格行，由于<TableRow>标签也是容器，因此在该标签中可以添加其他组件，每添加一个组件该行就增加一列，但表格布局并不会为每一行、每一列或每个单元格绘制边框。如果直接向 TableLayout 中添加组件，那么该组件也将直接占用一行。

在表格布局中，列的宽度由该列中最宽的那个组件决定，整个表格布局的宽度则取决于父容器的宽度。通过设置列的属性可以对列进行隐藏、伸展、收缩操作，从而可以充分利用屏幕空间，这 3 种属性如下。

- Collapsed，如果一列被标识为 collapsed，则该列将会被隐藏。
- Shrinkable，如果一列被标识为 shrinkable，则该列的宽度可以进行收缩，以使表格能够适应其父容器的大小。
- Stretchable，如果一列被标识为 stretchable，则该列的宽度可以进行拉伸，以使填满表格中空闲的空间。

TableLayout 继承自 LinearLayout 类，除了继承来自父类的属性和方法，还包含表格布局所特有的属性和方法。这些 XML 属性说明如表 5.3 所示。

表 5.3　　表格布局常用 XML 属性

XML 属性	说　明
android:collapseColumns	设置列为 Collapsed，列号从 0 开始，多个列号用“,”分隔
android:shrinkColumns	设置列为 Shrinkable，列号从 0 开始，多个列号用“,”分隔
android:stretchColumns	设置为 Stretchable，列号从 0 开始，多个列号用“,”分隔

setShrinkAllColumns() 和 setStretchAllColumns() 方法实现的功能是将表格中的所有列设置为 Shrinkable 或 Stretchable。

下面通过例 5.2 说明如何使用表格布局。

【例 5.2】　应用表格布局实现用户登录界面。

（1）修改主布局文件，将默认的布局管理器改为 TableLayout，将布局方向 android:gravity 设为 center_horizontal 水平居中，并通过定义表格属性 android:stretchColumns = "0,3"设置表格布局中的第 0 列，第 3 列自动拉伸。

```
<?xml version = "1.0" encoding = "utf-8"?>
<TableLayout xmlns:android = "http://schemas.android.com/apk/res/android"
    android:layout_width = "match_parent"
    android:layout_height = "match_parent"
    android:gravity = "center_horizontal"
    android:background = "#FFFAE4"
    android:stretchColumns = "0,3" >
    <!-- 待添加组件 -->
</TableLayout>
```

（2）在该布局管理器内添加一个 TextView 组件，该组件未放置于 TableRow 之内，则会单独占据一行，这时该组件中不需要设置 android:layout_span 跨行属性。极端情况下，如果表格布局管理器中没有 TableRow 组件，则实现类似线性管理器的纵向排列。

```
    <!-- 第一行，直接使用组件则独占一行 -->
    <ImageView
        android:layout_width = "match_parent"
```

```
        android:layout_height = "5px"
        android:gravity = "center_horizontal" />
```

（3）在该布局管理器内继续添加 TableRow 组件，表示占用一行，该 TableRow 内添加一个 ImageView 图片组件，设置宽和高都根据大小自动设定，位置水平居中，并通过设置属性 android:layout_span 值为 4，表示类似 HTML 应用，使该行可跨表格四列。

```
<!-- 第二行，用 TableRow 独占一行，需要设置跨 4 列 -->
<TableRow>
    <ImageView
        android:layout_width = "wrap_content"
        android:layout_height = "wrap_content"
        android:gravity = "center_horizontal"
        android:layout_span = "4"
        android:src = "@drawable/logo" />
</TableRow>
```

（4）在该布局管理器内继续添加 TableRow 组件，该组件内添加表示用户名的 TextView 组件，居右对齐，和用于输入用户名的 EditText 组件，通过属性 android:minWidth 设置最小宽度为 260 个像素，字体大小为 15sp。

```
<!-- 第三行 -->
<TableRow>
    <!-- 第一列，内容为空，用于自动伸展-->
    <TextView />
    <!-- 第二列，用户名文本框-->
    <TextView
        android:layout_width = "wrap_content"
        android:layout_height = "wrap_content"
        android:gravity = "right"
        android:text = "用户名    "
        android:textStyle = "bold" />
    <!-- 第三列，用户名输入框-->
    <EditText
        android:id = "@+id/username"
        android:layout_width = "wrap_content"
        android:layout_height = "wrap_content"
        android:minWidth = "260px"
        android:text = "admin"
        android:textSize = "15sp" />
    <!-- 第四列，内容为空，用于自动伸展 -->
    <TextView />
</TableRow>
```

（5）在该布局管理器内继续添加 TableRow 组件，该组件内添加表示密码的 TextView 组件，居右对齐，和用于输入密码的 EditText 组件，通过设置属性 android:inputType 的值为 textPassword，表示以加密形式显示所输入密码，字号 15sp。宽度虽然设置为 wrap_content 根据自身大小自动分配，但密码输入框所在列最小宽度为上一行的用户名输入框的 260 像素，所以本列宽度会自动拉伸到 260 像素。

```
<!-- 第四行 -->
```

```
<TableRow>
    <!-- 第一列，内容为空，用于自动伸展-->
    <TextView />
    <!-- 第二列，密码文本框-->
    <TextView
        android:layout_width = "wrap_content"
        android:layout_height = "wrap_content"
        android:gravity = "right"
        android:text = "密码    "
        android:textStyle = "bold" />
    <!-- 第三列，密码输入框-->
    <EditText
        android:layout_width = "wrap_content"
        android:layout_height = "wrap_content"
        android:inputType = "textPassword"
        android:text = "pass"
        android:textSize = "15sp" />
    <!-- 第四列，内容为空，用于自动伸展 -->
    <TextView />
</TableRow>
```

（6）在该布局管理器标签内最后添加 TableRow 组件，该组件内添加取消和登录按钮，这两个按钮的样式设置为 custom_button。

```
<!-- 第五行 -->
<TableRow>
    <TextView />
    <Button android:id = "@+id/cancel"
        android:layout_width = "wrap_content"
        android:layout_height = "wrap_content"
        android:background = "@drawable/custom_button"
        android:text = "取消"
        android:textSize = "14sp" />
    <Button android:id = "@+id/login"
        android:layout_width = "wrap_content"
        android:layout_height = "wrap_content"
        android:background = "@drawable/custom_button"
        android:text = "登录"
        android:textSize = "14sp" />
    <TextView />
</TableRow>
```

本例中第三行到第五行中一共添加了 6 个未设置属性的 TextView 组件，并且允许拉伸，用户名、密码提示和输入区域占据自身宽度外，剩余每行两侧空间通过 TextView 拉伸占满。这样设计是为了实现登录窗口居中显示，而文本组件不会显示到屏幕右边缘，更显美观。运行本实例，显示效果如图 5.4 所示。

如果某行不需要拉伸或收缩，设置其确定的宽度、高度即可。如第二行的图片组件不需要拉伸，可将其宽度 android:layout_width 按比例设置成 352px，而高度 android:layout_height 按比例设置成 131px，修改代码如下。

```
<ImageView
    android:layout_width = "352px"
    android:layout_height = "131px"
    android:gravity = "center_horizontal"
    android:layout_span = "4"
    android:src = "@drawable/logo" />
```

运行修改过的本实例，显示结果如图 5.5 所示。

图 5.4　logo 拉伸效果

图 5.5　logo 未拉伸效果

5.1.3　帧布局

帧布局也称为框架布局，使用<FrameLayout>标签来表示，它为每个加入其中的组件都创建一个空白的区域，通常称为一帧。每个组件对应的一帧都会被对齐到屏幕的左上角，即(0,0)坐标点开始布局。也就是说，把各个组件叠加在一起，而且所有的组件通过层叠的方式来进行显示。如果每个帧的大小都一样，gravity 属性也相同，则同一时刻只能看到最上面的帧，其他的则被其遮挡，这个特点在进行选项卡设计时会用到。帧布局常用的 XML 属性如表 5.4 所示。

表 5.4　　帧布局常用 XML 属性

XML 属性	说　明
android:foreground	设置该帧布局容器的前景图像
android:foregroundGravity	设置前景图像的 gravity 属性，即显示位置

（1）帧布局中没有 android:gravity 属性；

（2）帧布局中的组件如果没有设置对齐方式，组件会重叠在一起。因此，这些组件需要在属性 android:layout_gravity 中设置自己在帧布局中的位置。

下面将通过例 5.3 说明如何使用帧布局。

【例 5.3】应用帧布局实现渐进效果。

（1）修改主布局文件，将默认的布局管理器改为帧布局 FrameLayout，在布局中设置前景图片，以及前景图片显示的位置（右上方）。

```
<FrameLayout xmlns:android = "http://schemas.android.com/apk/res/android"
    android:layout_width = "match_parent"
    android:layout_height = "match_parent"
    android:foreground = "@drawable/logo"
    android:foregroundGravity = "top|right" >
    <!-- 待添加组件 -->
</FrameLayout>
```

（2）在该布局管理器内添加一个 TextView 组件，并为该组件设置宽度和高度，以及背景色，表示一个颜色区块。

```
    <!-- 最低层 -->
    <TextView
        android:layout_width = "wrap_content"
        android:layout_height = "wrap_content"
        android:background = "#DC143C"
        android:height = "410px"
        android:width = "700px" />
```

（3）在该布局管理器内再依次添加 6 个 TextView 组件，并为其指定不同的颜色，设置高度不变，宽度依次减小。后一个 TextView 组件会覆盖到前一个 TextView 组件上，实现一个渐进效果的层叠应用。

```
    <TextView
        android:layout_width = "wrap_content"
        android:layout_height = "wrap_content"
        android:background = "#FF6600"
        android:height = "410px"
        android:width = "600px" />
    <TextView
        android:layout_width = "wrap_content"
        android:layout_height = "wrap_content"
        android:background = "#FFFF00"
        android:height = "410px"
        android:width = "500px" />
    <TextView
        android:layout_width = "wrap_content"
        android:layout_height = "wrap_content"
        android:background = "#009966"
```

```
        android:height = "410px"
        android:width = "400px" />
    <TextView
        android:layout_width = "wrap_content"
        android:layout_height = "wrap_content"
        android:background = "#669900"
        android:height = "410px"
        android:width = "300px" />
    <TextView
        android:layout_width = "wrap_content"
        android:layout_height = "wrap_content"
        android:background = "#0000CC"
        android:height = "410px"
        android:width = "200px" />
    <!-- 最上层 -->
    <TextView
        android:layout_width = "wrap_content"
        android:layout_height = "wrap_content"
        android:background = "#993399"
        android:height = "410px"
        android:width = "100px" />
```

运行本实例，横屏显示结果如图 5.6 所示。

图 5.6 帧布局渐进色效果

5.1.4 相对布局

相对布局使用<RelativeLayout>标签来表示，是按照组件之间的相对位置来进行布局，是实际布局中最常用的方式之一。相对布局灵活性很大，属性较多，所以属性之间产生冲突的可能性也较大，但能够最大程度保证在各种屏幕类型的移动设备上正确显示界面布局。

相对布局支持的常用 XML 属性如表 5.5 所示。

表 5.5 相对布局常用 XML 属性

XML 属性	说　明
android:gravity	设置该布局容器内部各子组件的对齐方式
android:ignoreGravity	设置不受 gravity 组件影响的组件

在相对布局管理器中，除了上面介绍的两个属性之外，为了更好地控制该布局管理器中各个子组件的布局分布，RelativeLayout 类还提供了一个内部类 RelativeLayout.LayoutParams，通过这个类提供了较多的 XML 属性，其中值为 boolean 型的属性主要描述当前组件与父组件（布局管理器）的相对位置，如表 5.6 所示。

表 5.6　相对布局中取值为 boolean 型的属性

属 性 名 称	属 性 说 明
android:layout_centerHorizontal	当前组件位于父组件的横向中间位置
android:layout_centerVertical	当前组件位于父组件的纵向中间位置
android:layout_centerInParent	当前组件位于父组件的中央位置
android:layout_alignParentBottom	当前组件底端与父组件底端对齐
android:layout_alignParentLeft	当前组件左侧与父组件左侧对齐
android:layout_alignParentRight	当前组件右侧与父组件右侧对齐
android:layout_alignParentTop	当前组件顶端与父组件顶端对齐
android:layout_alignWithParentIfMissing	参照组件不存在或不可见时参照父组件

相对布局中除了通过表 5.6 介绍的当前组件与父组件位置之间的关系，来确定布局外，还有通过兄弟组件之间的相对位置来进行布局。这种方式需要指定组件属性值为其他组件的 id，此类属性如表 5.7 所示。

表 5.7　相对布局中组件之间的位置关系

属 性 名 称	属 性 说 明
android:layout_toRightOf	使当前组件位于给出 id 组件的右侧
android:layout_toLeftOf	使当前组件位于给出 id 组件的左侧
android:layout_above	使当前组件位于给出 id 组件的上方
android:layout_below	使当前组件位于给出 id 组件的下方
android:layout_alignTop	使当前组件的上边界与给出 id 组件的上边界对齐
android:layout_alignBottom	使当前组件的下边界与给出 id 组件的下边界对齐
android:layout_alignLeft	使当前组件的左边界与给出 id 组件的左边界对齐
android:layout_alignRight	使当前组件的右边界与给出 id 组件的右边界对齐

最后，相对布局中描述四周留白的属性，取值以像素为单位。其属性及说明如表 5.8 所示。

表 5.8　相对布局中取值为像素的属性及说明

属 性 名 称	属 性 说 明
android:layout_marginLeft	当前组件左侧的留白
android:layout_marginRight	当前组件右侧的留白
android:layout_marginTop	当前组件上方的留白
android:layout_marginBottom	当前组件下方的留白
android:layout_margin	当前组件上下左右四个方向的留白

注意

在进行相对布局时要注意以下规则。

（1）相对布局的子组件必须有唯一的 id 属性以使规则正确应用。

（2）当心循环规则，循环规则发生在两个组件具有互相指向时。

（3）保持相对布局规则最小化，能减小出现循环规则的机率并且使得布局更加可维护和灵活。

（4）测试布局设计在横屏和竖屏模式下，以及在不同的屏幕大小和解决方案下是不是符合预期。

（5）使用相对布局代替嵌套线性布局以改进程序性能和响应能力。

下面通过例 5.4 说明相对布局的使用。

【例 5.4】 应用相对布局实现一个瑜伽馆的介绍页面。

（1）修改主布局文件，将默认的布局管理器改为相对布局 RelativeLayout，并设置布局管理器的背景色。

```
<RelativeLayout xmlns:android = "http://schemas.android.com/apk/res/android"
    xmlns:tools = "http://schemas.android.com/tools"
    android:layout_width = "match_parent"
    android:layout_height = "match_parent"
    android:background = "#FFFAE4"
    tools:context = ".MainActivity" >
    <!-- 待添加组件 -->
</RelativeLayout>
```

（2）在此相对布局管理器中增加一个 ImageView 组件设置为瑜伽馆 logo，并设置该 ImageView 组件的属性 android:layout_alignParentTop 和属性 android:layout_centerHorizontal 为 true，表示当前组件居顶对齐，还保持水平居中，同时设置 android:paddingTop 为 0 像素，表示当前组件距界面顶部无间距。

```
<ImageView android:id = "@+id/logo"
    android:layout_width = "352px"
    android:layout_height = "131px"
    android:layout_alignParentTop = "true"
    android:layout_centerHorizontal = "true"
    android:paddingTop = "0px"
    android:src = "@drawable/logo" />
```

（3）在此相对布局管理器中增加一个 ImageView 组件作为瑜伽馆业务展示图片，设置 id 值为 spa，并设置 android:layout_centerInParent 为 true，表示该组件在布局中居中，即水平方向和垂直方向都居中。

```
<ImageView android:id = "@+id/spa"
    android:layout_width = "wrap_content"
    android:layout_height = "wrap_content"
    android:layout_centerInParent = "true"
    android:src = "@drawable/spa" />
```

（4）在此相对布局管理器中增加一个 id 值为 book 的 ImageView 组件，设置其属性 android:layout_below 和 android:layout_alignLeft 的值都为 spa，表示 book 组件位于 spa 组件下方，并和 spa 组件左对齐。

```
<ImageView android:id = "@+id/book"
    android:layout_width = "150px"
    android:layout_height = "54px"
    android:layout_alignLeft = "@id/spa"
    android:layout_below = "@id/spa"
    android:layout_marginTop = "5px"
    android:src = "@drawable/book" />
```

（5）在此相对布局管理器中增加一个 id 值为 gift 的 ImageView 组件，设置其属性 android:layout_below 和 android:layout_alignRight 的值都为 spa，表示 book 组件位于 spa 组件下方，并和 spa 组件右对齐。为了避免 book、gift 组件和 spa 太接近，所以在这两个组件中都设置了 android:layout_marginTop 为 5 个像素。

```
<ImageView android:id = "@+id/gift"
    android:layout_width = "150px"
    android:layout_height = "54px"
    android:layout_alignRight = "@id/spa"
    android:layout_below = "@id/spa"
    android:layout_marginTop = "5px"
    android:src = "@drawable/gift" />
```

（6）在此相对布局管理器中最后增加一个 TextView 作为版权说明信息，通过设置 android:layout_alignParentBottom 和 android:layout_centerHorizontal 的属性值为 true，使布局管理器居底对齐、水平居中对齐。

```
<TextView android:id = "@+id/copyright"
    android:layout_width = "wrap_content"
    android:layout_height = "wrap_content"
    android:layout_alignParentBottom = "true"
    android:layout_centerHorizontal = "true"
    android:layout_marginBottom = "5px"
    android:textSize = "12dp"
    android:text = "©Spa Treats 2012 | Design: Design themes" />
```

运行本实例，显示结果如图 5.7 所示。

为了界面美观可以隐藏标题栏，通过修改 MainActivity 类的 onCreate 方法，在 super.onCreate(savedInstanceState)语句之前，添加如下代码即可。

```
requestWindowFeature(Window.FEATURE_NO_TITLE);  //隐藏标题栏
```

为了实现交互应用，可在主 Activity 类的 onCreate 方法最后添加组件的处理事件。如获取 id 值为 book 的组件，对其添加单击事件监听器，在监听器中完成预定功能，代码如下所示。

```
ImageView book = (ImageView) findViewById(R.id.book);
book.setOnClickListener(new View.OnClickListener()
{
    public void onClick(View v)
    {
        //预定处理代码
        Toast.makeText(MainActivity.this, "预定成功，欢迎光临!",
                    Toast.LENGTH_LONG).show();
    }
});
```

运行修改过的实例，显示结果如图 5.8 所示。

图 5.7　相对布局未隐藏标题栏效果

图 5.8　相对布局隐藏标题栏效果

5.1.5　绝对布局

绝对布局使用<AbsoluteLayout>标签来表示，通过指定界面元素的坐标位置，来确定用户界面的整体布局，容器组件不再负责管理其子组件的位置。绝对布局自 Android 2.0 版本开始不再推荐使用，因为通过 x 坐标和 y 坐标确定界面元素位置后，Android 系统不能根据不同屏幕对界面元素的位置进行调整，降低了界面布局对不同类型和尺寸的屏幕的适应能力。鉴于此，对于绝对布局不再做过多介绍。

5.1.6　复用 XML 布局文件

在应用程序中，为了保持各窗口之间的风格的统一，在 UI 布局文件中，会用到很多相同的布局。如果针对每个 XML 布局文件，都把相同的布局实现一遍，会造成代码冗余，可读性很差，另外修改起来会比较繁琐，对后期的维护非常不利。所以，希望能实现包含多个组件布局的复用，这样仅需要将相同布局的代码单独作为一个模块，在其他布局中重用公共布局即可。Android 开发中，提供了<include>标签和<merge>标签，可实现布局的复用。下面分别予以介绍。

1.　<include>标签

<include>标签的使用格式如下，其中布局文件名为 layout 路径下布局文件对应在 R.layout 中生成的 id 值。另外可以覆盖 id 值，即重新定义 id 值。如果要覆盖布局的尺寸，就必须同时覆盖 android:layout_height 和 android:layout_width 的属性。若只覆盖其中一个，则无效。

```
<include
    android:id = "@+id/需引用布局文件的 id 值"
    android:layout_width = "match_parent"
    android:layout_height = "match_parent"
    layout = "@layout/布局文件名"/>
```

下面通过例 5.5 介绍如何重用布局文件。

【例 5.5】　应用相对布局引用一个页眉布局和一个页脚布局实现一个瑜伽馆的介绍页面。

（1）页眉布局 titlebar。

新建页眉布局文件，文件名为 titlebar.xml，并将该布局的 id 名设为 titlebar。该布局文件的布局管理器采用线性布局，布局方向为垂直。

```
<LinearLayout xmlns:android = "http://schemas.android.com/apk/res/android"
    android:id = "@+id/titlebar"
    android:layout_width = "match_parent"
    android:layout_height = "match_parent"
    android:orientation = "vertical" >
    <!-- 待添加组件 -->
</LinearLayout>
```

①在该布局管理器中，添加一个表示瑜伽馆标识的 ImageView 组件，id 值设为 logo，设置布局方向为水平居中，并将宽度、高度值设定为确定值，表示不允许拉伸。

```
    <ImageView android:id = "@+id/logo"
        android:layout_width = "352px"
        android:layout_height = "131px"
        android:layout_gravity = "center_horizontal"
        android:paddingTop = "0px"
        android:src = "@drawable/logo" />
```

②在该布局管理器中，再添加一个线性布局管理器。在添加的布局中放置 EditText 组件用于输入查询关键字，id 值为 key；一个 Button 组件，id 值为 search，用于单击搜索。

```
    <LinearLayout
        android:layout_width = "match_parent"
        android:layout_height = "wrap_content"
        android:orientation = "horizontal"
        android:gravity = "center_horizontal" >
        <EditText android:id = "@+id/key"
            android:layout_width = "250px"
            android:layout_height = "35sp"
            android:paddingTop = "2px"
            android:textSize = "14sp" />
        <Button android:id = "@+id/search"
            android:layout_width = "wrap_content"
            android:layout_height = "wrap_content"
            android:background = "@drawable/custom_button"
            android:paddingTop = "2px"
            android:text = "搜索"
            android:textSize = "14sp" />
    </LinearLayout>
```

该 titlebar 页眉布局显示效果，如图 5.9 所示。

（2）页脚布局 tailbar。

新建页脚布局文件，文件名为 tailbar.xml，布局管理器采用帧布局，并将该布局的 id 名设为 tailbar。

```
<?xml version = "1.0" encoding = "utf-8"?>
<FrameLayout xmlns:android = "http://schemas.android.com/apk/res/android"
    android:id = "@+id/tailbar"
    android:layout_width = "match_parent"
    android:layout_height = "match_parent" >
    <!-- 待添加组件 -->
</FrameLayout>
```

在该帧布局管理器中，添加一个排列方向为水平的线性布局管理器。新增的线性布局管理器，设置布局方向为垂直居中对齐，其内部组件为居顶对齐并水平居中。线性布局中水平方向依次放置一个表示预定功能的 ImageView 组件，id 值为 book；一个用于分割前后两个组件的 TextView，id 值为 blank，间隔为 5 个像素；一个表示赠送朋友功能的 ImageView 组件，id 值为 gift。

```
<LinearLayout
    android:layout_width = "match_parent"
    android:layout_height = "match_parent"
    android:layout_gravity = "center_vertical"
    android:gravity = "top|center_horizontal"
    android:orientation = "horizontal" >
    <ImageView android:id = "@+id/book"
        android:layout_width = "150px"
        android:layout_height = "54px"
        android:layout_marginTop = "5px"
        android:src = "@drawable/book" />
    <TextView android:id = "@+id/blank"
        android:layout_width = "5px"
        android:layout_height = "54px"
        android:layout_marginTop = "5px" />
    <ImageView android:id = "@+id/gift"
        android:layout_width = "150px"
        android:layout_height = "54px"
        android:layout_marginTop = "5px"
        android:src = "@drawable/gift" />
</LinearLayout>
```

在该帧布局管理器中，最后再添加一个表示版权信息的 TextView 组件，id 值为 copyright，布局设置为居底对齐、水平居中。

```
<TextView android:id = "@+id/copyright"
    android:layout_width = "wrap_content"
    android:layout_height = "wrap_content"
    android:layout_gravity = "bottom|center_horizontal"
    android:layout_marginBottom = "5px"
    android:text = "©Spa Treats 2012 | Design: Design themes"
    android:textSize = "12dp" />
```

该 tailbar 页脚布局显示结果，如图 5.10 所示。

图 5.9 需复用的页眉布局

图 5.10 需复用的页脚布局

（3）主布局 activity_main。

主布局文件名为 activity_main.xml，采用相对布局管理器。

```
<RelativeLayout xmlns:android = "http://schemas.android.com/apk/res/android"
    android:layout_width = "match_parent"
```

```
    android:layout_height = "match_parent"
    android:background = "#FFFAE4" >
    <!-- 待添加组件 -->
</RelativeLayout>
```

该相对布局管理器中，通过 include 标签的 layout 属性将 id 值为 titlebar 的页眉布局引入主布局文件中，即从工程 res/layout 目录下获取 titlebar.xml 布局文件。include 标签中未设置宽度和高度属性，说明宽高使用默认值。

```
    <include layout = "@layout/titlebar" />
```

该相对布局管理器中，再添加一项 id 值为 yoga 的业务介绍的 ImageView 组件，居中放置（水平和垂直方向都居中）。

```
    <ImageView android:id = "@+id/yoga"
        android:layout_width = "308px"
        android:layout_height = "363px"
        android:layout_centerInParent = "true"
        android:src = "@drawable/yoga" />
```

该相对布局管理器中，最后再次通过 include 标签将 id 值为 tailbar 的页脚布局引入主布局文件中，并将其设置位于 yoga 之下，居底对齐并且水平居中。这里对宽度和高度进行了重新设置，注意宽度、高度如需重新设置，必须都设置，不能只设置其中之一。

```
    <include
        android:layout_width = "match_parent"
        android:layout_height = "wrap_content"
        android:layout_below = "@id/yoga"
        android:layout_alignParentBottom = "true"
        android:layout_centerHorizontal = "true"
        layout = "@layout/tailbar" />
```

同样再设置一个布局文件，分别引入上述的页眉布局和页脚布局，修改中间的业务介绍的 ImageView 组件，代码类似不再赘述。运行此实例，依次打开两个窗体，显示结果如图 5.11、图 5.12 所示。

图 5.11　引入公共布局的窗体 1

图 5.12　引入公共布局的窗体 2

2. <merge>标签

当在布局文件中复用另外的布局时，<merge>标签能够在布局层次中消除多余的布局元素，增加 UI 效率。下面将通过一个使用<merge>标签的例子来说明如何重用布局，如例 5.6 所示。

【例 5.6】 应用线性布局复用实现浏览图片的功能。

（1）需复用的布局 pic。

新建布局文件 pic.xml，该布局文件中采用线性布局管理器，在管理器中添加了一个显示图片的 ImageView 组件，由于编译器的限制必须将这个 ImageView 置入一个布局管理器中。

```
<LinearLayout xmlns:android = "http://schemas.android.com/apk/res/android"
    android:layout_width = "match_parent"
    android:layout_height = "match_parent"
    android:orientation = "vertical" >
    <ImageView
        android:layout_width = "wrap_content"
        android:layout_height = "wrap_content"
        android:layout_margin = "5px"
        android:src = "@drawable/pic2" />
</LinearLayout>
```

（2）主布局 activity_main。

打开工程的布局文件 activity_main.xml，修改主布局为线性布局管理器，设置管理器中的组件位置为垂直排列。

```
<LinearLayout xmlns:android = "http://schemas.android.com/apk/res/android"
    xmlns:tools = "http://schemas.android.com/tools"
    android:layout_width = "match_parent"
    android:layout_height = "match_parent"
    android:orientation = "vertical"
    android:background = "#FFFAE4"
    tools:context = ".MainActivity" >
    <!-- 待添加组件 -->
</LinearLayout>
```

在该主布局管理器中，通过 include 标签引入 id 为 pic 的布局文件，并重新设定宽高属性值。

```
    <include
        android:layout_width = "wrap_content"
        android:layout_height = "wrap_content"
        layout = "@layout/pic" />
```

在该主布局中，再添加一个线性布局，设置居中放置。新添加的线性布局中水平居中放入两个按钮，用于打开上一张图片、下一张图片。

```
    <LinearLayout
        android:layout_width = "match_parent"
        android:layout_height = "match_parent"
        android:gravity = "center"
        android:orientation = "horizontal" >
        <Button
            android:layout_width = "wrap_content"
            android:layout_height = "wrap_content"
            android:background = "@drawable/custom_button"
            android:text = " 上一图 "
            android:textSize = "14sp" />
```

```
        <Button
            android:layout_width = "wrap_content"
            android:layout_height = "wrap_content"
            android:background = "@drawable/custom_button"
            android:text = "  下一图  "
            android:textSize = "14sp" />
    </LinearLayout>
```

从上面的代码可以看出，主布局复用了 pic 布局，即将 pic 布局的内容嵌入到了 include 标签所指定的位置。在主布局中就多嵌套了一层 pic 布局文件内的 LinearLayout 管理器，增加了视图层级结构，降低了代码的执行效率。运行本实例，显示结果如图 5.13 所示。

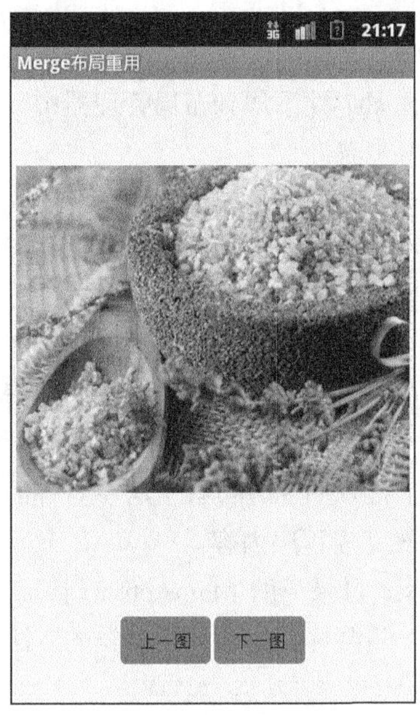

图 5.13　merge 标签实例

（3）使用 merge 标签修改的需复用的布局 pic，代码如下所示。

```
<merge xmlns:android = "http://schemas.android.com/apk/res/android" >
    <ImageView
        android:layout_width = "wrap_content"
        android:layout_height = "wrap_content"
        android:layout_margin = "5px"
        android:src = "@drawable/pic2" />
</merge>
```

将 pic 布局中的 LinearLayout 管理器取消，替换成 merge 标签，这样在主布局引入该 pic 布局时，仅将 merge 内的 ImageView 组件引入主布局，减少了嵌套，提高了代码效率。

在 RelativeLayout 布局管理器中使用 include 标签复用布局时，如发现 include 进来的组件无法用 layout_alignParentBottom = "true"之类的标签来调整位置，仅需要在 include 时同时重载 layout_width 和 layout_height 这两个属性即可。如果未重载，任何针对 include 的布局调整都是无效的。

5.2 控制视图界面的其他方法

Android 的视图界面可以通过上述的 XML 文件进行布局，优点是方便、快捷，但是该方法有失灵活。下面将分别介绍完全通过 Java 代码，以及使用 XML 和 Java 代码混合控制视图界面的方法，供用户根据实际需要进行选择。

5.2.1 代码控制视图界面

Android 开发中提供了类似 Java Swing 那样完全通过代码控制视图界面的方法。通过实例化组件类的方式创建，然后将这些视图组件添加到布局管理器中，从而实现界面设计。此方法控制视图界面，可以分为以下 3 个主要步骤。

（1）创建布局管理器，设置布局管理器的属性。

（2）创建视图组件，设置组件的布局和各种属性。

（3）将创建的具体组件添加到布局管理器中。

下面将通过例 5.7 来介绍如何使用 Java 代码控制视图界面。

【例 5.7】 应用完全通过代码控制视图界面的方法实现带导航栏的图片展示功能。

界面主布局是垂直排列的线性布局，其中包含一个实现了导航栏的帧布局，和业务说明的 ImageView 组件；帧布局中设置了背景图片，包含了居左对齐的返回按钮，和居中对齐的业务说明的 TextView。按照如下步骤依次完成程序内容。

（1）在新建的项目中，打开 src 目录下的 MainActivity.java 文件，删除方法 onCreate 中自动生成的代码。首先添加如下代码，隐藏标题栏和调用父类的方法 onCreate 进行初始化。

```
requestWindowFeature(Window.FEATURE_NO_TITLE);
super.onCreate(savedInstanceState);
```

（2）创建线性布局为主布局，通过其 setBackgroundColor 方法指定背景颜色，背景色以 16 进制数值表示，并设置布局方向为垂直方向，布局方向值通过 LinearLayout 的常量表示。

```
LinearLayout lLayout = new LinearLayout(this);  //创建线性布局
lLayout.setBackgroundColor(0xFFFFFAE4);  //设置布局颜色
lLayout.setOrientation(LinearLayout.VERTICAL);  //设置布局方向
```

（3）创建帧布局为子布局。

```
FrameLayout fLayout = new FrameLayout(this);  //实例化子布局
fLayout.setBackgroundResource(R.drawable.repeat_bg);  //设置背景资源
fLayout.setPadding(10, 0, 0, 0);  //设置间距，居左 10 个像素
```

（4）定义子布局中的布局参数类。Android 中 View 组件提供了丰富的属性设置方法，但是一些复杂设置包含权重，在布局中的位置等属性并没有提供。为了能够设置这些重要的属性，Android 中提供了布局参数类 LayoutParams 完成这个功能。本例中布局是帧布局，所以帧布局中组件的相关属性需要使用帧布局的内部类 LayoutParams 来设置。首先创建用于保存按钮的布局参数对象，在构造函数中指定按钮布局的宽度、高度，这里宽度设置为根据内容自动调整，高度为 35 个像素，并设置按钮对齐方式为居左对齐、垂直居中。

```
// 创建保存按钮布局参数的对象
FrameLayout.LayoutParams btnLp = new FrameLayout.LayoutParams(
                                LayoutParams.WRAP_CONTENT, 35);
btnLp.gravity = Gravity.LEFT | Gravity.CENTER_VERTICAL; //设置布局中的按钮对齐方式
```

（5）创建子布局中的按钮，设置文本、大小、背景等基本参数，并将上一步中建立的保存按钮参数的对象设置到按钮中。

```
Button btnReturn = new Button(this); //创建返回按钮
btnReturn.setText("返回"); //设置按钮显示文字
btnReturn.setTextSize(TypedValue.COMPLEX_UNIT_DIP, 11); //设置按钮的文字字号
btnReturn.setBackgroundDrawable(
this.getResources().getDrawable(R.drawable.custom_button)); //设置按钮的背景样式
btnReturn.setLayoutParams(btnLp); //设置按钮的参数对象
```

（6）与按钮类似，创建文本框 TextView 以及设置相关参数的代码如下所示。

```
// 创建保存文本框布局参数的对象
FrameLayout.LayoutParams textLp = new FrameLayout.LayoutParams(
LayoutParams.WRAP_CONTENT, LayoutParams.WRAP_CONTENT);
textLp.gravity = Gravity.CENTER; //设置布局中的文本框对齐方式，居中对齐
TextView tv = new TextView(this); //创建文本框
tv.setText("SPA生活馆"); //设置文本框显示文字
tv.setTextColor(Color.rgb(255, 255, 255)); //设置文本框中的字体颜色，用Color类指定
// tv.setTextColor(0xFFFFFFFF); //也可以用16进制形式指定颜色
tv.setTextSize(TypedValue.COMPLEX_UNIT_DIP, 18); //设置文本框的文字字号
tv.setLayoutParams(textLp); //设置文本框的参数对象
```

（7）定义主布局中的图片组件。由于该组件位于线性布局中，所以该组件使用的布局参数类为 LinearLayout 的内部类 LayoutParams，使用方法与前面类似。

```
// 创建保存图片布局参数的对象
LinearLayout.LayoutParams ivLp = new LinearLayout.LayoutParams(
LayoutParams.WRAP_CONTENT, LayoutParams.MATCH_PARENT);
// 设置图片对齐方式
ivLp.gravity = Gravity.CENTER_HORIZONTAL | Gravity.CENTER_VERTICAL;
ImageView iv = new ImageView(this); //创建图片对象
iv.setImageResource(R.drawable.spainfo); //设置图片资源
iv.setPadding(2, 2, 2, 2); //依次设置图片的左、上 、右、下间距
iv.setLayoutParams(ivLp); //设置图片参数对象
```

（8）将组件加入布局。

```
fLayout.addView(btnReturn); //在子布局中添加返回按钮
fLayout.addView(tv); //在子布局中添加文本框
lLayout.addView(fLayout); //在主布局增加子布局
lLayout.addView(iv); //在主布局增加图片组件
setContentView(lLayout); //将主布局加启动页面
```

运行本实例，竖屏、横屏的显示结果如图 5.14、图 5.15 所示。

图 5.14　竖屏效果

图 5.15　横屏效果

从本例可以看出，界面布局、添减组件、设置属性这些基本操作，用 Java 代码实现起来确实很繁琐，不推荐用户使用。

5.2.2　代码和 XML 联合控制视图界面

视图界面完全采用 Java 代码设计和控制的确比较灵活，可随时根据需要调整界面布局，但是代码量过大，样式调整复杂，不利于修改、解耦等。还有一种方法，就是结合 Java 代码和 XML 布局文件联合控制视图界面。这种方法通常将变化小、行为固定的组件设置在 XML 布局文件中，而把变化较多、行为控制比较复杂的组件通过 Java 代码来管理。

下面就通过例 5.8 来介绍如何使用 Java 代码和 XML 混合控制视图界面。

【例 5.8】　应用代码和 XML 联合控制视图界面的方法实现游戏列表显示。

（1）在新建的项目中，修改主布局文件。主布局为线性布局，设置 id 为 mainLayout，其中包含一个 id 为 flayout 的帧布局作为导航栏。帧布局中放置一个用户返回的 ImageView 组件，设置居左对齐，垂直居中；再放置一个显示“热门游戏”标题的 TextView 组件，设置颜色为白色，字号 18dip，居中对齐。

```
<LinearLayout xmlns:android = "http://schemas.android.com/apk/res/android"
    android:id = "@+id/mainLayout"
    android:layout_width = "match_parent"
    android:layout_height = "match_parent"
    android:background = "#FFFAE4"
    android:orientation = "vertical" >
    <!-- 第一行放置返回图标和题目 -->
    <FrameLayout
        android:id = "@+id/flayout"
        android:layout_width = "match_parent"
        android:layout_height = "wrap_content"
        android:background = "@drawable/titlebg"
        android:paddingLeft = "10px" >
```

```
        <!--左侧垂直居中位置放置返回图片-->
        <ImageView
            android:layout_width = "45px"
            android:layout_height = "35px"
            android:src = "@drawable/return1"
            android:layout_gravity = "left|center_vertical"  />
        <!--居中位置放置页面标题:热门游戏-->
        <TextView
            android:layout_width = "wrap_content"
            android:layout_height = "wrap_content"
            android:layout_gravity = "center"
            android:text = "热门游戏"
            android:textColor = "#FFFFFF"
            android:textSize = "18dip" />
    </FrameLayout>
    <!-- 待通过 Java 代码添加组件 -->
</LinearLayout>
```

（2）在主 Activity 文件中添加两个 private 的成员变量数组，其中 pics 数组存放游戏图片 id，titles 数组存放游戏简介。这两个数组值作为热门游戏的内容，代码如下所示。

```
//游戏图标数组
private int[] pics =
    { R.drawable.ico1, R.drawable.ico2, R.drawable.ico3, R.drawable.ico4,
      R.drawable.ico5, R.drawable.ico6 };
//游戏说明数组
private String[] titles =
    { "天天爱消除来自于萌萌星球，只要找出 3 个相同的动物手拉手排一起，它们就能 Happy 得打滚", "新的植物，新的僵尸，新的玩法，结合最先进的触摸屏技术，直接加入僵尸抵抗战", "免费手机社交应用，支持语音、文字消息、表情、图片、视频等方式交流", "绿猪逆袭啦！利用各种工具潜入小鸟巢穴，偷走蛋蛋", "Lep 孤军奋战，解救他的朋友们和家人。请帮助他获得更多强大的技能，打败巫师", "拖拽功夫虫躲避障碍，打破玻璃瓶营救小虫的益智游戏，让我们开始这场拯救行动吧！" };
```

（3）删除方法 onCreate 中自动生成的代码。添加如下代码，隐藏标题栏和调用父类的方法 onCreate 进行初始化，并设置 activity_main.xml 作为主布局文件。

```
requestWindowFeature(Window.FEATURE_NO_TITLE);
super.onCreate(savedInstanceState);
setContentView(R.layout.activity_main);
```

（4）通过 id 分别获得主布局和导航栏对应的布局，并设置导航栏布局的背景资源。

```
// 获得主布局
LinearLayout mainLayout = (LinearLayout) findViewById(R.id.mainLayout);
// 设置导航栏背景资源
FrameLayout fLayout = (FrameLayout) findViewById(R.id.flayout);
fLayout.setBackgroundResource(R.drawable.repeat_bg);
```

（5）在导航栏下，需放置多条游戏信息。每条信息包含一张图片和一个简单的文字说明，这就需要一个线性布局来包装该条信息，布局方向为水平。创建保存该线性布局参数的对象，设置宽度为与父窗体一致，高度为根据内容自动调整。同时，再创建一个保存图片布局参数的对象，设置图片的宽度、高度均为 100 像素。

```
//设置子布局的参数对象
LinearLayout.LayoutParams lps = new LinearLayout.LayoutParams(
LayoutParams.MATCH_PARENT, LayoutParams.WRAP_CONTENT);
//设置图片参数对象
LinearLayout.LayoutParams params = new LinearLayout.LayoutParams(100, 100);
```

（6）通过对图片数组和游戏简介数组的遍历，针对每个图片和游戏简介，建立一个线性布局存放，设置该线性布局的方向为水平，设置间距，以及上步创建的参数对象。创建游戏图片组件，设置其图片资源来自 pics 数组，以及创建游戏说明文本组件，设置内容来自 titles 数组。

```
//循环建立游戏图片和游戏说明,并加入布局中
for(int i = 0; i < pics.length; i + +) {
    //创建子布局存放图片和简介
    LinearLayout childLayout = new LinearLayout(this);
    childLayout.setOrientation(LinearLayout.HORIZONTAL);
    childLayout.setPadding(2, 5, 2, 5);
    childLayout.setLayoutParams(lps);
    ImageView iv = new ImageView(this);  //创建游戏图片组件
    iv.setImageResource(pics[i]);  //设置图片资源
    iv.setLayoutParams(params);  //设置图片组件参数对象
    TextView tv = new TextView(this);  //创建游戏说明文本框组件
    tv.setPadding(5, 5, 5, 5);  //设置文本间距
    tv.setText(titles[i]);  //设置文本框内容
```

（7）最后，将图片和游戏说明的文本加入到线性布局中，并将该线性布局加入到主布局中。

```
    // 子布局中加入图片和文字
    childLayout.addView(iv);
    childLayout.addView(tv);
    // 主布局中加入子布局
    mainLayout.addView(childLayout);
}
```

运行该实例，显示结果如图 5.16 所示。

从本例可以看出，结合 XML 布局文件和 Java 代码混合控制视图界面，既具有了灵活性，又具有了便捷性。

图 5.16　代码和 XML 联合控制视图界面

5.3 多界面的使用

在 Android 应用中，经常会需要在多个界面之间交换数据。本节将介绍如何使用 Intent 和 Bundle 在 Activity 间交换数据，以及调用另一个界面并返回结果。

在多界面场合传递数据，可以通过 Intent 来实现，通常 Intent 可被称为两个界面之间的信使。对于需传递的数据，有两种封装方式。一种是使用 Intent 封装数据，另一种是使用 Bundle 封装数据。

界面间传递数据一般使用以下 4 个步骤。

（1）创建 Intent 实例。

实例化 Intent，通常采用如下 3 种方式。

- 在 Intent 构造方法中指定当前 Activity 和需打开的 Activity。

```
Intent intent = new Intent(A.class, B.class);
```

其中 A.class 表示当前 Activity 类，B.class 表示需打开的 Activity 类。

- 通过 Intent 的 setClass()方法指定当前 Activity 和需打开的 Activity。

```
Intent intent = new Intent();
intent.setClass(A.class, B.class);
```

- 通过 Intent 的 setClassName()方法指定当前 Activity 和需打开的 Activity。

```
Intent intent = new Intent();
intent.setClassName(A.this, "package.B");
```

第一个参数为当前 Activity 类名，第二个参数为“包名.类名”。

或：

```
Intent intent = new Intent();
intent.setClassName("package- ", "package.B");
```

第一个参数是 Activity 类所在的包名，第二个参数为“包名.类名”。

（2）封装数据。

封装要传递到下一界面的数据，一般有两种方式，分别由 Intent 或 Bundle 对象来封装，将由下面小节介绍。

（3）打开下一 Activity 界面。

打开下一 Activity 界面通常有两种方式：

- startActivity(Intent intent)

该方法直接转入下一界面中。

- startActivityForResult(Intent intent, int requestCode)

该方法转入下一界面，下一界面关闭后会自动返回当前页。其中 requestCode 必须大于等于 0，是请求编码。

（4）在打开的下一界面中接收数据。

首先通过 getIntent()方法获得 Intent 实例，再通过 getStringExtra()方法获取各项数据。例如：

```
Intent intent = getIntent();
String name = intent.getStringExtra("name");
```

下面分别对使用 Intent、Bundle 封装数据和获取另一个界面返回结果予以简单的介绍。

5.3.1 使用 Intent 封装数据

将数据封装在 Intent 中，通常使用 putExtra(String name, XXX value)方法，第一个参数表示封装数据名称，第二个参数表示各种类型的值。例如：

```
intent.putExtra("name", "张三");
intent.putExtra("age", 20);
```

下面通过例 5.9 来介绍使用 Intent 封装数据传递到下一界面。

【例 5.9】 使用 Intent 封装数据，预定瑜伽馆的服务。

（1）新建 Android 项目，修改主布局文件。将主布局改为垂直排列的线性布局，依次添加瑜伽馆 logo、输入姓名文本框、预定日期文本框、选择服务项目单选按钮、选择技师文本框和预定按钮。代码如下所示。

```
<?xml version = "1.0" encoding = "utf-8"?>
<LinearLayout xmlns:android = "http://schemas.android.com/apk/res/android"
    android:layout_width = "match_parent"
    android:layout_height = "match_parent"
    android:background = "#FFFAE4"
    android:gravity = "center_horizontal"
    android:orientation = "vertical" >
    <ImageView
        android:layout_width = "352px"
        android:layout_height = "131px"
        android:gravity = "center_horizontal"
        android:src = "@drawable/logo" />
    <EditText android:id = "@+id/name"
        android:layout_width = "wrap_content"
        android:layout_height = "wrap_content"
        android:hint = "请输入姓名"
        android:minWidth = "400px"
        android:textSize = "20dp" />
    <EditText android:id = "@+id/date"
        android:layout_width = "wrap_content"
        android:layout_height = "wrap_content"
        android:hint = "请输入日期"
        android:minWidth = "400px"
        android:onClick = "chooseDate"
        android:textSize = "20dp" />
    <RadioGroup android:id = "@+id/service"
        android:layout_width = "wrap_content"
        android:layout_height = "wrap_content"
        android:minWidth = "400px"
        android:orientation = "horizontal" >
        <RadioButton
            android:layout_width = "wrap_content"
            android:layout_height = "wrap_content"
            android:textSize = "20dp"
            android:text = "Spa" />
        <RadioButton
            android:layout_width = "wrap_content"
            android:layout_height = "wrap_content"
```

```
            android:textSize = "20dp"
            android:text = "Yoga" />
        <RadioButton
            android:layout_width = "wrap_content"
            android:layout_height = "wrap_content"
            android:textSize = "20dp"
            android:text = "Detox" />
    </RadioGroup>
    <EditText android:id = "@+id/receiver"
        android:layout_width = "wrap_content"
        android:layout_height = "wrap_content"
        android:hint = "请选择技师"
        android:minWidth = "400px"
        android:textSize = "20dp" />
    <Button
        android:layout_width = "400px"
        android:layout_height = "wrap_content"
        android:background = "@drawable/custom_button"
        android:onClick = "book"
        android:text = "预定"
        android:textSize = "20dp" />
</LinearLayout>
```

（2）修改主 Activity 文件 BookActivity.java，代码如下所示。

```
public class BookActivity extends Activity {
    private EditText name, date, receiver;
    private RadioGroup service;
    private int year, month, day;
    private String serviceName;
    @Override
    protected void onCreate(Bundle savedInstanceState) {
        super.onCreate(savedInstanceState);
        setContentView(R.layout.main);
        name = (EditText) findViewById(R.id.name);
        date = (EditText) findViewById(R.id.date);
        receiver = (EditText) findViewById(R.id.receiver);
        service = (RadioGroup) findViewById(R.id.service);
    }
    //通过日期对话框组件选择日期
    public void chooseDate(View view) {
        DatePickerDialog datepd = new DatePickerDialog(this, new DatePickerDialog.
OnDateSetListener(){
            @Override
            public void onDateSet(DatePicker dp, int year, int month, int day) {
                BookActivity.this.year = year;
                BookActivity.this.month = month;
                BookActivity.this.day = day;
                date.setText(year + "-" + (month + 1) + "-" + day); //写入到日期框
            }
        }, 2013, 8, 20); //默认为2013-8-20日
        datepd.setMessage("请选择日期");
        datepd.show();
```

```
    }
    //预定
    public void book(View view) {
        for(int i = 0; i < service.getChildCount(); i + +) {
            RadioButton r = (RadioButton) service.getChildAt(i);
            if(r.isChecked()) {
                serviceName = r.getText().toString();
                break;
            }
        }
        Intent intent = new Intent();
        // 以下三种方法均可指定要打开的 Activity 界面
        // intent.setClassName("com.example.book",
        //             "com.example.book.BookResultActivity");
        // intent.setClassName(BookActivity.this,
        //             "com.example.book.BookResultActivity");
        intent.setClass(BookActivity.this, BookResultActivity.class);
        intent.putExtra("name", name.getText().toString());
        intent.putExtra("date", date.getText().toString());
        intent.putExtra("receiver", receiver.getText().toString());
        intent.putExtra("serviceName", serviceName);
        startActivity(intent);
    }
}
```

上述代码中在单击日期文本框时，会调用 chooseDate()方法，该方法使用日期对话框组件 DatePickerDialog 选择日期。日期对话框实例化需要五个参数，分别是当前上下文环境 Context，DatePickerDialog.OnDateSetListener 监听器对象，默认年、月、日的值。通过 Intent 类的 putExtra()方法保存姓名、日期、选择服务和技师，并通过 startActivity()方法打开下一 Activity 界面。

（3）创建接收预定的界面布局文件 bookresult.xml，代码如下所示。

```
<?xml version = "1.0" encoding = "utf-8"?>
<LinearLayout xmlns:android = "http://schemas.android.com/apk/res/android"
    android:layout_width = "match_parent"
    android:layout_height = "match_parent"
    android:background = "#FFFAE4"
    android:orientation = "vertical" >
    <TextView
        android:layout_width = "match_parent"
        android:layout_height = "wrap_content"
        android:text = "您的预订信息是:"
        android:textSize = "20dp" />
    <TextView android:id = "@+id/result"
        android:layout_width = "match_parent"
        android:layout_height = "wrap_content"
        android:textSize = "20dp" />
</LinearLayout>
```

（4）创建接收预定的界面 Activity 文件 BookResultActivity.java，代码如下所示。

```
public class BookResultActivity extends Activity {
    @Override
```

```
    protected void onCreate(Bundle savedInstanceState) {
        super.onCreate(savedInstanceState);
        setContentView(R.layout.bookresult);
        TextView result = (TextView) findViewById(R.id.result);
        Intent intent = getIntent(); //获得 Intent 对象
        String bookItem = "姓名：" + intent.getStringExtra("name") + "\n";
        bookItem += "日期：" + intent.getStringExtra("date") + "\n";
        bookItem += "技师：" + intent.getStringExtra("receiver") + "\n";
        bookItem += "选择项目：" + intent.getStringExtra("serviceName");
        result.setText(bookItem);
    }
}
```

（5）在 AndroidManifest.xml 文件中的<application>标记内加入第二个界面的<activity>标记，代码如下所示。

```
<activity
    android:name = "com.example.book.BookResultActivity"
    android:label = "@string/booktitle" >
</activity>
```

运行本实例，结果分别如图 5.17、图 5.18、图 5.19 所示。

图 5.17　预定首页

图 5.18　选择日期对话框

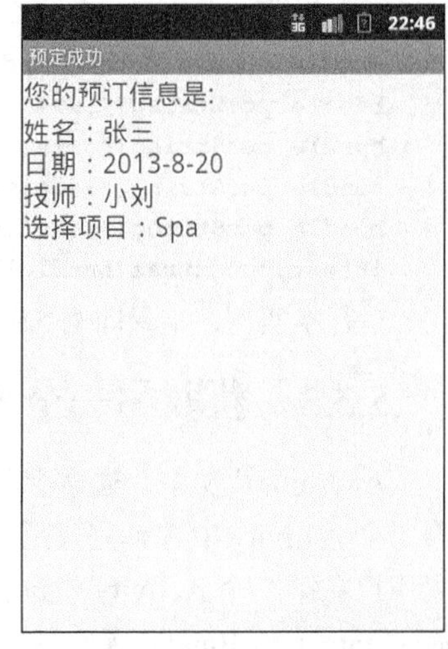

图 5.19　第二页接收数据

5.3.2　使用 Bundle 封装数据

Bundle 是用于不同 Activity 之间的数据传递类，其常用方法如表 5.9 所示。

表 5.9　Bundle 常用方法

属　性　值	说　　明
clear()	清除此 Bundle 映射中的所有保存的数据
clone()	克隆当前 Bundle
containsKey(String key)	返回指定 key 的值

续表

属 性 值	说 明
getString(String key)	返回指定 key 的字符
hasFileDescriptors()	指示是否包含任何捆绑打包文件描述符
isEmpty()	如果这个捆绑映射为空，则返回 true
putXXX(String key, XXX value)	保存给定 key 的值，XXX 表示各种类型
remove(String key)	移除指定 key 的值

Bundle 封装数据的一般过程为先获取 Bundle 实例，再通过 putXXX(String key, XXX value)方法封装数据，最后使用 Intent 对象的 putExtras()方法将 Bundle 对象加入 Intent 中。例如：

```
Bundle bundle = new Bundle();
bundle.putString("name", "张三");
bundle.putInt("age", 20);
intent.putExtras(bundle);
```

下面通过例 5.10 来介绍使用 Bundle 封装数据。

【例 5.10】 使用 Bundle 封装数据，预定瑜伽馆的服务。

（1）使用例 5.9 项目的各文件。

（2）修改主 Activity 文件 BookActivity.java，将 Intent 封装数据的代码替换为如下所示代码。

```
Bundle bundle = new Bundle();
bundle.putString("name", name.getText().toString());
bundle.putString("date", date.getText().toString());
bundle.putString("receiver", receiver.getText().toString());
bundle.putString("serviceName", serviceName);
intent.putExtras(bundle);
```

运行本实例，结果同例 5.9 的结果。

5.3.3 获取另一个界面返回结果

Android 开发中，通常会存在这种应用：第一个 Activity 打开第二个 Activity，当用户在第二个 Activity 中选择完成后，程序会自动返回到第一个 Activity 中，第一个 Activity 可以获取用户在第二个 Activity 中选择的结果，如程序中常出现的“返回上一步”功能。这时也可以通过 Intent 和 Bundle 来实现，与前例所介绍的两个 Activity 之间交互数据不同的是，此时需要通过 startActivityForResult()方法来打开第二个 Activity，同时第一个 Activity 中还需要完成 onActivityResult()方法用于回调。

下面通过例 5.11 来介绍如何获取另一个界面返回结果。

【例 5.11】 通过获取另一个界面返回结果，选择瑜伽馆技师。

（1）使用例 5.9 项目的各文件。

（2）修改主布局文件，在 id 为 receiver 的技师文本框中，添加如下属性。

```
android:onClick = "chooseReceiver"
```

即单击该文本框时，调用 chooseReceiver()方法。

（3）修改主 Activity 文件 BookActivity.java，添加两个方法，chooseReceiver() 和 onActivityResult()。代码如下所示。

```
// 打开选择技师 Activity
```

```
public void chooseReceiver(View view) {
    Intent intent = new Intent();
    intent.setClass(BookActivity.this, ReceiverListActivity.class);
    startActivityForResult(intent, 0x111);
}
// 回调方法
@Override
protected void onActivityResult(int requestCode, int resultCode, Intent data) {
    super.onActivityResult(requestCode, resultCode, data);
    if(requestCode == 0x111 && resultCode == 0x111) {
        Bundle bundle = data.getExtras();
        receiver.setText(bundle.getString("receiverName"));
    }
}
```

其中在 chooseReceiver()中创建一个 Intent 实例，指定打开的 Activity 界面类，最后调用 startActivityForResult()方法打开指定界面。startActivityForResult()方法与 startActivity()都是打开界面的方法，区别在于，当前界面需要获取打开界面的数据。本例中新创建的 startActivityForResult()方法用于获得打开界面返回值，是个回调方法，即打开的界面关闭后会主动调用第一个界面的该方法。当请求码和结果码（requestCode 和 resultCode）都与打开第二个界面时指定的请求码的值（0x111）相同的话，从 bundle 中获取返回的数据。

（4）创建技师选择 Activity 的布局文件 receiverlist.xml，将布局设置为垂直排列的线性布局，放置一个提示文本框和一个表格视图组件。代码如下所示。

```
<?xml version = "1.0" encoding = "utf-8"?>
<LinearLayout xmlns:android = "http://schemas.android.com/apk/res/android"
    android:layout_width = "match_parent"
    android:layout_height = "match_parent"
    android:background = "#FFFAE4"
    android:gravity = "center_horizontal"
    android:orientation = "vertical" >
    <TextView
        android:layout_width = "match_parent"
        android:layout_height = "wrap_content"
        android:text = "请选择技师："
        android:textSize = "20dp" />
    <GridView android:id = "@+id/gridView"
        android:layout_width = "match_parent"
        android:layout_height = "match_parent"
        android:layout_margin = "5dp"
        android:horizontalSpacing = "5px"
        android:verticalSpacing = "5px"
        android:numColumns = "2" />
</LinearLayout>
```

（5）创建技师列表 Activity 文件 ReceiverListActivity.java，代码如下所示。

```
public class ReceiverListActivity extends Activity {
    private GridView gridView;
    // img 和 receiverName 分别为技师照片和名字数组
    private int[] img= { R.drawable.receiver1, R.drawable.receiver2,
```

```
            R.drawable.receiver3, R.drawable.receiver4 };
    private String[] receiverName= { "小王", "小徐", "小黄", "小薛" };
    @Override
    protected void onCreate(Bundle savedInstanceState) {
        super.onCreate(savedInstanceState);
        setContentView(R.layout.receiverlist);
        gridView = (GridView) findViewById(R.id.gridView);
        // 创建适配器
        BaseAdapter adapter = new BaseAdapter() {
            @Override
            public View getView(int position, View view, ViewGroup parent) {
                ImageView imageView;
                if(view == null) {
                    imageView = new ImageView(ReceiverListActivity.this);
                    imageView.setAdjustViewBounds(true);
                    imageView.setMaxWidth(200);
                    imageView.setMaxHeight(200);
                    imageView.setPadding(5, 5, 5, 5);
                } else
                    imageView = (ImageView) view;
                imageView.setImageResource(img[position]);
                return imageView;
            }
            @Override
            public long getItemId(int position) {
                return position;
            }
            @Override
            public Object getItem(int position) {
                return position;
            }
            @Override
            public int getCount() {
                return img.length;
            }
        };
        gridView.setAdapter(adapter); // 表格视图组件设置适配器
        // 表格视图组件添加监听器
        gridView.setOnItemClickListener(new AdapterView.OnItemClickListener() {
            @Override
            public void onItemClick(AdapterView<?> parent, View view, int position, long id) {
                Intent intent = getIntent();
                Bundle bundle = new Bundle();
                bundle.putString("receiverName", receiverName[position]);
                intent.putExtras(bundle);
                setResult(0x111, intent);
                finish();
            }
        });
```

```
        }
    }
```

上述代码中，分别定义了保存技师照片和名字的 img 和 receiverName 数组，并创建 BaseAdapter 适配器与表格视图组件关联，最后给表格视图组件添加监听器，当单击某张照片后将对应的技师名封装到 Bundle 中返回，通过 finish()方法关闭这个新打开的界面。

（6）在 AndroidManifest.xml 文件中的<application>标记内加入技师界面的<activity>标记，代码如下所示。

```
<activity
    android:name = "com.example.book.ReceiverListActivity"
    android:label = "@string/chooseteceiver" >
</activity>
```

运行修改过的工程，打开技师界面之前如图 5.20 所示，打开技师界面如图 5.21 所示，关闭技师界面如图 5.22 所示。

图 5.20　打开技师界面之前

图 5.21　打开技师界面

图 5.22　关闭技师界面之后

5.4　小　　结

本章主要介绍了用户界面设计方面的相关内容，包括了 Android 中控制界面的四种常用的布局管理器（绝对布局已经废弃，不再推荐使用）。使用布局管理器进行视图界面控制的思路是将程序的表现层和控制层分离，在后期修改用户界面时无需更改程序的源代码，用户还能通过可视化工具方便快捷地查看所设计的界面，便于加快界面设计的过程，体现了 MVC 的设计思想，推荐用户使用。

随后又讲解了控制视图界面的其他方法，首先介绍了使用纯代码控制界面的方法，解决了在 XML 布局文件中控制视图界面的缺点，就是不够灵活的问题，但也带来了样式调整复杂，过于繁琐的问题。接着介绍了使用 XML 布局文件和 Java 代码共同控制视图界面的方法，具有了灵活性和快捷性，能够达到很好的效果。

最后介绍了界面之间进行数据交互的方法，数据的封装可以使用 Intent，也可以使用 Bundle，

Bundle 应用范围更为广泛。另外讲解了打开界面的两种方法，其中 startActivity()仅仅是打开界面而不做其他操作，而 startActivityForResult()方法打开界面等待下个界面传递过来的数据，通过重载方法 onActivityResult()来接收数据。

通过本章的学习，用户可以了解到设计良好的布局对于应用程序的性能很重要。若需要提高布局性能，可以使用像 monitor 和 lint 之类的 Android SDK 工具来调试和优化应用程序，这方面内容可自行查找相关资料学习。

练　习

1. 使用线性布局显示一个带图片和文字说明的歌曲列表。
2. 使用表格布局设计一个用户注册界面。
3. 使用帧布局设计一个带有标题、内容和页脚的界面。
4. 使用相对布局设计一个图片按十字型排列的界面。
5. 使用代码控制视图界面的方法设计一个登录界面。
6. 设计一个照片墙应用，可单击上一页、下一页进行翻页。

第 6 章 Android 数据存储与共享

本章介绍 Android 应用程序中如何将数据存储在手机内存或 SD 卡上，以供后续处理使用。常用的数据存储方式有首选项信息存储、数据文件存储、SQLite 数据库存储。此外本章还将介绍 ContentProvider 和 ContentResolver，这是 Android 平台中提供的跨应用程序数据共享的方式。

6.1 数据存储与共享方式概述

Android 平台提供了多种不同的方式来实现应用程序数据的存储与共享，同时能够读写和操作手机内存及 SD 卡上的各种文件的数据。智能手机机身内存容量有限，除了 Android 系统本身占用的内存空间外，可供用户使用的存储空间并不多，所以这部分空间仅仅用来存放一些小规模的数据，比如用户首选项信息，应用程序的配置信息等。Android 提供了 SharedPreferences 类及相关的一系列方法来操作和处理这些数据信息，该部分的内容将在 6.2 节中进行介绍。

如果需要按照特定的格式来保存和读取少量数据，通常使用数据文件的形式。数据文件可以保存在机身内存或者 SD 卡中。Android 数据文件的读写操作采用了 Java API 中的 FileInputStream 类和 FileOutputStream 类，以及相关的一系列方法。如果用户熟悉 Java 中文件的读写操作，则可以将其直接应用到 Android 应用程序中来，这部分内容将在 6.3 节中进行介绍。

如果应用程序中有大量的数据需要频繁地进行增、删、改、查的操作，那么使用数据库存储和操作数据将是最佳的选择。Android 系统中集成了 SQLite 数据库，并且为数据库的操作提供了相关的类和方法，便于没有数据库开发经验的开发者编写程序。SQLite 数据库介绍以及详细的数据库操作将在 6.4 节中进行介绍。

本章最后的 6.5 节将介绍 Android 平台中利用 Content Provider 机制来实现跨应用程序数据共享的方式。一个应用程序可以通过 Content Provider 来发布自己的数据，但是并不会泄露这些数据在其应用程序内部的组织和存储方式。其他的应用程序可以通过 Content Resolver 来获取自己感兴趣的某个应用程序发布出来的共享数据。

6.2 首选项信息

为了实现更好的用户体验和人性化的操作界面，大部分应用程序提供了可定制的设置选项或菜单，用户可以按照自己的喜好来更改程序的界面风格、操作习惯、常用列表等数据信息。通常把这些信息叫做首选项信息，这些信息基本都是以键值对（key-value）的形式来存储的。Android 提供了 SharedPreferences 类及相关的一系列方法来操作和处理这些首选项信息。这样用户就无需自己再建立数据文件和指定数据格式，可以直接利用 API 提供的相关接口非常方便地进行数据的存储操作。

需要强调的是虽然用户不需要自己去建立数据文件，但是系统仍然会通过创建一个 XML 格式数据文件的方式来保存首选项信息的数据。一个应用程序中可以通过指定不同的 XML 文件名来创建多个首选项信息数据文件，除此以外，SharedPreferences 类还可以控制首选项信息的保护和公开模式。有些首选项信息仅仅是对本应用程序私有的，需要保护。而有些首选项信息则是可以公开的，或者需要公开的，以便其他应用程序可以通过这些信息获取所需的数据。根据首选项信息数据的保护及不同共享方式的需要，SharedPreferences 类提供 3 种不同的数据权限管理模式：

- MODE_PRIVATE：私有模式
- MODE_WORLD_READABLE：全局可读模式
- MODE_WORLD_WRITEABLE：全局可写模式

以下就分私有首选项信息数据存储和公有首选项信息数据存储与共享两个方面来进行介绍。

6.2.1 私有数据存储

利用 SharedPreferences 方式来存储和操作私有首选项信息数据的步骤如下。

1. SharedPreferences 实例的获取

通过 Context 类的 getSharedPreferences()方法来获取 SharedPreferences 类的实例，该方法的 API 为：

- public SharedPreferences getSharedPreferences(String name, int mode)

该方法需要两个参数，并返回一个 SharedPreferences 类的实例。两个参数分别为字符串类型的首选项信息数据文件名，和整数类型的访问权限模式。

首选项信息数据文件唯一的存储格式为 XML 格式，所以在第一个参数指定数据文件名时无需加.xml 的后缀；第二个参数即为上面所列出的三种数据权限管理模式之一。

- public SharedPreferences getSharedPreference()

如果仅需要一个共享文件，则可以使用该方法。因为只有一个文件，它不需要提供名称。

下例为获取 SharedPreferences 对象的方法，这里对于私有数据使用 MODE_PRIVATE 模式。

```
SharedPreferences sp = getSharedPreferences("myShare", MODE_PRIVATE);
```

2. SharedPreferences.Editor 实例的获取

SharedPreferences 对象本身只提供读取数据的方法，而不能进行数据的存储和修改操作。这些操作需要通过 SharedPreferences 类的嵌套类 SharedPreferences. Editor 类所提供的方法来实现。Editor 的实例可以通过 SharedPreferences 类的 edit()方法获得，该方法不需要任何参数，如下例所示。

```
Editor editor = sp.edit();
```

3. 数据的保存

Editor 类提供了五个方法来分别保存五种基本类型的数据（boolean、float、int、long、String），这也是首选项信息数据文件中所支持的几种数据类型。所对应的方法分别是：

- putBoolean(String key, boolean value)
- putFloat(String key, float value)
- putInt(String key, int value)
- putLong(String key, long value)
- putString(String key, String value)

可以看到这些方法都需要两个参数，第一个都是字符串类型的 key 键，第二个分别是对应每个方法的基本数据类型。

除了以上保存数据的方法，Editor 类还提供了删除和保存提交数据的方法：

- remove(String key)

该方法用来删除由参数 key 指定的已经保存的数据。

- commit()

该方法用来提交所有更新过的数据。如果想要保存用户提交的姓名、年龄、分数等数据，可以编写如下例所示代码。

```
editor.putString("name", "angel");
editor.putInt("age", 20);
editor.putFloat("score", 98.5);
editor.commit();
```

4. 数据的获取

对应的，SharedPreferences 类提供了 5 个方法来获取已存储的首选项信息数据，分别是：

- boolean getBoolean(String key, boolean defValue)
- float getFloat(String key, float defValue)
- int getInt(String key, int defValue)
- long getLong(String key, long defValue)
- String getString(String key, String defValue)

可以看到这些方法都需要两个参数，并且返回所对应数据类型的数据。第一个参数都是字符串类型的 key 键，通过该值查找并返回所对应的数据。第二个参数是默认值，当首选项信息数据文件中找不到由第一个参数指定的键所对应的值时，方法返回这个默认值作为返回值。

除此以外，SharedPreferences 类还提供 getAll()方法来返回所有的首选项信息数据，和 contains(String key)方法来检查是否存在某个特定的键值信息。这两个方法的使用请用户自行查阅 API 文档，此处不再赘述。

下面通过例 6.1 来介绍如何使用 SharedPreferences 类和 Editor 类来存储和读取首选项信息数据。

【例 6.1】　SharedPreferences 中私有数据的存取应用。

（1）创建 Android 工程，工程名为 SP_Share，主 Activity 类名为 SPShare Activity，位于 com.example.sp 包内。

（2）修改主布局文件，将默认的布局管理器改为 LinearLayout 线性布局，垂直排列。在主布局中再放置三个水平线性布局，第一个线性布局中放入文本为 key 的文本框和文本输入框，第二个线性布局中放入文本为 value 的文本框和文本输入框，第三个线性布局中放入保存和读取按钮，代码如下所示。

```
<?xml version = "1.0" encoding = "utf-8"?>
<LinearLayout
   xmlns:android = "http://schemas.android.com/apk/res/android"
   android:layout_width = "match_parent"
   android:layout_height = "match_parent"
   android:background = "#FFFAE4"
   android:orientation = "vertical" >
   <LinearLayout
      android:layout_width = "match_parent"
      android:layout_height = "wrap_content"
      android:orientation = "horizontal" >
      <TextView
         android:layout_width = "wrap_content"
         android:layout_height = "wrap_content"
         android:text = "Key    " />
      <EditText android:id = "@+id/key"
         android:layout_width = "match_parent"
         android:layout_height = "wrap_content"  />
   </LinearLayout>
   <LinearLayout
      android:layout_width = "match_parent"
      android:layout_height = "wrap_content"
      android:orientation = "horizontal" >
      <TextView
         android:layout_width = "wrap_content"
         android:layout_height = "wrap_content"
         android:text = "Value" />
      <EditText android:id = "@+id/value"
         android:layout_width = "match_parent"
         android:layout_height = "wrap_content" />
   </LinearLayout>
   <LinearLayout
      android:layout_width = "match_parent"
      android:layout_height = "wrap_content"
      android:gravity = "center"
      android:orientation = "horizontal" >
      <Button android:id = "@+id/save"
         android:layout_width = "wrap_content"
         android:layout_height = "wrap_content"
         android:text = "保存" />
      <Button android:id = "@+id/read"
         android:layout_width = "wrap_content"
         android:layout_height = "wrap_content"
      android:text = "读取" />
   </LinearLayout>
</LinearLayout>
```

（3）修改主 Activity 文件 SPShareActivity.java，单击保存按钮可以将 key 和 value 值保存在首选项信息数据文件中。该文件位于系统文件目录“/data/data/<package_name>/shared_prefs”下，

文件名（SPShare.xml）由程序中 getSharedPreferences()方法的参数指定，如图 6.1 所示。当用户只输入 key 的值，然后单击读取按钮，可以从数据文件中获取相应的 value 值，并显示在文本框内。代码如下所示。

```
public class SPShareActivity extends Activity {
    @Override
    public void onCreate(Bundle savedInstanceState) {
        super.onCreate(savedInstanceState);
        setContentView(R.layout.main);
        final EditText key = (EditText) findViewById(R.id.key);
        final EditText value = (EditText) findViewById(R.id.value);
        final Button save = (Button) findViewById(R.id.save);
        final Button read = (Button) findViewById(R.id.read);
        final SharedPreferences sp = getSharedPreferences("SPShare",
                                MODE_PRIVATE); //私有模式,不对外公开
        final SharedPreferences.Editor editor = sp.edit();
        //监听器中保存键值对
        save.setOnClickListener(new View.OnClickListener() {
            @Override
            public void onClick(View arg0) {
                String k = key.getText().toString();
                String v = value.getText().toString();
                editor.putString(k, v);
                editor.commit();
                Toast.makeText(SPShareActivity.this, "保存成功!",
                                Toast.LENGTH_LONG).show();
            }
        });
        //监听器中根据键取值
        read.setOnClickListener(new View.OnClickListener() {
            @Override
            public void onClick(View arg0) {
                String k = key.getText().toString();
                String v = sp.getString(k, "NO_VALUE");
                value.setText(v);
            }
        });
    }
}
```

运行本实例，结果如图 6.2 所示。

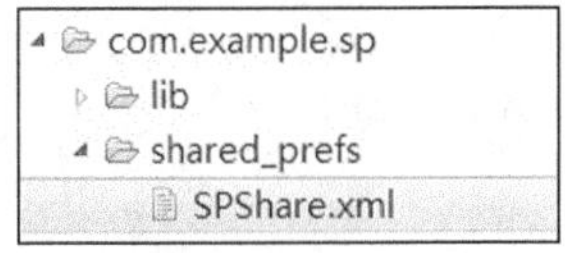

图 6.1　首选项文件位置

图 6.2　首选项信息数据存储

6.2.2 公有数据存储与共享

上面的示例程序例 6.1 中，创建首选项信息数据文件时使用的模式为 MODE _PRIVATE，表明该数据文件是私有的，不对其他的应用程序公开。而访问权限模式改为 MODE_WORLD_READABLE，则可以对外部应用公开。

下面通过例 6.2 介绍如何访问其他应用中的 SharedPreferences 数据。

【例 6.2】 SharedPreferences 中公有数据的存储与共享应用。

（1）将例 6.1 中的访问权限模式改为 MODE_WORLD_READABLE，并运行填入一组数据，这样创建的数据文件可以被其他应用程序读取。

（2）创建 Android 工程 SP_Test，主 Activity 类名为 ReadSPDataActivity，位于 com.example.sp2 包内。

（3）修改本工程主布局文件内容同例 6.1 中的主布局文件。

（4）修改主 Activity 文件 ReadSPDataActivity.java，代码如下所示。

```
public class ReadSPDataActivity extends Activity {
    @Override
    public void onCreate(Bundle savedInstanceState) {
        super.onCreate(savedInstanceState);
        setContentView(R.layout.main);
        final EditText key = (EditText) findViewById(R.id.key);
        final EditText value = (EditText) findViewById(R.id.value);
        final Button save = (Button) findViewById(R.id.save);
        final Button read = (Button) findViewById(R.id.read);
        Context context = null;
        try {
            context = createPackageContext("com.example.sp", Context.CONTEXT_IGNORE_
SECURITY); //指定例 6.1 中的包名(共享首选项在包内)
        }
        catch(NameNotFoundException e) {
            e.printStackTrace();
        }
        final SharedPreferences sp = context.getSharedPreferences("SPShare", MODE_WORLD_
READABLE);
        final SharedPreferences.Editor editor = sp.edit();
        save.setOnClickListener(new View.OnClickListener() {
            @Override
            public void onClick(View view) {
                String k = key.getText().toString();
                String v = value.getText().toString();
                editor.putString(k, v);
                editor.commit(); //因为例 6.1 提供只读，不能写入
                Toast.makeText(ReadSPDataActivity.this, "保存成功!",
                        Toast.LENGTH_LONG).show();
            }
        });
        read.setOnClickListener(new View.OnClickListener() {
            @Override
            public void onClick(View view) {
                String k = key.getText().toString();
```

```
                String v = sp.getString(k, "NO_VALUE");
                value.setText(v);
            }
        });
    }
}
```

运行本实例，其显示的界面与例 6.1 完全一样。输入一个 key 值，单击 Read 按钮，可以读出在例 6.1 程序中保存的相应 value 值。如果尝试修改 value 值，则提示错误，因为首选项信息数据文件设定的访问权限为 MODE_WORLD_ READABLE，为只读的权限。仔细观察例 6.2 和例 6.1 的不同之处，可以看到要想从一个应用程序中读取另外一个应用程序公开的首选项信息文件，需要首先创建一个对应该应用程序的 Context 对象，使用 createPackageContext()方法，将所要访问的包名作为参数传入即可。然后使用该 Context 对象的 getSharedPre ferences()方法获取 SharedPreferences 对象，方法与前面所讲的完全一样。后面的读取首选项信息数据和存储首选项信息数据的方法也和前面所讲完全一样，只是操作别的应用程序的首选项信息数据文件前需要确保能获得相应的操作权限。

6.3　数 据 文 件

通过 SharedPreferences 类存储的首选项信息数据只能是以键值对的形式处理。如果需要按照特定的格式来保存和读取少量数据，通常使用数据文件的形式。Android 数据文件的读写操作采用了 Java API 中的 FileInputStream 类和 FileOutputStream 类，以及相关的一系列方法。数据文件可以保存在机身内存中或者 SD 卡中。将数据文件保存在机身内存中具有操作便捷、读写速度快、稳定以及卸载 SD 卡后不影响程序正常运行的优点，但缺点是机身内存容量有限，不能保存大量数据。如果数据量较大，则需要将数据文件存储在 SD 卡中。以下分别来进行介绍如何在机身内存和 SD 卡中创建和操作数据文件。

6.3.1　内存数据文件

与文件操作相关的两个主要类是 FileInputStream 和 FileOutputStream。这两个类的 API 与 Java API 中完全一样，此处不再赘述。在 Android 中，每个数据文件都是和创建它的应用程序所绑定的，当删除应用程序时，其在内存中所关联的数据文件也将被同时删除。利用 Context 类提供的 openFileInput()和 openFile Output()方法可以获取本应用程序相应的文件输入和输出流。这两个类的常用方法有：

- FileInputStream openFileInput(Stringname) throws FileNotFoundException

该方法通过参数指定要读取的文件名，返回文件输入流。

- FileOutputStream openFileOutput(String name,int mode) throws FileNotFoundException

该方法需要两个参数，即文件名和文件操作模式，返回文件输出流。文件操作模式为四个整数类型的常量：

- MODE_PRIVATE：默认输出模式
- MODE_APPEND：追加模式
- MODE_WORLD_READABLE：全局只读模式

- MODE_WORLD_WRITEABLE：全局只写模式

另外还需要注意这两个方法都需要处理 FileNotFoundException 的异常。下面通过例 6.3 来介绍如何在应用程序中读写机身内存数据的文件。

【例 6.3】 读写机身内存数据的应用。

（1）修改主布局文件，依次添加文本框用于输入文件名，保存和读取按钮，最后添加文本框用于显示读取文件内容，代码如下所示。

```
<LinearLayout xmlns:android = "http://schemas.android.com/apk/res/android"
    android:layout_width = "match_parent"
    android:layout_height = "match_parent"
    android:orientation = "vertical" >
    <EditText android:id = "@+id/title"
        android:layout_width = "match_parent"
        android:layout_height = "wrap_content"
        android:hint = "请输入文件名" />
    <LinearLayout
        android:layout_width = "match_parent"
        android:layout_height = "wrap_content"
        android:gravity = "center"
        android:orientation = "horizontal" >
        <Button android:id = "@+id/save"
            android:layout_width = "wrap_content"
            android:layout_height = "wrap_content"
            android:text = "保存" />
        <Button android:id = "@+id/read"
            android:layout_width = "wrap_content"
            android:layout_height = "wrap_content"
            android:text = "读取" />
    </LinearLayout>
    <EditText android:id = "@+id/text"
        android:layout_width = "match_parent"
        android:layout_height = "match_parent" />
</LinearLayout>
```

（2）修改主 Activity 文件 ReadMemActivity，单击保存按钮可以按照指定的文件名将 id 名为 text 中的所输入的文本保存，单击读取按钮可以按照指定的文件名读取文件并将内容显示在 id 名为 text 的文本框内，代码如下所示。

```
public class ReadMemActivity extends Activity {
    @Override
    public void onCreate(Bundle savedInstanceState) {
        super.onCreate(savedInstanceState);
        setContentView(R.layout.main);
        final EditText title = (EditText) findViewById(R.id.title);
        final EditText text = (EditText) findViewById(R.id.text);
        final Button save = (Button) findViewById(R.id.save);
        final Button read = (Button) findViewById(R.id.read);
        save.setOnClickListener(new OnClickListener() {
            @Override
            public void onClick(View arg0) {
                String t = title.getText().toString();
```

```
                String v = text.getText().toString();
                try {
                    FileOutputStream out = openFileOutput(t, MODE_PRIVATE);
                    PrintStream ps = new PrintStream(out);
                    ps.println(v);
                    ps.close();
                }
                catch (Exception e) {
                    e.printStackTrace();
                }
            }
        });
        read.setOnClickListener(new OnClickListener() {
            @Override
            public void onClick(View view) {
                String t = title.getText().toString();
                try {
                    FileInputStream in = openFileInput(t);
                    byte[] buff = new byte[1024];
                    int hasRead = 0;
                    StringBuilder sb = new StringBuilder("");
                    while ((hasRead = in.read(buff)) > 0) {
                        sb.append(new String(buff, 0, hasRead));
                    }
                    text.setText(sb.toString());
                }
                catch (Exception e) {
                    e.printStackTrace();
                }
            }
        });
    }
}
```

运行本实例，结果如图 6.3 所示。由程序创建的数据文件通常被保存在“/data/data/<package_name>/files”目录下，如图 6.4 所示。

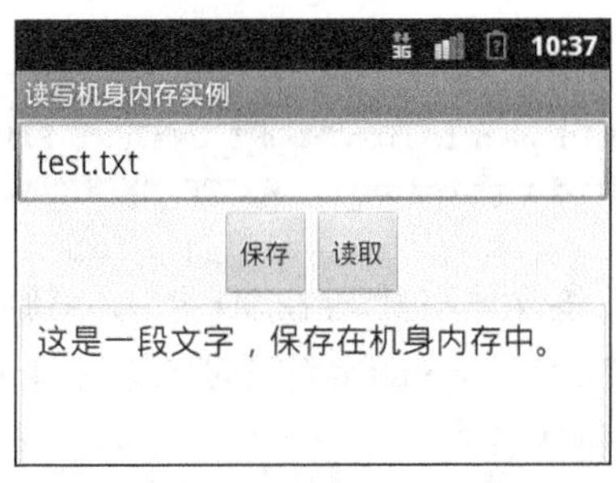

图 6.3　数据文件存储

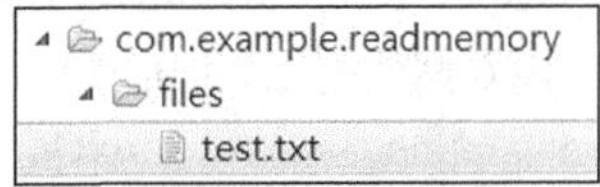

图 6.4　机身内存文件位置

除了以上的基本文件读写操作，Context 类还提供了一些方法进行其他的文件操作，主要有：

- deleteFile()：删除文件
- getDir()：创建/获取一个文件目录

- fileList()：列出本应用程序创建的所有文件
- getFileDir()：获取已创建文件的路径

用户可以查阅 API 文档获取这些方法的详细使用说明。

6.3.2 SD 卡数据文件

在 SD 卡上进行数据文件的各种操作方式与在机身内存中的文件操作基本类似。所不同的主要有两点，一是要在 AndroidManifest.xml 文件中申明对 SD 卡（外部存储空间）的使用权限，二是要在程序中编写检查 SD 卡是否可用的程序。因为程序运行过程中有可能出现 SD 卡暂时不可用的情况。

下面通过例 6.4 来介绍如何在 SD 卡上读写数据。

【例 6.4】 读写 SD 卡数据文件的应用。

（1）参照例 6.3，创建主布局文件，保持程序界面不变。

（2）修改 AndroidManifest.xml 文件，添加允许 SD 卡写入权限的两个<uses-permission>标记，代码如下所示。

```
<?xml version = "1.0" encoding = "utf-8"?>
<manifest xmlns:android = "http://schemas.android.com/apk/res/android"
    package = "com.example.readsdcard"
    android:versionCode = "1"
    android:versionName = "1.0" >
    <uses-sdk
        android:minSdkVersion = "8"
        android:targetSdkVersion = "17" />
    <application
        android:icon = "@drawable/ic_launcher"
        android:label = "@string/app_name" >
        <activity
            android:name = ".ReadSDActivity"
            android:label = "@string/app_name" >
            <intent-filter>
                <action android:name = "android.intent.action.MAIN" />
                <category android:name = "android.intent.category.LAUNCHER" />
            </intent-filter>
        </activity>
    </application>
    <uses-permission android:name = "android.permission.MOUNT_UNMOUNT_FILESYSTEMS"/>
    <uses-permission android:name = "android.permission.WRITE_EXTERNAL_STORAGE"/>
</manifest>
```

（3）修改主 Activity 文件 ReadSDActivity.java，将数据文件保存在 SD 卡中，增加编写检查 SD 卡是否可用的程序，以及修改创建 FileInputStream 和 FileOutput Stream 对象的方式。代码如下所示。

```
public class ReadSDActivity extends Activity {
    @Override
    public void onCreate(Bundle savedInstanceState) {
        super.onCreate(savedInstanceState);
        setContentView(R.layout.main);
        final EditText title = (EditText) findViewById(R.id.title);
        final EditText text = (EditText) findViewById(R.id.text);
        final Button save = (Button) findViewById(R.id.save);
```

```
        final Button read = (Button) findViewById(R.id.read);
        save.setOnClickListener(new View.OnClickListener() {
            @Override
            public void onClick(View view) {
                if(Environment.getExternalStorageState().equals(Environment.MEDIA_MOUNTED)){
                    //获得 SD 卡路径
                    File sdCardDir = Environment.getExternalStorageDirectory();
                    String t = title.getText().toString();
                    String v = text.getText().toString();
                    try {
                        FileOutputStream out = new FileOutputStream(
                                sdCardDir.getCanonicalPath() + "/" + t);
                        PrintStream ps = new PrintStream(out);
                        ps.println(v);
                        ps.close();
                    }
                    catch (Exception e) {
                        e.printStackTrace();
                    }
                }
            }
        });
        read.setOnClickListener(new View.OnClickListener() {
            @Override
            public void onClick(View view) {
                if (Environment.getExternalStorageState().equals(Environment.MEDIA_MOUNTED)) {
                    File sdCardDir = Environment.getExternalStorageDirectory();
                    String t = title.getText().toString();
                    try {
                        FileInputStream in = new
                        FileInputStream(sdCardDir.getCanonicalPath() + "/" + t);
                        byte[] buff = new byte[1024];
                        int hasRead = 0;
                        StringBuilder sb = new StringBuilder("");
                        while ((hasRead = in.read(buff)) > 0) {
                            sb.append(new String(buff, 0, hasRead));
                        }
                        text.setText(sb.toString());
                    }
                    catch (Exception e) {
                        e.printStackTrace();
                    }
                }
            }
        });
    }
}
```

运行本实例，结果同例 6.3，区别在于文件保存在 SD 卡上。

6.4　SQLite 数据库

SQLite 是一种轻量级的基于文件的数据库管理系统，是由 C 语言编写，实现了标准 SQL 中的 CRUD 操作。由于 SQLite 小巧、高效，所以特别适合用于手机等嵌入式设备中来进行大量数据的存储和各种操作。Android 平台提供了对 SQLite 的良好支持，尤其是提供了实现各种数据库操作的 API，方便开发者编写应用程序。

SQLite 内部支持以下几种数据类型：

- NULL：空值
- INTEGER：有符号整数，根据值的大小存储在 1/2/3/4/6/8 字节的空间中
- REAL：浮点数，存储在 8 个字节的空间
- TEXT：文本字符串，以数据库的编码方式存储文本
- BLOB：二进制字节数据

6.4.1 SQLite 基本操作

SQLiteDatabase 是操作 SQLite 数据库的主要类，它提供了一系列方法来进行数据库的各种操作。以下介绍一些常用的方法，更多方法请用户可以查看 API 文档。

- public static SQLiteDatabase openOrCreateDatabase(String path,SQLiteDatabase.CursorFactory factory)

顾名思义，openOrCreateDatabase()方法用来打开或创建一个数据库。这是一个静态方法，其返回值为已创建或打开的数据库对象。如果不能打开或创建数据库，则抛出 SQLiteException 异常。该方法需要两个参数，第一个为数据库文件路径（包括文件名），第二个参数可以指定一个 CursorFactory 对象，也可以直接传入 null 来使用默认的 CursorFactory 对象。当从数据库进行查询操作时，返回的游标对象由该 CursorFactory 对象创建。

- public void execSQL(String sql)

该方法执行一个 SQL 语句，参数为要执行的 SQL 语句。该 SQL 语句不能是查询语句，可以是其他的如创建表、插入数据、删除数据等。该方法没有返回值，如果 SQL 语句执行有错误发生，则抛出 SQLiteException 异常。

- public Cursor rawQuery(String sql, String[] selectionArgs)

该方法执行一个查询 SQL 语句，第一个参数即为该语句。如果所有的查询条件都写在第一个参数里，则第二个参数可以为 null。或者第一个参数 SQL 语句中用“？”保留一些位置，其值由第二个参数字符串数组中的对应值进行替换，该方法返回一个 Cursor 对象。后面将介绍 Cursor 接口的常用方法。

- public void close()

该方法关闭数据库。

Cursor 是一个接口，提供了对于数据库查询结果集进行随机读写的一系列方法。以下列举出一些常用的方法，更多方法请用户查看相关 API 文档。

- boolean moveToNext()

该方法将游标向后移动一次。当一个查询语句返回一个 Cursor 对象时，游标位置停留在第一条数据项的前面，向后移动一次可以到达第一条数据。如果查询结果为空，或者向后移动游标后已经超出了数据集的最后一条数据，该方法返回 false，否则返回 true。

- boolean moveToPrevious()

游标向前移动一位。

- boolean moveToFirst()

游标移动到最前。

- boolean moveToLast()

游标移动到最后。

- boolean moveToPosition(int position)

游标移动到 position 位置处。

- boolean move(int offset)

游标移动 offset 位。

- boolean isFirst()

判断游标是否在第一行。

- boolean isLast()

判断游标是否在最后一行。

- int getPosition()

获取当前游标在整个结果集中的位置。

- int getCount()

获取整个结果集中所有记录项的总数。

- int getColumnCount()

获取结果集中每条记录的字段总数。

- int getInt(int columnIndex)，long getLong(int columnIndex)，float getFloat(int columnIndex)，double getDouble(int columnIndex)，short getShort(int column Index)，String getString(int columnIndex)

这些方法都用于从指定的字段位置获取相对应数据类型的数据。从方法的名字可以看出其作用，此处不再一一介绍。

下面通过例 6.5 说明 SQLite 的使用。

【例 6.5】　通过 SQLite 数据库保存学生成绩信息，并予以显示。

（1）创建工程 SQLite1，修改布局文件，将默认的布局管理器改为 LinearLayout 线性布局，垂直排列。其中界面上方的三个文本框用来输入每个学生的信息，包括学号、姓名、成绩、中间是插入和读取按钮，下方放置一个用于显示已输入学生信息的文本框。代码如下所示。

```
<LinearLayout xmlns:android = "http://schemas.android.com/apk/res/android"
    android:layout_width = "match_parent"
    android:layout_height = "match_parent"
    android:orientation = "vertical" >
    <LinearLayout
        android:layout_width = "match_parent"
        android:layout_height = "wrap_content"
        android:orientation = "horizontal" >
        <TextView
            android:layout_width = "wrap_content"
            android:layout_height = "wrap_content"
            android:gravity = "right"
            android:text = "学号" />
        <EditText android:id = "@+id/id"
            android:layout_width = "wrap_content"
            android:layout_height = "wrap_content"
            android:inputType = "number" />
    </LinearLayout>
```

```
    <LinearLayout
        android:layout_width = "match_parent"
        android:layout_height = "wrap_content"
        android:orientation = "horizontal" >
        <TextView
            android:layout_width = "wrap_content"
            android:layout_height = "wrap_content"
            android:gravity = "right"
            android:text = "姓名" />
        <EditText android:id = "@+id/name"
            android:layout_width = "wrap_content"
            android:layout_height = "wrap_content" />
    </LinearLayout>
    <LinearLayout
        android:layout_width = "match_parent"
        android:layout_height = "wrap_content"
        android:orientation = "horizontal" >
        <TextView
            android:layout_width = "wrap_content"
            android:layout_height = "wrap_content"
            android:gravity = "right"
            android:text = "成绩" />
        <EditText android:id = "@+id/score"
            android:layout_width = "wrap_content"
            android:layout_height = "wrap_content"
            android:inputType = "number" />
    </LinearLayout>
    <LinearLayout
        android:layout_width = "match_parent"
        android:layout_height = "wrap_content"
        android:gravity = "center_horizontal"
        android:orientation = "horizontal" >
        <Button android:id = "@+id/save"
            android:layout_width = "wrap_content"
            android:layout_height = "wrap_content"
            android:text = "插入" />
        <Button android:id = "@+id/read"
            android:layout_width = "wrap_content"
            android:layout_height = "wrap_content"
            android:singleLine = "false"
            android:text = "读取" />
    </LinearLayout>
    <EditText android:id = "@+id/list"
        android:layout_width = "match_parent"
        android:layout_height = "match_parent" />
</LinearLayout>
```

（2）修改主 Activity 文件 SQLite1Activity.java，建立数据库 stu.db3，在该库中建立 student 表。当用户单击插入按钮时，这条学生记录被插入数据库中的 student 表中。用户可通过单击读取按钮，将 student 表中现有的数据全部读取出来，并显示在下方的文本框内。代码如下所示。

```
public class SQLite1Activity extends Activity {
```

```
    @Override
    public void onCreate(Bundle savedInstanceState) {
        super.onCreate(savedInstanceState);
        setContentView(R.layout.main);
        final EditText id = (EditText) findViewById(R.id.id);
        final EditText name = (EditText) findViewById(R.id.name);
        final EditText score = (EditText) findViewById(R.id.score);
        final Button save = (Button) findViewById(R.id.save);
        final Button read = (Button) findViewById(R.id.read);
        final EditText list = (EditText) findViewById(R.id.list);
        final SQLiteDatabase db = SQLiteDatabase.openOrCreateDatabase(this.getFilesDir().
                                  toString() + "/stu.db3", null);
        try {
            db.execSQL("create table student(id integer, name varchar(50), score
                       integer)");
        }
        catch(Exception e) {
            e.printStackTrace();
        }
        save.setOnClickListener(new View.OnClickListener() {
            @Override
            public void onClick(View view) {
                String i = id.getText().toString();
                String n = name.getText().toString();
                String s = score.getText().toString();
                db.execSQL("insert into student values(?, ?, ?)",
                        new String[] { i, n, s });
                Toast.makeText(SQLite1Activity.this, "保存成功!", Toast.LENGTH_SHORT).
                              show();
            }
        });
        read.setOnClickListener(new View.OnClickListener() {
            @Override
            public void onClick(View view) {
                list.setText("");// 清空原有数据
                Cursor cursor = db.rawQuery("select * from student", null);
                while(true) {
                    if(cursor.moveToNext() == false)
                        break;
                    int i = cursor.getInt(0);
                    String n = cursor.getString(1);
                    int s = cursor.getInt(2);
                    String tmp = list.getText().toString();
                    list.setText(tmp + "\n" + i + "    " + n + "    " + s);
                }
            }
        });
    }
}
```

运行本实例，结果如图 6.5 所示。所创建的数据库文件保存在文件系统“/data/data/<package_name>/files/stu.db3”处，如图 6.6 所示。

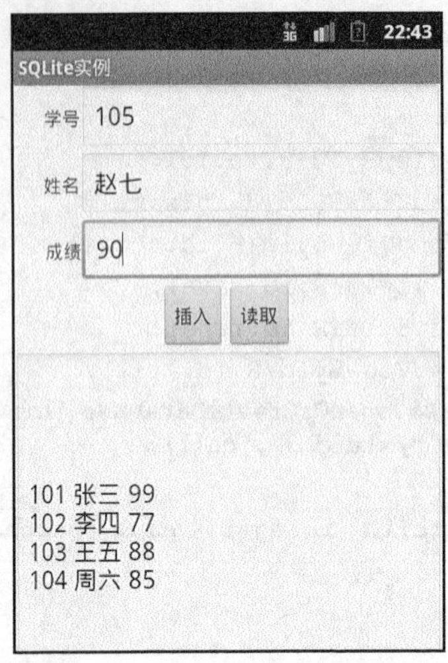

图 6.5 SQLite 数据库存储

图 6.6 数据库存储位置

上面的示例程序中演示了如何使用数据库管理应用程序数据。程序中先后使用了打开/创建数据库、创建表、插入数据、读取数据等数据库操作。其他的如更新数据、删除数据等操作与之类似，用户可以自行编写相应代码进行试验。

6.4.2 SQLiteOpenHelper

类 SQLiteOpenHelper 是一个帮助类，提供了方法来辅助创建和打开数据库，管理数据库的不同版本。使用时需要创建一个类，继承自类 SQLiteOpenHelper，并重写其相对应的方法来实现创建、打开、更新数据库的操作。这几个方法如下。

- void onCreate(SQLiteDatabase db)

指定的数据库被创建时调用。

- void onOpen(SQLiteDatabase db)

指定的数据库被打开时调用。

- void onUpgrade(SQLiteDatabase db, int oldVersion, int newVersion)

由第一个参数指定的数据库被更新时调用。第二个和第三个参数分别用来指定需要被更新的旧的数据库版本号和更新后新的数据库版本号。

除了以上三个回调方法外，类 SQLiteOpenHelper 还提供了以下三个常用方法。

- SQLiteDatabase getWritableDatabase()

第一次调用时会创建一个新的可供读写的数据库，并打开，以后每次调用将只打开数据库。

- SQLiteDatabase getReadableDatabase()

与上述方法类似，只是以只读的形式打开数据库。

- void close()

关闭用上述两个方法打开的数据库。

下面通过例 6.6 说明 SQLiteOpenHelper 的用法。

【例 6.6】 使用 SQLiteOpenHelper 及相关类实现学生数据的保存和读取。

创建工程 SQLite2，本例使用例 6.5 的布局文件，实现效果也同例 6.5，但本工程使用 SQLiteOpenHelper 及相关类来实现学生数据的保存和读取。

（1）建立学生数据类，其中包含学生编号、姓名、成绩三个变量和它们的 getter()、settter() 方法。

```
public class Student {
    private int id;
    private String name;
    private int score;
    public int getId() {
        return id;
    }
    public void setId(int id) {
        this.id = id;
    }
    public String getName() {
        return name;
    }
    public void setName(String name) {
        this.name = name;
    }
    public int getScore() {
        return score;
    }
    public void setScore(int score) {
        this.score = score;
    }
}
```

（2）创建 DBOpenHelper 类，该类继承自 SQLiteOpenHelper，能够根据用户指定的数据库名、版本号、表名和 sql 语句，创建数据库和表。

```
public class DBOpenHelper extends SQLiteOpenHelper {
    private String CREATE_TABLE = "";
    private String tableName = "";
    public DBOpenHelper(Context context, String dbName, int dbVersion, String tableName, 
String sql) {
        super(context, dbName, null, dbVersion);
        CREATE_TABLE = sql;
        this.tableName = tableName;
    }
    @Override
    public void onCreate(SQLiteDatabase db) {
        db.execSQL(CREATE_TABLE);
    }
    @Override
    public void onUpgrade(SQLiteDatabase db, int oleVersion, int newVersion) {
        db.execSQL("drop table if exists " + tableName);
        onCreate(db);
    }
}
```

（3）创建 DBHelper 类，该类提供各个对数据库访问的方法，这里完成写入和读取方法，用户可在此基础上添加删除、修改等方法。代码如下所示。

```
public class DBHelper {
    private static final String DATABASE_NAME = "mydb";
    private static final int DATABASE_VERSION = 1;
    private static final String TABLE_NAME = "student"; //数据库表名
    private static final String[] COLUMNS= { "id", "name", "score" }; //数据库表字段名
```

```
    private String sql = "";
    private DBOpenHelper helper;
    private SQLiteDatabase db;
    public DBHelper(Context context) {
        sql = "create table " + TABLE_NAME + " (" + COLUMNS[0] + " integer primary key,
" + COLUMNS[1] + " varchar(50)," + COLUMNS[2] + " integer);";
        helper = new DBOpenHelper(context, DATABASE_NAME,
                            DATABASE_VERSION, TABLE_NAME, sql);
        db = helper.getWritableDatabase();
    }
    public void insert(Student data) {
        ContentValues values = new ContentValues();
        values.put(COLUMNS[0], data.getId());
        values.put(COLUMNS[1], data.getName());
        values.put(COLUMNS[2], data.getScore());
        db.insert(TABLE_NAME, null, values);
    }
    public ArrayList<Student> find() {
        ArrayList<Student> list = new ArrayList<Student>();
        Student stu = null;
        Cursor cursor = db.query(TABLE_NAME, COLUMNS, null, null, null, null, null);
        while(cursor.moveToNext()) {
            stu = new Student();
            stu.setId(cursor.getInt(0));
            stu.setName(cursor.getString(1));
            stu.setScore(cursor.getInt(2));
            list.add(stu);
        }
        cursor.close(); //关闭游标
        return list;
    }
}
```

（4）修改主 Activity 类 SQLite2Activity.java 文件，完成插入、读取按钮事件的处理，代码如下所示。

```
public class SQLite2Activity extends Activity {
    private EditText id, name, score, list;
    private Button save, read;
    @Override
    protected void onCreate(Bundle savedInstanceState) {
        super.onCreate(savedInstanceState);
        setContentView(R.layout.main);
        id = (EditText) findViewById(R.id.id);
        name = (EditText) findViewById(R.id.name);
        score = (EditText) findViewById(R.id.score);
        save = (Button) findViewById(R.id.save);
        read = (Button) findViewById(R.id.read);
        list = (EditText) findViewById(R.id.list);
        save.setOnClickListener(new View.OnClickListener() {
            @Override
            public void onClick(View arg0) {
                Student info = null;
                DBHelper helper = new DBHelper(SQLite2Activity.this);
                info = new Student();
                info.setId(Integer.parseInt(id.getText().toString()));
                info.setName(name.getText().toString());
```

```
                info.setScore(Integer.parseInt(score.getText().toString()));
                helper.insert(info);
                id.setText(""); //清空 id、姓名、成绩文本框
                name.setText("");
                score.setText("");
                Toast.makeText(SQLite2Activity.this, "保存成功!",
                        Toast.LENGTH_SHORT).show();
            }
        });
        read.setOnClickListener(new View.OnClickListener() {
            @Override
            public void onClick(View arg0) {
                list.setText(""); //清空原有数据
                DBHelper helper = new DBHelper(SQLite2Activity.this);
                //将所有学生数据写入动态数组
                ArrayList<Student> stuList = helper.find();
                //遍历动态数组，输出学生数据
                for(Student info : stuList) {
                    list.setText(list.getText().toString() + info.getId() + " "
                            + info.getName() + " " + info.getScore() + "\n");
                }
            }
        });
    }
}
```

运行本实例，结果如图 6.5 所示。

6.5　Content Provider

Content Provider 是 Android 平台中提供的跨应用程序数据共享的方式。一个应用程序可以通过 Content Provider 来发布自己的数据，但是并不会泄露这些数据在其应用程序内部的组织和存储方式。其他的应用程序可以通过 Content Resolver 来获取自己感兴趣的某个应用程序发布出来的共享数据。通信录、通话记录、短信记录、相片库、铃声库等系统应用程序也通过 Content Provider 来共享自己的数据。图 6.7 描述了 Content Provider 和 Content Resolver 以及应用程序三者之间的关系。

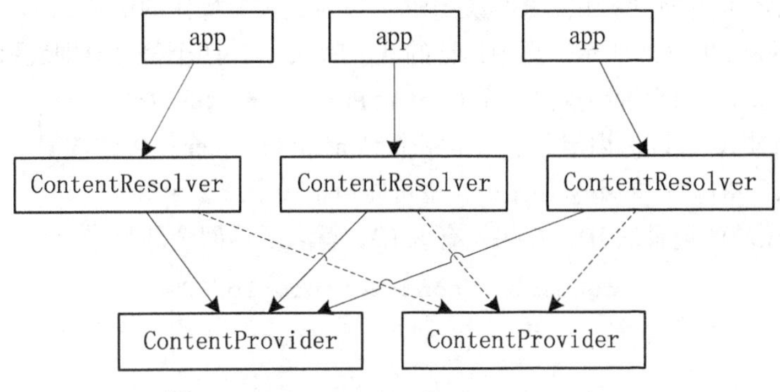

图 6.7　Content Provider 和 Content Resolver

6.5.1 使用 Content Provider 发布数据

在应用程序通过 Content Provider 发布数据分为以下两步。

- 从类 ContentProvider 派生出一个子类，即自定义一个 ContentProvider，并根据需要重写其中的方法。
- 在 AndroidManifest.xml 中注册这个自定义的 ContentProvider。

Content Provider 提供的数据以类似于数据库表的二维结构来组织。每一行表示一条记录，每一行包含若干列。通过调用类 ContentProvider 提供的方法，其他应用程序可以对 Content Provider 提供的数据进行查询、更新、插入、删除等操作。类 ContentProvider 中提供的，需要重写的主要方法如下。

- boolean onCreate()

启动 ContentProvider 时回调的方法。如果 ContentProvider 启动成功，返回 true，否则返回 false。

- Cursor query(Uri uri,String[]projection,String selection, String[] selectionArgs, String sortOrder)

查询数据的方法，有五个参数，返回值是一个 Cursor 对象。第一个参数是 URI 标签，后面将详细介绍。第二个参数是返回结果中需要包含的字段，如果为 null 表示包含所有字段。第三个参数是返回结果中需要包含哪些行，类似于 SQL 语句中的查询子句，如果为 null 表示包含所有行。第四个参数为前一个参数中查询子句需要包含的参数值。第五个参数为返回结果排序的方式，如果为 null 表示可以按照任意方式排序。

- int update(Uri uri,ContentValues values,String selection,String[]selectionArgs)

更新数据的方法，有四个参数，返回更新操作所影响的数据记录个数。四个参数中的第 1、3、4 三个参数与上述 query()方法中的参数作用相同。第 2 个参数为更新操作中对应于某个字段的需要更新的字段/值对。

- Uri insert(Uri uri, ContentValues values)

插入数据的方法，有两个参数，第一个是插入操作的 URI 标签，第二个是插入操作中对应于某个字段的需要更新的字段/值对。返回值是新插入的记录的 URI 标签。

- int delete(Uri uri, String selection, String[] selectionArgs)

删除数据的方法，有两个参数，分别指定需要删除数据的查询子句，以及相应的查询子句中需要包含的参数值，返回删除掉的记录个数。

- String getType(Uri uri)

返回指定 URI 的数据的 MIME 类型。如果没有类型，返回 null。

上述方法参数中的 uri 是 Content Provider 提供的访问数据的标签，它利用 URI 的格式来描述要操作的数据由哪个应用程序提供，路径是什么，以及具体要操作的记录。

URI 的一个示例如图 6.8 所示。图中标记的 A、B、C、D 四个符号分别表示：

- A：标准前缀，用来说明这是一个 Content Provider 控制的数据。
- B：URI 的标识，该标识定义了这是哪个 Content Provider 提供的数据。
- C：路径，URI 中可能不包括路径，也可能包括一个或多个。
- D：需要获取的记录的 ID；如果没有该 ID，就表示返回全部数据。

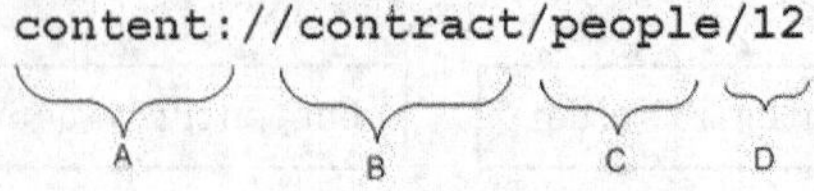

图 6.8　URI 示例

下面通过例 6.7 介绍如何使用 Content Provider 发布数据。

【例 6.7】 使用 Content Provider 发布数据。

（1）在例 6.5 的工程上创建类 StuProvider，继承自类 ContentProvider。用于向其他应用程序共享数据，为了简便起见，这里只介绍向其他应用程序提供查询的 ContentProvider 接口，其他更新、插入、删除等接口作为练习，请用户自行添加。StuProvider 类，代码如下所示。

```
public class StuProvider extends ContentProvider {
    SQLiteDatabase db;
    @Override
    public int delete(Uri arg0, String arg1, String[] arg2) {
        return 0;
    }
    @Override
    public String getType(Uri uri) {
        return null;
    }
    @Override
    public Uri insert(Uri uri, ContentValues values) {
        return null;
    }
    @Override
    public boolean onCreate() {
        try {
            db = SQLiteDatabase.openOrCreateDatabase(
                    this.getContext().getFilesDir().toString() + "/stu.db3", null);
        }
        catch(Exception e) {
            return false;
        }
        return true;
    }
    @Override
    public Cursor query(Uri uri, String[] projection, String selection,
                    String[] selectionArgs, String sortOrder) {
        Cursor cursor = db.rawQuery("select * from student", null);
        return cursor;
    }
    @Override
    public int update(Uri uri, ContentValues values, String selection,
                String[] selectionArgs) {
        return 0;
    }
}
```

（2）在例 6.5 的 AndroidManifest.xml 文件中添加<provider>标记，注册上面定义的 StuProvider 类，代码如下所示。

```
<manifest xmlns:android = "http://schemas.android.com/apk/res/android"
    package = "com.example.sqlite1"
    android:versionCode = "1"
    android:versionName = "1.0" >
    <uses-sdk
        android:minSdkVersion = "8"
```

```
        android:targetSdkVersion = "17" />
    <application
        android:allowBackup = "true"
        android:icon = "@drawable/ic_launcher"
        android:label = "@string/app_name"
        android:theme = "@style/AppTheme" >
        <activity
            android:name = "com.example.sqlite1.SQLite1Activity"
            android:label = "@string/app_name" >
            <intent-filter>
                <action android:name = "android.intent.action.MAIN" />
                <category android:name = "android.intent.category.LAUNCHER" />
            </intent-filter>
        </activity>
        <provider android:name = ".StuProvider"
             android:authorities = "com.example.sqlite1.StuProvider" />
    </application>
</manifest>
```

6.5.2 使用 Content Resolver 获取数据

要获取和操作其他应用程序通过 Content Provider 提供的数据，需要使用 Content Resolver。通过类 Content 的 getContentResolver()方法可以获得一个类 ContentResolver 的实例。类 ContentResolver 的常用方法有：

- Cursor query(Uri uri,String[]projection,String selection,String[]selectionArgs, String sortOrder)
- int update(Uri uri, ContentValues values, String where, String[]selectionArgs)
- Uri insert(Uri url, ContentValues values)
- int delete(Uri url, String where, String[] selectionArgs)

这四个方法分别用于对给定的 URI 进行数据查询、更新、插入和删除操作。这些方法的参数的作用与前述 ContentProvider 类中相对应方法的参数的作用完全一样。

下面通过例 6.8 介绍如何利用 ContentResolver 来获取例 6.7 中利用 ContentProvider 发布的数据。

【例 6.8】 使用 Content Resolver 获取数据。

（1）创建 Android 工程，工程名为 ReadContent，修改主布局文件，将默认的布局管理器改为 LinearLayout 线性布局，垂直排列。在该布局中依次放置按钮，和文本数据框。代码如下所示。

```
<LinearLayout xmlns:android = "http://schemas.android.com/apk/res/android"
    android:layout_width = "match_parent"
    android:layout_height = "match_parent"
    android:orientation = "vertical" >
    <Button android:id = "@+id/getData"
        android:layout_width = "match_parent"
        android:layout_height = "wrap_content"
        android:text = "获取 Content 数据" />
    <EditText android:id = "@+id/list"
        android:layout_width = "match_parent"
```

```
    android:layout_height = "wrap_content" />
</LinearLayout>
```

（2）修改主 Activity 文件，在单击“获取 Content 数据”事件处理中通过 ContentResolver 类提供的 query()方法（在 ContentResolver 子类 StuProvider 中已实现该方法），获取 URI 为“content://com.example.sqlite1.StuProvider”处共享的数据，代码如下所示。

```
public class ReadContentActivity extends Activity {
    @Override
    protected void onCreate(Bundle savedInstanceState) {
        super.onCreate(savedInstanceState);
        setContentView(R.layout.main);
        Button btn = (Button) findViewById(R.id.getData);
        final EditText list = (EditText) findViewById(R.id.list);
        btn.setOnClickListener(new View.OnClickListener() {
            @Override
            public void onClick(View view) {
                ContentResolver resolver = getContentResolver();
                Uri uri = Uri.parse("content://com.example.sqlite1.StuProvider");
                Cursor cursor = resolver.query(uri, null, null, null, null);
                while(true) {
                    if(cursor.moveToNext() == false)
                        break;
                    String i = cursor.getString(0);
                    String n = cursor.getString(1);
                    int s = cursor.getInt(2);
                    String tmp = list.getText().toString();
                    list.setText(tmp + "\n" + i + " " + n + " " + s);
                }
            }
        });
    }
}
```

运行本实例，查询结果如图 6.9 所示。

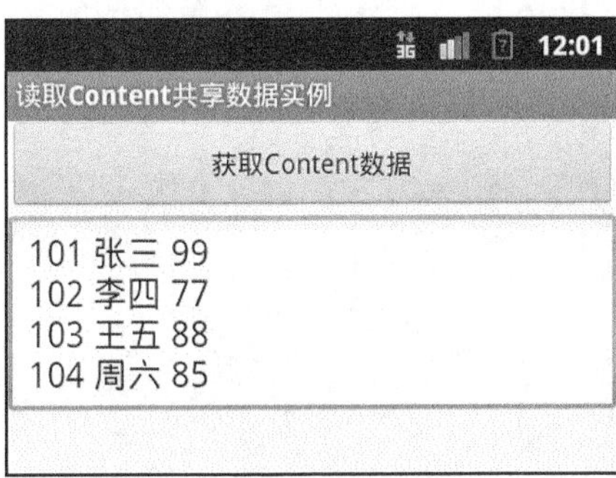

图 6.9　ContentResolver 获取数据

6.6 小　　结

本章介绍 Android 应用程序中如何将数据存储在手机内存或 SD 卡上，以供后续处理使用。常用的数据存储方式有首选项信息存储、数据文件存储、SQLite 数据库存储。此外本章还介绍了

ContentProvider 和 ContentResolver，这是 Android 平台中提供的跨应用程序数据共享的方式。

练　习

1. SharedPreferences 类如何保存用户的首选项信息数据？
2. Android 中对数据文件的读写操作模式有哪些？
3. 在 SD 卡上进行数据文件的操作与在机身内存中的文件操作有何不同？
4. 简述 SQLite 进行数据存储、查询、更新等操作的主要步骤。
5. 简述 Content Provider 提供数据共享以及 Content Resolver 获取共享数据的方式。
6. 编写一个用户登录的程序，让用户输入用户名和密码，进行验证后登录。验证过程可以由程序模拟进行（如假设用户名为 aaa，密码为 bbb）。然后增加一个让用户自动登录的功能。需要将用户信息进行保存和读取。尝试使用本章讲过的各种数据存储方法实现这个功能。

第 7 章 多线程及消息处理

使用多线程编程是很多应用程序所必需的，如游戏、多媒体程序、有网络连接的程序、进行耗时计算的程序等。Android 中可以通过继承 Thread 类或者实现 Runnable 接口创建线程，并提供了 Handler 类和 AsyncTask 类进行线程间通信及消息处理。Timer 定时器也是常用的实现多线程程序的方式，用于实现定时的后台任务。本章最后还介绍了 Android 多线程通信机制。

7.1 Android 多线程概述

Android 平台支持多线程编程，其采用了与 Java 中相同的创建和操作线程的方式。本小节将介绍如何通过继承 Thread 类和实现 Runnable 接口两种方式来创建线程，并通过相关的方法来操作线程。

7.1.1 创建线程

Thread 类提供了若干构造方法和成员方法用于创建和操作线程，一个 Thread 对象就代表了一个具体的线程。Runnable 接口是一个内容非常简单的接口，仅仅包含了一个抽象方法 run()，而 Thread 类本身就实现了 Runnable 接口。不论是通过继承 Thread 类来创建线程，还是通过实现 Runnable 接口来创建线程，都需要重写 run()方法，将需要这个线程所要进行的工作的代码放在其中。下面的两段代码分别表示如何通过继承 Thread 类或者实现 Runnable 接口来创建线程。

（1）通过继承 Thread 类来创建线程。

```
class MyThread extends Thread{
    public void run(){
        // 线程需要完成的工作
    }
    public static void main(){
        Thread thread1 = new MyThread(); // 创建线程
    }
}
```

（2）通过实现 Runnable 接口来创建线程。

```
class MyThread implements Runnable{
    public void run(){
        // 线程需要完成的工作
```

```
    }
    public static void main(){
        Thread thread2 = new Thread(new MyThread()); // 创建线程
    }
}
```

Thread 类提供了多个构造方法用来创建线程，上述两段代码中分别使用了不用参数的构造方法和使用一个 Runnable 对象的构造方法。除此之外，还可以提供一些其他的参数，如线程的名字、线程组等来使用构造方法创建线程。下面列举出了 Thread 类常用的构造方法。

- Thread()
- Thread(Runnable runnable)
- Thread(String threadName)
- Thread(Runnable runnable, String threadName)
- Thread(ThreadGroup group, String threadName)
- Thread(ThreadGroup group, Runnable runnable)
- Thread(ThreadGroup group, Runnable runnable, String threadName)

前面两段创建线程的程序仅仅用于演示语法，故比较简单。实际程序在自定义线程类时往往会通过使用构造方法参数的方式来进行数据的传递。此外，程序中也经常会使用内部类的方式来自定义线程类，以简化代码并提高内聚性。本章后续的小节中有相关的示例程序。

7.1.2 操作线程

线程对象创建后，并不会马上就自动运行，而是需要通过调用线程的 start()方法来启动线程。线程被启动后，也不一定会马上就能够运行，因为可能在同一个时刻有多个线程都在等待运行，这就需要由虚拟机的线程控制器根据线程的优先级，线程调度策略等来决定运行哪一个线程。线程从创建，到启动，运行，直到最后的停止，会经历若干个状态的变化。图 7.1 描述了这些状态的变化以及影响状态变化的条件。

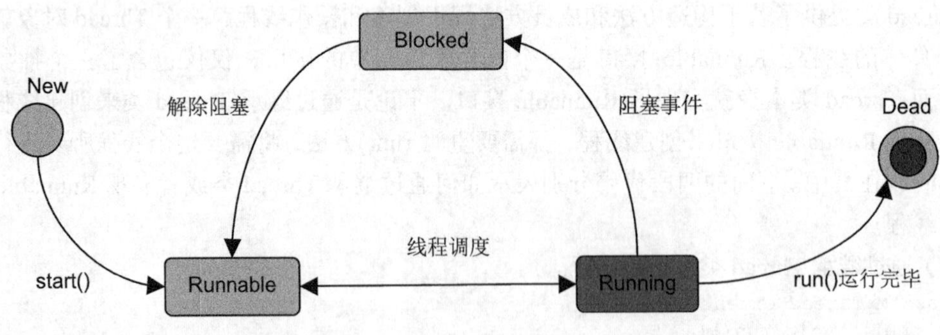

图 7.1 线程状态

当新建了一个 Thread 对象后，线程的状态为 New，此时线程尚未开始运行。通过调用线程的 start()方法来启动线程。线程被启动后，其状态变为 Runnable，等待线程控制器的调度。如果该线程被调度运行，则其状态变为 Running。线程的工作完成后，即重写的 run()方法中代码执行完毕，则线程的状态变为 Dead。线程在运行过程中，有可能尚未运行完毕就被重新调度回到 Runnable 状态，等待下次调度运行。线程在运行过程中，遇到阻塞事件不能继续运行时会进入 Blocked 状态，当阻塞解除后，线程回到 Runnable 状态等待下次调度运行。

下面列举出了常用的操作线程的方法，具体的使用可以参考 API 文档。注意 wait()、notify()、notifyAll()等方法为 Object 类的方法，其余方法为 Thread 类的方法。

- start()：启动新建的线程，进入 Runnable 状态
- run()：线程进入 Running 状态后运行
- sleep()：线程睡眠指定的一段时间
- join()：调用这个方法的主线程，会等待加入的子线程完成
- yield()：线程放弃执行，使其他优先级不低于此线程的线程有机会运行
- isAlive()：测试线程是否存活
- getPriority()/setPriority()：获取/设置线程的优先级
- wait()：线程进入等待状态，等待被唤醒
- notify()/notifyAll()：唤醒其他/所有线程

7.2　UI 线程与非 UI 线程

每个 Android 程序启动时都会创建一个进程，同时在该进程中创建一个主线程。如果不再启动新的线程，则该程序就是只有这一个线程的单线程程序。主线程主要负责处理与 UI 相关的事件，如用户的按键事件，用户接触屏幕的事件以及屏幕绘图事件，并把相关的事件分发到对应的 UI 组件进行处理。所以主线程通常又被叫作 UI 线程。从 UI 线程中操作 UI 组件是安全的，而且 Android 中的 UI 操作必须要在 UI 线程中执行。如果创建了其他新的线程，则不允许在其他线程直接进行 UI 操作。也就是说 Android 中在非 UI 线程中进行 UI 操作不是线程安全的，必须通过 Android 提供的消息处理机制，由非 UI 线程向 UI 线程发出请求消息，由 UI 线程处理这些消息，并进行相关的 UI 操作。以下通过例 7.1 来介绍创建一个新的非 UI 线程，并尝试进行 UI 操作时会出现的异常。

【例 7.1】　非 UI 线程进行 UI 操作。

创建一个 Android 项目，项目名为 HelloDemo。前面章节中讲过，这个新建的项目，即使不进行任何修改，不添加任何代码，也能正常运行，运行的效果如图 7.2 所示。

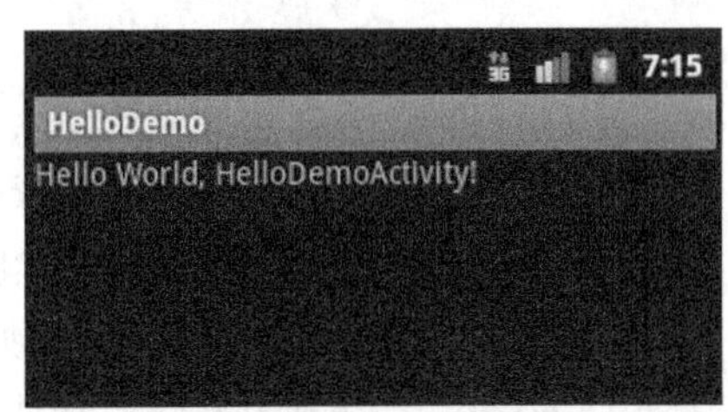

图 7.2　HelloDemo 程序界面

（1）下面修改这个程序，在主布局文件中为 TextView 组件添加一个 id，代码如下所示。

```
<RelativeLayout xmlns:android = "http://schemas.android.com/apk/res/android"
   xmlns:tools = "http://schemas.android.com/tools"
   android:layout_width = "match_parent"
   android:layout_height = "match_parent"
   android:paddingBottom = "@dimen/activity_vertical_margin"
   android:paddingLeft = "@dimen/activity_horizontal_margin"
   android:paddingRight = "@dimen/activity_horizontal_margin"
   android:paddingTop = "@dimen/activity_vertical_margin"
   tools:context = ".MainActivity" >
   <TextView android:id = "@+id/text"
      android:layout_width = "wrap_content"
      android:layout_height = "wrap_content"
      android:text = "@string/hello_world" />
</RelativeLayout>
```

（2）修改主 Activity 文件，代码如下所示。

```
public class HelloDemoActivity extends Activity {
    private TextView text;
    @Override
    public void onCreate(Bundle savedInstanceState) {
        super.onCreate(savedInstanceState);
        setContentView(R.layout.main);
        text = (TextView) findViewById(R.id.text);
        Thread t = new Thread() {
            public void run() {
                try {
                    Thread.sleep(3000);  //睡眠 3 秒
                } catch (InterruptedException e) {
                    e.printStackTrace();
                }
                text.setText("UI Thread Test.");
            }
        };
        t.start(); //启动线程
    }
}
```

从程序中可以看出，通过一个匿名内部类创建一个新的线程 t，并运行。该线程暂停 3 秒钟后，尝试将程序界面上的文字修改为“UI Thread Test.”。运行该程序，首先会出现如图 7.2 所示的界面，3 秒钟后，并没有按照预期出现文字的变化，而是发生了异常。界面如图 7.3 所示。在 LogCat 中出现的异常描述是：“android.view.ViewRoot$CalledFromWrongThreadException: Only the original thread that created a view hierarchy can touch its views.”。这就说明只有创建 View 的 UI 线程才能进行 UI 操作，其他线程不能进行 UI 操作。

图 7.3　程序异常终止界面

同时，即使不创建新的线程，而是在 UI 线程中调用 Thread.sleep()方法，如果涉及了 UI 操作，也不能保证会按照预期的设计出现结果。以下通过修改上述例子来进行介绍。

（3）修改主 Activity 文件，代码如下所示。

```
public class HelloDemoActivity extends Activity {
    @Override
    public void onCreate(Bundle savedInstanceState) {
        super.onCreate(savedInstanceState);
        setContentView(R.layout.main);
        TextView text = (TextView) findViewById(R.id.text);
        text.setText("Sleep start.");
        try {
            Thread.sleep(3000); //睡眠 3 秒
        }
        catch (InterruptedException e) {
           e.printStackTrace();
        }
        text.setText("Sleep end.");
    }
}
```

按照以上的代码，程序运行时界面上应该首先显示“Sleep start.”的文字，然后等待 3 秒钟时间，文字变成“Sleep end.”。在模拟器中运行该程序，可以看到程序运行后，界面上开始是空白的，既没有之前的“Hello World, HelloDemoActivity!”字样，也没有“Sleep start.”字样，等待 3 秒钟后，界面上显示“Sleep end.”。可见由于在主线程中使用 Thread.sleep()方法而造成了 UI 操作的不安全问题。下面的小节中将介绍如何消除这种问题。

7.3 多线程中的常用类

Android 中提供了 Handler 类和 AsyncTask 类进行线程间通信及消息处理。Handler 类可以处理较为复杂的线程间通信及消息处理，AsyncTask 类提供的一个轻量级的基于多线程的进行后台异步工作处理的类。另外 Timer 定时器也是常用的实现多线程程序的方式。下面分别予以介绍。

7.3.1 Handler 类

如前所述，Android 中新创建的线程不能直接进行 UI 操作，必须通过 Android 提供的消息处理机制，由非 UI 线程向 UI 线程发出请求消息，由 UI 线程处理这些消息，并进行相关的 UI 操作。这个发送消息和处理消息的过程由 Handler 类来协助进行处理。Handler 类有一个供其子类重写的方法 handleMessage(Message msg)。通常在 UI 线程中创建 Handler 类的子类，并通过重写这个方法来实现 UI 操作。当有其他线程向这个自定义子类的对象发送消息时，handleMessage()方法会被调用，通过识别参数 msg 的值，可以进行相应的 UI 操作。

Message 类用来封装所发送消息的值，这些值可以是各种数据类型的，可以通过 Bundle 类来封装这些不同类型的值。这里简单介绍一下 Message 类和 Bundle 类的基本使用方式。

Bundle 类用于提供从字符串到某个具体值的映射。一个 Bundle 对象可以封装多个键值对，键只能是字符串类型的，值可以是各种数据类型的，包括基本数据类型和引用类型。Bundle 类提供了若干个 putXXX()方法用于添加键值对，如 void putInt(String key, int value)用于添加一个值为 int 类型的键值对。相应的 Bundle 类有一系列个 getXXX()方法用于根据键来获取键值对中的值，如 int getInt(String key)。

Message 类的方法 void setData(Bundle data)用于将一个 Bundle 对象放入消息中，而 Bundle getData()方法用于获取放入的消息。Message 类还提供了两个 int 类型的域变量和一个 Object 类型的域变量作为额外的存储空间，如果程序中只需要将一两个整数或者一个对象引用作为消息传递的值，可以不用 Bundle 对象而直接使用这些域变量。需要注意的是要创建一个 Message 对象，可以直接使用这个类的构造方法，但是最好尽可能使用消息池中现有的可重复使用的消息，而不去创建新的消息。有两个方法可以从消息池中获取一个可供使用的消息，Message.obtain()方法和 Handler.obtainMessage()方法。

Handler 类提供了多个发送消息的方法，最常用的有以下两个。

- boolean sendMessage(Message msg)

该方法首先将信息值封装在一个 Message 对象中，然后以此为参数调用。sendMessage()方法发出消息。

● boolean sendEmptyMessage(int what)

该方法适用于更简单的情况，不需要 Message 对象，只是以一个整数作为参数。不同的整数值代表不同的需要处理情况的编码，通过 Message 类的 what 属性可以获得这个值，从而确定需要进行怎样的处理。

除了上面列出的两个方法，Handler 类还有其他的方法用于获取和判断消息，处理消息等等。

下面通过例 7.2 介绍 Handler 类如何向主线程发送消息。

【例 7.2】 UI 线程中接收 Handler 类发送的消息。

本例在程序中创建一个新的线程，该线程处理一些耗时的工作（用 Thread.sleep()方法暂停 3 秒代替），然后通过 Handler 类的对象向主线程发送消息。主线程接受并处理这个消息，在用户界面上显示出新线程发送的数据。

（1）使用例 7.1 的主布局文件。

（2）修改主 Activity 文件，代码如下所示。

```
public class HelloDemoActivity extends Activity {
    @Override
    public void onCreate(Bundle savedInstanceState) {
        super.onCreate(savedInstanceState);
        setContentView(R.layout.main);
        final TextView text = (TextView) findViewById(R.id.text);
        final Handler handler = new Handler() {
            public void handleMessage(Message msg) { //接收消息
                super.handleMessage(msg);
                text.setText("Message Code: " + msg.what);
            }
        };
        Thread t = new Thread() {
            public void run() {
                try {
                    Thread.sleep(3000);
                }
                catch (InterruptedException e) {
                    e.printStackTrace();
                }
                handler.sendEmptyMessage(1); //发送消息
            }
        };
        t.start(); //启动线程
    }
}
```

该程序运行后，会首先显示如图 7.2 所示的界面，3 秒钟后，显示如图 7.4 所示的界面，其中显示的代码 1 就是调用 sendEmptyMessage()方法时传递的参数 1。

下面通过例 7.3 来介绍一般多线程编程的过程。

【例 7.3】 使用多线程模拟秒表应用。

本例是一个模拟秒表的程序，在初始界面单击开始按钮后，数字时钟开始计时。这时单击记录按钮，可以将当前的时间记录在界面下方的文本框内，可以根据需

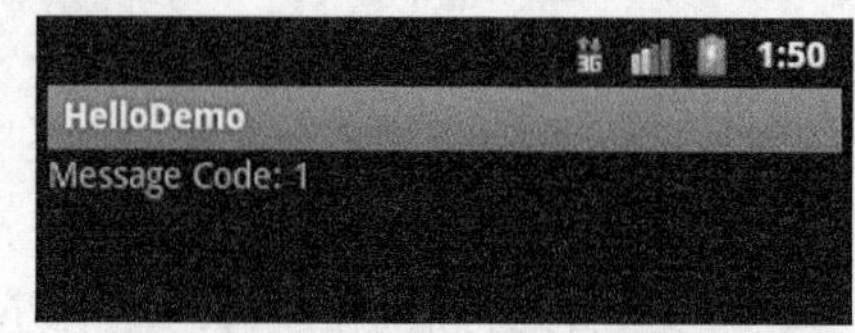

图 7.4 调用 sendEmptyMessage()方法

要记录多个时间。单击停止按钮，时钟停止计时。时钟的控制通过一个独立的线程进行，由于该线程不能直接操作 UI 组件，需要使用前面介绍过的通过使用 Handler 类来从非 UI 线程中发送消息，再由 UI 线程处理消息并进行 UI 组件的控制。

（1）创建 Android 工程，修改主布局文件的布局管理器为线性布局，其中放入计时的文本框、一个线性布局管理器和记录时间的文本框。其中的线性布局中水平放置开始按钮和记录按钮。代码如下所示。

```
<?xml version = "1.0" encoding = "utf-8"?>
<LinearLayout xmlns:android = "http://schemas.android.com/apk/res/android"
   android:layout_width = "match_parent"
   android:layout_height = "match_parent"
   android:gravity = "center"
   android:orientation = "vertical" >
   <TextView android:id = "@+id/timer"
      android:layout_width = "match_parent"
      android:layout_height = "wrap_content"
      android:text = "0:0.0"
      android:gravity = "center"
      android:textSize = "20pt" />
   <LinearLayout
      xmlns:android = "http://schemas.android.com/apk/res/android"
      android:layout_width = "match_parent"
      android:layout_height = "wrap_content"
      android:gravity = "center"
      android:orientation = "horizontal" >
      <Button android:id = "@+id/start"
         android:layout_width = "wrap_content"
         android:layout_height = "wrap_content"
         android:layout_margin = "5pt"
         android:text = " 开始 "
         android:textSize = "10pt" />
      <Button android:id = "@+id/record"
         android:layout_width = "wrap_content"
         android:layout_height = "wrap_content"
         android:layout_margin = "5pt"
         android:text = " 记录 "
         android:textSize = "10pt" />
   </LinearLayout>
   <EditText android:id = "@+id/show"
      android:layout_width = "match_parent"
      android:layout_height = "match_parent"
      android:editable = "false"
      android:textSize = "12pt" />
</LinearLayout>
```

（2）修改主 Activity 文件，单击开始按钮开始计时，并将开始按钮的文本改为停止，每隔 100 毫秒更新一次计时文本框的值，代码如下所示。

```
public class TimerActivity extends Activity {
   Boolean isRunning = false;
   TextView timer;
   Button start, record;
   EditText show;
   Handler handler;
   @Override
   public void onCreate(Bundle savedInstanceState) {
```

```
        super.onCreate(savedInstanceState);
        setContentView(R.layout.main);
        timer = (TextView)findViewById(R.id.timer);
        start = (Button)findViewById(R.id.start);
        record = (Button)findViewById(R.id.record);
        show = (EditText)findViewById(R.id.show);
        start.setOnClickListener(new View.OnClickListener() {
            public void onClick(View view) {
                if(isRunning){
                    isRunning = false;
                    start.setText(" 开始 ");
                }else{
                    isRunning = true;
                    start.setText(" 停止 ");
                    show.setText("");
                    new TimerThread().start();
                }
            }
        });
        record.setOnClickListener(new View.OnClickListener() {
            public void onClick(View view) {
                if(isRunning){
                    show.append(timer.getText() + "\n");
                }
            }
        });
        handler = new Handler() {
            public void handleMessage(Message msg) {
                super.handleMessage(msg);
                timer.setText((String)msg.obj);
            }
        };
    }
    class TimerThread extends Thread{
        public void run(){
            int i = 0;
            int j = 0;
            int k = 0;
            String time = "";
            while(isRunning){
                k + +;
                if(k = =10){
                    k = 0;
                    j + +;
                }
                if(j = =60){
                    j = 0;
                    i + +;
                }
                time = i + ":" + j + "." + k;
                Message msg = new Message();
                msg.obj = time;
                handler.sendMessage(msg);
                time = "";
                try {
```

```
                        Thread.sleep(100);
                    }
                    catch (InterruptedException e) {
                        e.printStackTrace();
                    }
                }
            }
        }
    }
```

运行本实例，显示结果如图 7.5、图 7.6 所示。

需要注意的是程序中通过定义一个 Boolean 型的变量 isRunning 来作为一个标志变量。这个变量的值既用来控制按钮进行不同的操作，也用来控制新线程中时钟的运行与停止。

图 7.5　起始界面

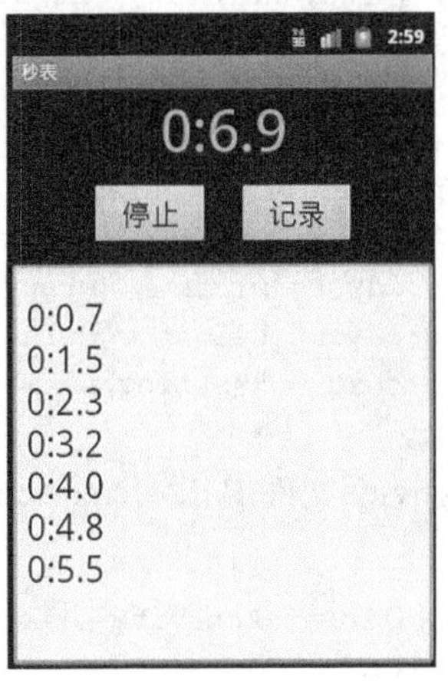

图 7.6　操作结果

7.3.2　AsyncTask 类

AsyncTask 类是 Android 提供的一个轻量级的基于多线程的进行后台异步工作处理的类。在后台工作比较简单，只需要向 UI 线程传递一些简单数据，可以不用像前面一节讲的那样使用 Thread 类和 Handler 类，只使用 AsyncTask 类即可。使用 AsyncTask 类的方法是创建一个子类，重写其相关的方法，然后在 UI 线程中使用 execute()方法运行这个自定义类即可。

定义 AsyncTask 类的子类时，需要指定该类所需的 3 个泛型，分别是 AsyncTask<Params, Progress, Result>。比如：

```
class MyAsyncTask extends AsyncTask<String, Integer, String>
{
    //重写方法
}
```

这三个类型对应的分别是执行后台任务所需的参数的类型，后台任务运行过程中向 UI 线程传递状态的类型，以及后台任务执行完毕后返回的结果的类型。

AsyncTask 类中需要子类重写的方法有：

- onPreExecute()：当后台任务使用 execute()方法运行后，会立即由 UI 线程调用，一般用于初始化操作和在界面上显示后台任务的初始状态。
- doInBackground(Params...)：当 onPreExecute()方法调用结束后，由后台线程调用该方法，开始执行后台任务。在后台任务执行过程中，可以调用 publishProgress(Progress...)方法向 UI 线程发布当前后台任务执行的状态，以便程序界面进行相应的更新。

- onProgressUpdate(Progress...)：当后台线程调用 publishProgress(Progress...)方法后，由 UI 线程调用该方法进行 UI 操作。
- onPostExecute(Result)：当后台任务全部执行完毕后，由 UI 线程调用该方法，更新界面的显示，或进行其他的后续操作。

下面通过例 7.4 来说明 AsyncTask 类的使用。

【例 7.4】 使用 AsyncTask 类修改 UI 组件值。

（1）创建 Android 工程，修改主布局文件的布局管理器为线性布局，其中放入一个文本框。代码如下所示。

```
<LinearLayout xmlns:android = "http://schemas.android.com/apk/res/android"
    android:layout_width = "match_parent"
    android:layout_height = "match_parent"
    android:background = "#FFFAE4"
    android:orientation = "vertical">
    <TextView android:id = "@+id/text"
        android:layout_width = "wrap_content"
        android:layout_height = "wrap_content"
        android:text = "@string/hello_world" />
</LinearLayout>
```

（2）修改主 Activity 文件 HelloDemoActivity.java，通过 AsyncTask 类来修改文本框的值。代码如下所示。

```
public class HelloDemoActivity extends Activity {
    TextView text;
    @Override
    public void onCreate(Bundle savedInstanceState) {
        super.onCreate(savedInstanceState);
        setContentView(R.layout.main);
        text = (TextView) findViewById(R.id.text);
        new MyAsyncTask().execute(null);
    }
    class MyAsyncTask extends AsyncTask<String, Integer, String> {
        public String doInBackground(String... params) {
            try {
                Thread.sleep(3000);
                publishProgress(50);
            } catch (InterruptedException e) {
                e.printStackTrace();
            }
            return "";
        }
        public void onProgressUpdate(Integer... values){
            text.setText("value: " + values[0]);
        }
    }
}
```

程序运行结果首先显示如图 7.2 所示的界面，3 秒钟后，后台任务通过 publish Progress(50)方法发布一次消息，UI 线程接收到后调用 onProgressUpdate()方法将程序界面进行更新（本例中即修改文本框的值），如图 7.7 所示。

很多手机应用程序中需要在一段时间内进行一些后台工作，如从数据库获取数据、初始化程序数据、进行网络连接、进行数据搜索等，这些后台工作都应该放在一个单独的非 UI 线程中来完成。下面通过例 7.5 介绍如何通过 AsyncTask 类和 ProgressBar 来配合显示后台工作的完成比例。

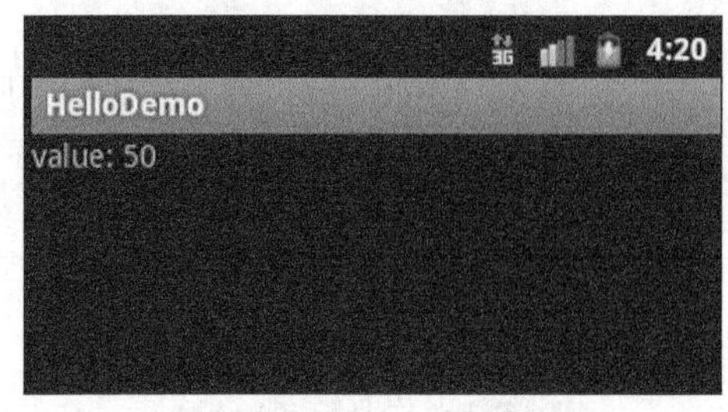

图 7.7　使用 AsyncTask 类的界面

【例 7.5】　使用 AsyncTask 类模拟进度变化。

本例为了简便起见，程序中并没有进行真正的后台工作，而是用了 Thread.sleep(100)来进行模拟。第 4 章介绍进度条时使用 Handler 类完成模拟下载任务，本例采用 AsyncTask 类来模拟获取数据过程。

（1）创建 Android 工程，修改主布局文件，代码如下所示。

```
<?xml version = "1.0" encoding = "utf-8"?>
<LinearLayout xmlns:android = "http://schemas.android.com/apk/res/android"
    android:layout_width = "match_parent"
    android:layout_height = "match_parent"
    android:orientation = "vertical" >
    <Button
        android:id = "@+id/get"
        android:layout_width = "match_parent"
        android:layout_height = "wrap_content"
        android:layout_margin = "10dp"
        android:text = "  点击开始获取数据  " />
    <ProgressBar
        android:id = "@+id/bar1"
        style = "@android:style/Widget.ProgressBar.Horizontal"
        android:layout_margin = "5dp"
        android:layout_width = "match_parent"
        android:layout_height = "wrap_content" />
    <LinearLayout xmlns:android = "http://schemas.android.com/apk/res/android"
    android:layout_width = "match_parent"
    android:layout_height = "match_parent"
    android:layout_margin = "10dp"
    android:orientation = "horizontal" >
    <ProgressBar
        android:id = "@+id/bar2"
        style = "?android:attr/progressBarStyleLarge"
        android:layout_width = "wrap_content"
        android:layout_height = "wrap_content" />
    <TextView
        android:id = "@+id/show"
        android:gravity = "center"
        android:layout_margin = "15dp"
        android:textSize = "30dp"
        android:text = "0%"
        android:layout_width = "wrap_content"
        android:layout_height = "wrap_content" />
```

```
    </LinearLayout>
</LinearLayout>
```

（2）修改主 Activity 文件，代码如下所示。

```
public class ProgressBarTestActivity extends Activity {
    @Override
    public void onCreate(Bundle savedInstanceState) {
        super.onCreate(savedInstanceState);
        setContentView(R.layout.main);
        final Button button = (Button) findViewById(R.id.get);
        final ProgressBar bar1 = (ProgressBar) findViewById(R.id.bar1);
        final ProgressBar bar2 = (ProgressBar) findViewById(R.id.bar2);
        final TextView show = (TextView) findViewById(R.id.show);
        bar1.setVisibility(View.INVISIBLE);
        bar1.setMax(100);
        bar2.setVisibility(View.INVISIBLE);
        show.setVisibility(View.INVISIBLE);
        class MyAsyncTask extends AsyncTask<String, Integer, String> {
            public String doInBackground(String... params) {
                try {
                    for (int i = 0; i <= 100; i + +) {
                        Thread.sleep(100);
                        publishProgress(i);
                    }
                } catch (InterruptedException e) {
                    e.printStackTrace();
                }
                return "";
            }
            public void onProgressUpdate(Integer... values) {
                show.setText(values[0] + "%");
                bar1.setProgress(values[0]);
                if(values[0] = =100)
                    bar2.setVisibility(View.INVISIBLE);
            }
        }
        button.setOnClickListener(new View.OnClickListener() {
            public void onClick(View v) {
                bar1.setVisibility(View.VISIBLE);
                bar2.setVisibility(View.VISIBLE);
                show.setVisibility(View.VISIBLE);
                new MyAsyncTask().execute(null);
            }
        });
    }
}
```

运行本实例，首先显示如图 7.8 所示的界面。单击按钮后，界面如图 7.9 所示，有一个水平条状的 ProgressBar 和一个圆圈状的 ProgressBar 出现，同时有一个数字的比例，这三者一起显示后台程序的运行状态。

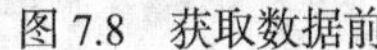
图 7.8　获取数据前

图 7.9　获取数据中

7.3.3　Timer 定时器

在 Android 中可以使用 Java API 的 Timer 类和 TimerTask 类来实现定时器的功能。但是在定时器中也不能直接进行 UI 操作，需要使用以下几种方式之一来间接地进行 UI 操作：

- 通过 Handler 类处理
- 通过 Activity.runOnUiThread(Runnable)方法
- 通过 View.post(Runnable)方法

通过 Handler 类来实现从定时器中向 UI 线程发送请求消息，由 Handler 来进行消息处理和 UI 操作的过程与前述从一个新线程中发送请求消息，由 Handler 来进行处理的方式完全一样。这里重点介绍后两种方式。

Activity 类提供了 runOnUiThread()方法，用来在 UI 线程之上启动另外一个线程，其 API 为：

```
public void runOnUiThread(Runnable action)
```

该方法被调用后，如果当前线程是 UI 线程，则立即启动由参数 action 所指定的线程。如果该方法被调用时的当前线程不是 UI 线程，则由参数 action 所指定的线程会被放入 UI 线程的事件队列中等待处理。通常程序调用 runOnUi Thread()方法是在非 UI 线程中进行。

下面修改例 7.4 的主 Activity 文件 HelloDemoActivity.java，使用定时器来完成对 UI 组件值的修改。

```
public class HelloDemoActivity extends Activity {
    TextView text;
    @Override
    public void onCreate(Bundle savedInstanceState) {
        super.onCreate(savedInstanceState);
        setContentView(R.layout.main);
        text = (TextView) findViewById(R.id.text);
        TimerTask task = new TimerTask() {
            public void run() {
                Thread t = new Thread() {
                    public void run() {
                        text.setText("runOnUiThread()");
                    }
                };
                runOnUiThread(t);
            }
```

```
        };
        Timer timer = new Timer();
        timer.schedule(task, 3000);
    }
}
```

本例中创建定时器任务 task，该任务创建一个新的线程 t，在新线程中将界面中文本框组件的文字修改为“runOnUiThread()”。前面讲过，如果直接使用 t.start()方法启动该线程，则程序会出现异常。这里使用了 runOnUiThread(t)方法来启动该线程。接着，一个定时器对象 timer 被创建，通过调用 timer.schedule(task, 3000)方法来设定等待 3 秒钟后运行定时器任务 task。运行程序后，程序界面首先如图 7.2 所示，3 秒钟后，界面文字被修改为“runOnUiThread()”，如图 7.10 所示，实现了预期的目标。

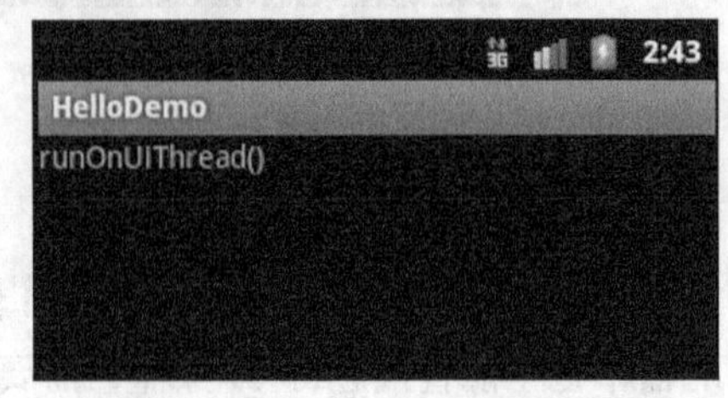

图 7.10　调用 runOnUiThread()方法

另外 View 类提供了 post()方法，用来在 UI 线程中启动一个新线程，该方法的 API 是：

```
public boolean post(Runnable action)
```

该方法的参数与上述的 runOnUiThread()方法一样，都是待启动的线程。两者的不同之处是 runOnUiThread()方法会将该线程放入 UI 线程的事件队列中等待处理，而 post()方法是将该线程放入某个 View 对象的消息队列中等待处理。如果能将线程成功放入 View 对象的消息队列中，post()方法返回 true，否则返回 false。post()方法的使用与 runOnUiThread()方法类似，下面举例说明。同样修改例 7.4 的主 Activity 文件 HelloDemoActivity.java，代码如下所示。

```
public class HelloDemoActivity extends Activity {
    TextView text;
    @Override
    public void onCreate(Bundle savedInstanceState) {
        super.onCreate(savedInstanceState);
        setContentView(R.layout.main);
        text = (TextView) findViewById(R.id.text);
        TimerTask task = new TimerTask() {
            public void run() {
                Thread t = new Thread() {
                    public void run() {
                        text.setText("view.post()");
                    }
                };
                text.post(t);
            }
        };
        Timer timer = new Timer();
        timer.schedule(task, 3000);
    }
}
```

与前面的程序类似，一个定时器任务 task 被创建，其工作是创建一个新的线程 t，在该线程中尝试将程序界面的文字修改为“view.post()”，并调用 text.post(t)方法来启动该线程。接着，一个定时器对象 timer 被创建，通过调用 timer.schedule(task, 3000)方法来设定等待 3 秒钟后运行定时器任务 task。运行程序后，程序界面首先如图 7.2 所示，3 秒钟后，界面文字被修改为“view.post()”，如图 7.11 所示，实现了对 UI 组件的修改。

下面通过例 7.6 手机内存监视器程序来演示 Timer 定时器的使用，以及定时器任务如何请求 UI 线程进行 UI 操作。程序运行后，每隔一秒钟会读取当前新的内存信息并显示在界面上。

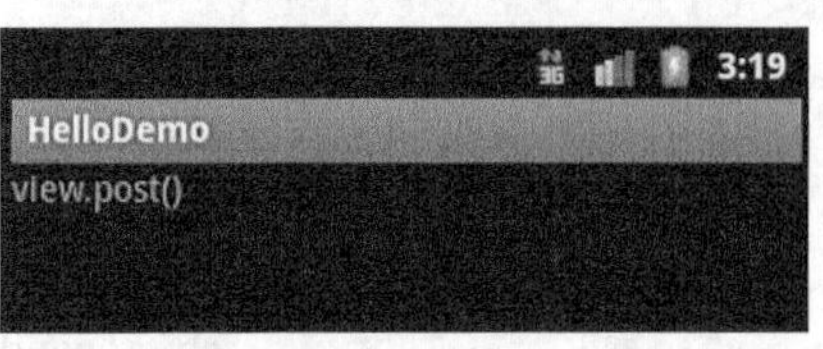

图 7.11　使用 view 的 post 方法修改 UI 组件

【例 7.6】 使用 Timer 定时器实现手机内存监视器。

（1）创建 Android 项目，项目名为 MemoryMonitor。修改主布局文件，代码如下所示。

```
<?xml version = "1.0" encoding = "utf-8"?>
<LinearLayout xmlns:android = "http://schemas.android.com/apk/res/android"
    android:layout_width = "match_parent"
    android:layout_height = "match_parent"
    android:orientation = "vertical" >
    <TextView android:id = "@+id/show"
        android:layout_width = "match_parent"
        android:textSize = "10pt"
        android:text = "正在获取内存信息..."
        android:layout_height = "wrap_content" />
</LinearLayout>
```

（2）修改主 Activity 文件，代码如下所示。

```
public class MemoryMonitorActivity extends Activity {
    @Override
    public void onCreate(Bundle savedInstanceState) {
        super.onCreate(savedInstanceState);
        setContentView(R.layout.main);
        final TextView show = (TextView) findViewById(R.id.show);
        TimerTask task = new TimerTask() {
            public void run() {
                Thread t = new Thread() {
                    public void run() {
                        ActivityManager.MemoryInfo info =
                                new ActivityManager.MemoryInfo();
                        ActivityManager manager = (ActivityManager)
                            getSystemService(Context.ACTIVITY_SERVICE);
                        manager.getMemoryInfo(info);
                        StringBuffer sb = new StringBuffer();
                        sb.append("当前总的可用内存: "
                                + ((info.availMem) >> 10) + "KB\n");
                        MemoryInfo memInfo = new MemoryInfo();
                        Debug.getMemoryInfo(memInfo);
                        sb.append("dalvik 虚拟机 PD 内存: "
                                + memInfo.dalvikPrivateDirty + "KB\n");
                        sb.append("dalvik 虚拟机 PSS 内存: "
                                + memInfo.dalvikPss + "KB\n");
                        sb.append("dalvik 虚拟机 SD 内存: "
                                + memInfo.dalvikSharedDirty + "KB\n");
                        sb.append("本地堆 PD 内存: "
                                + memInfo.nativePrivateDirty + "KB\n");
                        sb.append("本地堆 PSS 内存: "
                                + memInfo.nativePss + "KB\n");
```

```
                    sb.append("本地堆 SD 内存: "
                            + memInfo.nativeSharedDirty + "KB\n");
                    sb.append("其他 PD 内存: "
                            + memInfo.otherPrivateDirty + "KB\n");
                    sb.append("其他 PSS 内存: " + memInfo.otherPss + "KB\n");
                    sb.append("其他 SD 内存: "
                            + memInfo.otherSharedDirty + "KB");
                    show.setText(sb.toString());
                }
            };
            show.post(t);
        }
    };
    Timer timer = new Timer();
    timer.scheduleAtFixedRate(task, 1000, 1000);
  }
}
```

运行本实例，显示结果如图 7.12 所示。

图 7.12　内存监视器

上述代码中，通过语句 timer.scheduleAtFixedRate(task, 1000, 1000)定义了一个定时间隔为 1 秒的定时器来实现定时刷新内存信息的功能。在定时器任务中，定义了一个后台线程来获取内存信息，并通过 show.post(t)语句发布这个线程。内存信息分为两类，分别通过两个类来获取。ActivityManager.MemoryInfo 类的实例通过其 availMem 属性可以获得当前系统可用内存。在类 android. os.Debug. MemoryInfo 中可以获取与调试相关的内存信息，如 dalvik 虚拟机的内存使用状况，本地堆的内存使用状况等。这些数据都是通过 MemoryInfo 类的属性来获得的。读者可以查阅 API 文档获取相关的信息。

7.4　Android 多线程通信机制

前面介绍过，Android 中的多线程通信，尤其是非 UI 线程与 UI 线程之间的通信，一般使用 Handler 类通过发送与处理消息的形式来进行。消息由类 Message 来进行封装。本小节介绍另外两个与多线程编程相关的类：MessageQueue 和 Looper，并介绍 Android 中的多线程通信机制。

顾名思义，MessageQueue 类是消息队列的类。消息队列中可以包含多个消息，这些消息按照先进先出的原则依次被进行处理。MessageQueue 不需要由用户来进行创建和管理，而是通过 Looper 类用来管理。Android 中的每个线程都可以启动一个自己的 Looper，当 Looper 启动后，其会自动创建和管理 MessageQueue。每个线程最多只能有一个 Looper 和一个 MessageQueue。UI 线程启动时，系统会自动启动其 Looper。其他由用户新创建的线程不会自动启动 Looper，需要时可以手动启动。启动 Looper 通过调用静态方法 Looper.prepare()来实现。一个线程只有启动了 Looper，才能在其中创建 Handler 的对象，进行消息处理，否则程序会出现异常。下面通过例 7.7 来进行介绍。

【例 7.7】 Looper 类的使用。

（1）创建 Android 工程，工程名为 LooperTest。修改主 Activity 文件，代码如下所示。

```
public class LooperTestActivity extends Activity {
    @Override
    public void onCreate(Bundle savedInstanceState) {
        super.onCreate(savedInstanceState);
        setContentView(R.layout.main);
        new MyThread().start();
    }
    class MyThread extends Thread{
        public void run(){
            Handler handler = new Handler();
        }
    }
}
```

在类 LooperTestActivity 中定义了一个内部类 MyThread，继承自 Thread 类，作为一个新的线程，并在其 run()方法中创建一个 Handler 对象。运行程序，启动新创建的线程时，程序报错，抛出一个异常，异常的描述为：

```
java.lang.RuntimeException: Can't create handler inside thread that has not called
Looper.prepare()
```

该异常说明必须要首先调用 Looper.prepare()方法，然后才能在线程中创建 Handler 对象。将语句 Looper.prepare()加入上述代码中的语句 Handler handler = new Handler()之前，再次运行程序，没有异常发生，程序可以正常运行。

一个线程中只能有一个 Looper 和一个 MessageQueue，但是可以有多个 Handler 对象。程序中可以使用不同的 Handler 对象处理不同的消息。下面通过修改 LooperTestActivity.java 文件中 MyThread 类的定义来介绍 Looper 和 MessageQueue 的用法。

（2）重新修改主 Activity 文件，代码如下所示。

```
public class LooperTestActivity extends Activity {
    @Override
    public void onCreate(Bundle savedInstanceState) {
        super.onCreate(savedInstanceState);
        setContentView(R.layout.main);
        new MyThread().start();
    }
    class MyThread extends Thread {
        public void run() {
            Looper.prepare();
            Handler handler1 = new Handler() {
                public void handleMessage(Message msg) {
                    System.out.println("handler1 handle message " + msg.what);
                }
            };
            Handler handler2 = new Handler() {
                public void handleMessage(Message msg) {
                    System.out.println("handler2 handle message " + msg.what);
                }
            };
            Message msg1 = Message.obtain();
```

```
                Message msg2 = Message.obtain();
                msg1.what = 1;
                msg2.what = 2;
                handler1.sendMessage(msg1);
                handler2.sendMessage(msg2);
                Looper.loop();
            }
        }
    }
```

可以看到这个线程内部类中定义了两个 Handler 对象，handler1 和 handler2。然后通过 Message.obtain()方法创建了两个 Message 对象，msg1 和 msg2。给这两个 Message 对象分别赋值 1 和 2 以作区分。接着调用 sendMessage()方法发送这两个消息。注意程序最后一行的 Looper.loop() 方法，用来启动消息队列循环，处理 MessageQueue 中所有的消息。运行实例，可以看到输出的控制台信息中显示了两条消息被处理的结果。

```
handler1 handle message 1
handler2 handle message 2
```

7.5 小　　结

本章介绍了 Android 平台的多线程编程。Android 中可以通过继承 Thread 类或者实现 Runnable 接口创建线程。Thread 类提供了一系列方法用于控制线程。Android 中不能直接从非 UI 线程访问 UI 组件，需要通过传递消息和处理消息的形式来进行。Android 中提供了 Handler 类和 AsyncTask 类进行线程间通信及消息处理。Handler 类可以处理较为复杂的线程间通信及消息处理。AsyncTask 类提供的一个轻量级的基于多线程的进行后台异步工作处理的类。在后台工作比较简单，只需要向 UI 线程传递一些简单数据时可以使用 AsyncTask 类。Timer 定时器也是常用的实现多线程程序的方式。但是在定时器中也不能直接进行 UI 操作，需要通过 Handler 类，Activity.runOnUiThread (Runnable)方法，或者 View.post(Runnable)方法等方式来间接地进行 UI 操作。本章最后还介绍了 Android 多线程通信机制，多线程的消息传递与处理涉及 Message、MessageQueue 和 Looper 等类。Android 中的每个线程都可以启动一个自己的 Looper，当 Looper 启动后，其会自动创建和管理 MessageQueue。每个线程最多只能有一个 Looper 和一个 MessageQueue，但是可以有多个 Handler 对象进行消息处理。

练　　习

1. 通过继承 Thread 类创建线程和通过实现 Runnable 接口创建线程有何异同?
2. Android 中若从一个非 UI 线程操作 UI 组件，有哪些不同的方法?
3. 编写程序，在界面上添加一个按钮“计算”，单击按钮后，启动一个线程，计算从 2～200 之间所有的素数和，最后将结果显示在界面上。
4. 通过定时器控制界面背景色，实现每隔一秒自动更换一次背景色。
5. 简述 Android 多线程通信机制。

第 8 章 网络通信

网络通信功能是很多手机应用程序必不可少的部分，通过与服务器或其他客户端程序通信可以进行数据传输，实现所需功能。Android 平台中提供了众多的进行网络通信编程的接口，本章将分别介绍 WebView 组件、HttpURLConnection、Socket 编程和 Web Service 访问等网络通信方式。此外，本章还将介绍 JSON 数据处理和 XML 数据处理的内容，这是两种网络编程中经常要处理的数据格式。

8.1 通过 HTTP 访问网络

通过 HTTP 访问网络资源是目前最为广泛的应用之一，通常用于从各种网站获取需要的信息。本节首先介绍如何编写一个简单的用于测试的 Web 服务器应用，供后续各小节的 Android 示例程序使用。

8.1.1 测试用 Web 服务器

这里使用 Tomcat 6.0 作为 Web 服务器，编写一些静态 HTML 页面、动态的 JSP 页面和相应的 Servlet 程序供测试使用。本章中使用 Eclipse for Java EE 版，也可使用 MyEclipse 等其他 Web 开发环境。对于熟悉这部分内容的用户，也可自行进行设计。

在 Eclipse 中新建一个 “Dynamic Web Project”，工程名称为 TestServer，其他的配置选项均使用默认选项。创建工程完毕后的目录结构如图 8.1 所示。

```
TestServer
  JAX-WS Web Services
  Deployment Descriptor: TestServer
  Java Resources
    src
    Libraries
  JavaScript Resources
  build
  WebContent
    META-INF
    WEB-INF
      lib
      web.xml
```

图 8.1 Web 工程目录结构

1. 创建静态页面

在工程的 WebContent 目录下新建一个静态 HTML 文件 index.html。该文件代码如下所示。

```
<html>
    <head>
        <meta http-equiv = "Content-Type" content = "text/html; charset = UTF-8">
        <title>Test Page</title>
    </head>
    <body>
        <h1>测试 Hello World.</h1>
    </body>
</html>
```

2. 创建 Servlet：Login

在工程的 src 文件夹中新建一个包 com.example.web，在包中创建一个 Servlet 类 Login。如果是利用向导创建，则会在 web.xml 中自动进行 Servlet 信息注册。如果是手动创建的 servlet，则需要用户手动注册。Login 是一个很简单的 Servlet，通过 doGet()或 dotPost()方法接收用户提交的 Web 表单中的 name 和 pass 两个字段值，如果这两个字段的内容相同，则认为验证通过，向用户返回标识 1，否则返回标识 0。Login.java 代码如下所示。

```
package com.example.web;
import java.io.IOException;
import javax.servlet.ServletException;
import javax.servlet.ServletOutputStream;
import javax.servlet.http.HttpServlet;
import javax.servlet.http.HttpServletRequest;
import javax.servlet.http.HttpServletResponse;
public class Login extends HttpServlet {
    private static final long serialVersionUID = 1L;
    public Login() {
        super();
    }
    //接收 get 请求
    protected void doGet(HttpServletRequest request, HttpServletResponse response)
        throws ServletException, IOException {
        String name = request.getParameter("name");
        String pass = request.getParameter("pass");
        ServletOutputStream out = response.getOutputStream();
        if ( name! = null && pass! = null && name.equals(pass)) {
            out.print(1);
        }
        else
            out.print(0);
    }
    //接收 post 请求
    protected void doPost( HttpServletRequest request, HttpServletResponse response)
        throws ServletException, IOException {
        doGet(request,response);  //调用 doGet()，统一处理
    }
}
```

3. 创建 Servlet：Json

创建 Servlet 类 Json，用于后面介绍 JSON 数据处理时提供 JSON 数据。Json.java 代码如下所示。

```
package com.example.web;
```

```
import java.io.IOException;
import javax.servlet.ServletException;
import javax.servlet.ServletOutputStream;
import javax.servlet.http.HttpServlet;
import javax.servlet.http.HttpServletRequest;
import javax.servlet.http.HttpServletResponse;
public class Json extends HttpServlet {
    private static final long serialVersionUID = 1L;
    public Json() {
        super();
    }
    protected void doGet(HttpServletRequest request, HttpServletResponse response)
        throws ServletException, IOException {
        ServletOutputStream out = response.getOutputStream();
        out.print("[{\"city\":\"beijing\",\"code\":\"010\"}"
             + ",{\"city\":\"shanghai\",\"code\":\"021\"}"
             + ",{\"city\":\"xian\",\"code\":\"029\"}]");
    }
    protected void doPost(HttpServletRequest request, HttpServletResponse response)
        throws ServletException, IOException {
        doGet(request, response);
    }
}
```

完成以上工作后，将 TestServer 工程加入 Tomcat6.0 服务器，并启动服务器。

8.1.2　WebView 组件

WebView 是一个浏览器 UI 组件，使用了 Webkit 引擎用以显示 Web 页面。类似于其他的视图组件，可以将 WebView 放在界面中合适的位置，然后通过调用相应的方法来实现对其的操作。

WebView 类提供的常用方法如下。

- public void loadUrl(String url)

装载由参数指定的 URL。

- public void loadData(String data, String mimeType, String encoding)

装载数据，第一个参数一般是 HTML 代码，第二个参数是装载的数据的格式，如 text/html、image/jpeg 等，第三个参数是页面编码类型，如 utf-8、base64 等。

- public void loadDataWithBaseURL(String baseUrl, String data, String mime Type, String encoding, String failUrl)

与上述 loadData 方法类似，多了两个参数。第一个参数用来提供有相对路径的 baseUrl，最后一个参数表明如果加载页面失败，可以转向的 URL 位置。

- public void goBack()

返回前一个页面。

- public void goForward()

前往下一个页面。

- public boolean zoomIn()

放大当前页面，如果成功返回 true，否则返回 false。

- public boolean zoomOut()

缩小当前页面，如果成功返回 true，否则返回 false。

下面通过例 8.1 来介绍 WebView 的使用。

【例 8.1】 通过 WebView 组件访问网页。

（1）新建 Android 工程，工程名称为 WebViewTest。修改主布局文件，将默认的布局管理器改为 LinearLayout 线性布局，垂直排列。在该布局中添加 WebView 组件，布满整个界面，代码如下所示。

```
<?xml version = "1.0" encoding = "utf-8"?>
<LinearLayout xmlns:android = "http://schemas.android.com/apk/res/android"
    android:layout_width = "match_parent"
    android:layout_height = "match_parent"
    android:orientation = "vertical" >
    <WebView android:id = "@+id/view"
        android:layout_width = "match_parent"
        android:layout_height = "match_parent" />
</LinearLayout>
```

（2）修改主 Activity 类 WebViewTestActivity.java，代码如下所示。

```
public class WebViewTestActivity extends Activity {
    @Override
    public void onCreate(Bundle savedInstanceState) {
        super.onCreate(savedInstanceState);
        setContentView(R.layout.main);
        WebView v = (WebView)findViewById(R.id.view);
        v.setWebViewClient(new WebViewClient() {
            public boolean shouldOverrideUrlLoading(WebView view, String url) {
                view.loadUrl(url);
                return super.shouldOverrideUrlLoading(view, url);
            }
        });
        v.loadUrl("http://10.0.2.2:8888/TestServer");  //加载网址
    }
}
```

上例中通过使用 WebView 类的 loadUrl()方法加载网址，其中的参数 http://10.0.2.2:8888/TestServer 是从 Android 模拟器中访问本机 Tomcat 服务器的 URL 地址，8888 为 Tomcat 中设置的端口号。由于之前编写了 index.html 页面，该页面将作为 Tomcat 服务器的默认初始页面，也就是程序运行后，WebView 组件将装载并显示 index.html 页面。代码中 setWebViewClient()方法用于指定 loadUrl()方法打开页面的载体。如果无此行代码，程序也能运行，但是 index.html 页面会由 WebView 组件转交给 Android 系统的默认浏览器打开。加上此行代码后，页面将由 WebView 组件本身打开。

在 Android 程序中进行各种联网操作，需要在 AndroidManifest.xml 文件中添加如下的授权声明，后续的各小节中不再单独进行说明。

```
<uses-permission
android:name = "android.permission.INTERNET"/>
```

运行本实例，显示结果如图 8.2 所示。

通过 WebView 组件访问和显示 Web 页面非常简单，用户可以在添加控制按钮，如向前、返回、刷新等，和 URL 地址输入栏等之后，使用上述介绍的 WebView 类中的其他方法对所显示的页面进行控制。

图 8.2　WebView 加载网页

8.1.3 HttpURLConnection

HttpURLConnection 类位于 java.net 内，提供了基于 HTTP 的网络访问方法。Android API 中也包含了这个类，可以按照和 Java 中相同的使用方法来进行 Web 访问。使用 HttpURLConnection 访问网络，主要的操作步骤为：

- 利用 URL 地址实例化 URL 类
- 由 URL 类创建 HttpURLConnection 对象
- 以 GET/POST 方式向服务器发送请求
- 接收服务器响应

下面通过例 8.2 来介绍如何使用 HttpURLConnection。

【例 8.2】 使用 HttpURLConnection 访问 Servlet 验证用户。

该例将访问 8.1.1 中建立的 TestServer 工程的 Login Servlet。

（1）新建 Android 工程，工程名为 HttpTest。修改主布局文件，添加两个输入框和一个按钮，代码如下所示。

```
<?xml version = "1.0" encoding = "utf-8"?>
<LinearLayout xmlns:android = "http://schemas.android.com/apk/res/android"
    android:layout_width = "match_parent"
    android:layout_height = "match_parent"
    android:gravity = "center_horizontal"
    android:orientation = "vertical" >
    <EditText android:id = "@+id/name"
        android:layout_width = "match_parent"
        android:layout_height = "wrap_content"
        android:textSize = "25sp"
        android:layout_margin = "5sp"
        android:hint = "输入用户名" />
    <EditText android:id = "@+id/pass"
        android:layout_width = "match_parent"
        android:layout_height = "wrap_content"
        android:textSize = "25sp"
        android:layout_margin = "5sp"
        android:password = "true"
        android:hint = "输入密码" />
    <Button android:id = "@+id/login"
        android:layout_width = "wrap_content"
        android:layout_height = "wrap_content"
        android:textSize = "25sp"
        android:layout_margin = "5sp"
        android:text = "    登  录    " />
</LinearLayout>
```

（2）修改主 Activity 类 HttpTestActivity.java，实现通过服务器进行登录验证的功能。两个输入框分别用来让用户输入用户名和密码，当单击“登录”按钮，将用户名和密码发送到服务器请求验证，如果验证通过，将进入下一个欢迎界面。代码如下所示。

```
public class HttpTestActivity extends Activity {
    @Override
    public void onCreate(Bundle savedInstanceState) {
```

```
        super.onCreate(savedInstanceState);
        setContentView(R.layout.main);
        final EditText name = (EditText)findViewById(R.id.name);
        final EditText pass = (EditText)findViewById(R.id.pass);
        final Button button = (Button) findViewById(R.id.login);
        button.setOnClickListener(new View.OnClickListener() {
            public void onClick(View v) {
                if(name.getText().toString().equals("")) {
                    Toast toast = Toast.makeText(HttpTestActivity.this, "输入用户名",
                                    Toast.LENGTH_LONG);
                    toast.show();
                }
                else if(pass.getText().toString().equals("")) {
                    Toast toast = Toast.makeText(HttpTestActivity.this, "输入密码",
                                    Toast.LENGTH_LONG);
                    toast.show();
                }
                else {
                    try {
                        String n = name.getText().toString();
                        String p = pass.getText().toString();
                        URL url = new URL("http://10.0.2.2:8888/TestServer/Login?name=
                                    "+n+"&pass="+p);
                        HttpURLConnection uc = (HttpURLConnection)url.openConnection();
                        InputStream out = uc.getInputStream();
                        String result = String.valueOf(out.read());
                        if(result.equals("1")) {
                            Intent intent = new Intent(HttpTestActivity.this, WelcomeActivity.
                                            class);
                            startActivity(intent);
                        }
                        else {
                            Toast toast = Toast.makeText(HttpTestActivity.this,
                              "用户名或密码错误", Toast.LENGTH_LONG);
                            toast.show();
                        }
                    } catch (IOException e) {
                        e.printStackTrace();
                    }
                }
            }
        });
    }
}
```

（3）欢迎界面，该界面仅显示简单的欢迎字符，其所对应的布局文件是 main2.xml，代码如下所示。

```
<LinearLayout xmlns:android = "http://schemas.android.com/apk/res/android"
    android:layout_width = "match_parent"
    android:layout_height = "match_parent"
    android:orientation = "vertical" >
    <TextView android:id = "@+id/text"
```

```
        android:layout_width = "match_parent"
        android:layout_height = "wrap_content"
        android:textSize = "15pt"
        android:text = "Welcome" />
</LinearLayout>
```

（4）欢迎界面的 Activity 程序为 WelcomeActivity.java，未做修改，代码如下所示。

```
public class WelcomeActivity extends Activity {
    public void onCreate(Bundle savedInstanceState) {
        super.onCreate(savedInstanceState);
        setContentView(R.layout.main2);
    }
}
```

当用户单击“登录”按钮后，首先判断是否已经输入了用户名与密码，如果没有输入，则通过 Toast 提示用户进行输入。接着把用户名与密码拼接成以 get 方式向服务器发送请求的 URL 字符串，即：

```
http://10.0.2.2:8888/TestServer/Login?name = " + n + "&pass = " + p
```

以该字符串为参数创建 URL 对象 url，使用 url 的 openConnection()方法获得 HttpURLConnection 对象 uc。再通过 getInputStream()方法获得携带服务器返回信息的输入流，从流中读取出服务器返回的信息。如果返回的是 1，则表示验证通过，进入 welcome 界面，否则通过 Toast 提示用户验证未通过。

在运行程序之前，还需要在 AndroidManifest.xml 文件中注册 Welcoe Activity，以及添加访问网络的权限声明。运行本实例，结果如图 8.3、图 8.4 所示。

图 8.3　登录界面

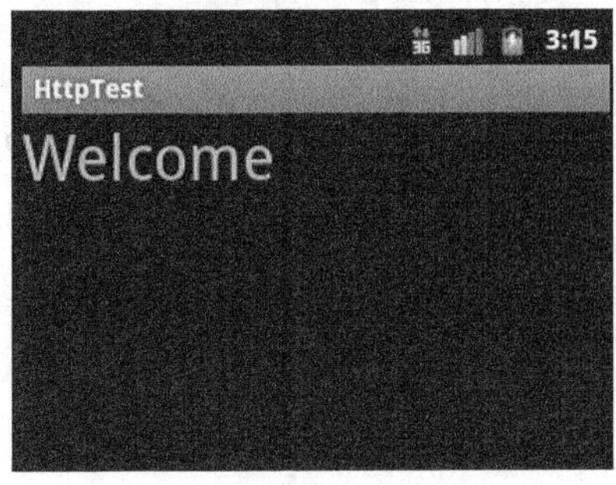

图 8.4　验证通过打开界面

8.2　Socket 编程

Android 支持 Java 的 ServerSocket 和 Socket 类，可以实现 TCP/IP 通信；也支持 DatagramSocket，实现 UDP 通信。可以利用这些类实现手机间的通信，如聊天、传送数据等。下面通过例 8.3 来介绍如何使用 ServerSocket 和 Socket 类。该例实现一个简单的多人聊天软件，包括一个 PC 上的服务器程序和一个 Android 的客户端程序。多个客户端可以同时连接服务器，每个客户端发送的聊天信息由服务器转发给其他客户端，实现了多人进行聊天的功能。

【例 8.3】　使用 Socket 实现多人聊天应用。

（1）首先编写服务器端程序。由于服务器程序要处理多个客户端程序发送的聊天信息，因此需要使用多线程技术，为每一个客户端启动一个单独的线程来进行处理。程序基本的思路如下：主线程（对应于 ChatServer 类）负责等待客户端的连接请求，一旦有客户端请求连接，则建立连接并将该客户端加入一个 ArrayList 中进行管理。然后为该客户端启动一个独立的聊天线程（对应于 ClientThread 类），该线程接受该客户端发送来的聊天信息，然后向 ArrayList 中现有的所有客户端进行转发。ChatServer 类和 ClientThread 类可以作为两个单独的类，本例子中采用了将 ClientThread 类作为 ChatServer 类的内部类的方式来实现。因为这两个类关联紧密，且 ClientThread 的实例只被 ChatServer 类使用，这样处理更加合理和便捷。代码如下所示。

```
public class ChatServer {
    public List<Socket> clients = new ArrayList<Socket>();
    private void startServer() {
        ServerSocket server;
        try {
            server = new ServerSocket(12345);
            while (true) {
                System.out.println("waiting...");
                Socket s = server.accept();
                clients.add(s);
                this.new ClientThread(s).start();
            }
        }
        catch (IOException e) {
            e.printStackTrace();
        }
    }
    public class ClientThread extends Thread {
        Socket s;
        BufferedReader in;
        public ClientThread(Socket client) {
            s = client;
            try {
                in = new BufferedReader(new InputStreamReader(s.getInputStream(), "utf-8"));
            }
            catch (Exception e) {
                e.printStackTrace();
            }
        }
        public void run() {
            try {
                String sentence = null;
                while ((sentence = in.readLine()) != null) {
                    for (Socket s : clients) {
                    try {
                        s.getOutputStream().write(
                            (sentence + "\n").getBytes("utf-8"));
                    }
                    catch (IOException e) {
                        e.printStackTrace();
                    }
                }
```

```
                }
            }
            catch (Exception e) {
                e.printStackTrace();
            }
        }
    }
    public static void main(String[] args) throws IOException {
        new ChatServer().startServer();
    }
}
```

ServerSocket类用于创建服务器端的套接字对象，构造方法只要指定一个用于通信的端口号即可。本程序中使用了端口号“12345”。ServerSocket类的accept()用于等待和接受客户端的连接。如果一直没有客户端请求进行连接，则该方法会阻塞，直到有了客户端的请求。该方法的返回值是一个Socket对象，即可以与客户端进行通信的对象，这个对象会被加入ArrayList，并以该对象为参数启动新的线程处理与该对象的通信。在线程ClientThread类中，通过一个无限循环不断地处理客户端发送来的聊天信息。Socket类的getInputStream()方法可以获得输入流，用于从中读取客户端发送的聊天信息。本例子中将这个输入流包装成为一个BufferedReader对象以便处理。readLine()方法也是一个阻塞方法，如果读取不到客户端的信息，就一直等待。如果读取到，就向ArrayList中所保存的所有的客户端转发聊天信息。完成此程序，在服务器（或PC机）上启动即可。

（2）编写Android客户端程序。创建一个Android工程，名称为chat。在布局文件中添加两个文本框，分别用于输入用户名与聊天信息，再添加一个文本框设定为只读，用来显式所有的聊天信息。代码如下所示。

```
<?xml version = "1.0" encoding = "utf-8"?>
<LinearLayout xmlns:android = "http://schemas.android.com/apk/res/android"
    android:layout_width = "match_parent"
    android:layout_height = "match_parent"
    android:orientation = "vertical" >
    <EditText android:id = "@+id/name"
        android:layout_width = "match_parent"
        android:layout_height = "wrap_content"
        android:layout_margin = "2pt"
        android:hint = " 输入用户名 "
        android:textSize = "8pt" />
    <EditText android:id = "@+id/sentence"
        android:layout_width = "match_parent"
        android:layout_height = "wrap_content"
        android:layout_margin = "2pt"
        android:hint = " 输入聊天内容 "
        android:textSize = "8pt" />
    <Button android:id = "@+id/send"
        android:layout_width = "wrap_content"
        android:layout_height = "wrap_content"
        android:layout_margin = "2pt"
        android:text = "    发 送    "
        android:textSize = "8pt" />
```

```
    <EditText android:id = "@+id/show"
        android:layout_width = "match_parent"
        android:layout_height = "match_parent"
        android:layout_margin = "2pt"
        android:gravity = "top"
        android:editable = "false"
        android:textSize = "8pt" />
</LinearLayout>
```

（3）在主 Activity 程序中需要完成两方面的工作。一是当用户单击“发送”按钮时，将聊天信息发送至服务器，二是接收服务器发送来的新的聊天信息，并在界面上显示。由于接收服务器的信息需要等待，不能直接将这个处理过程写在主线程中，所以也需要通过启动一个独立的线程来实现这个功能，当接收到服务器发送来的消息后，由该线程向主线程发送消息，由主线程将内容显示在界面上。本例中使用到了之前章节中所讲到的 Android 中基于 Handler 类的多线程通信处理方法。ChatActivity 类的代码如下所示，其中作为线程的 ChatThread 类也是以内部类的形式封装在 ChatActivity 类之中。

```
public class ChatActivity extends Activity {
    Socket socket;
    Handler handler;
    @Override
    public void onCreate(Bundle savedInstanceState) {
        super.onCreate(savedInstanceState);
        setContentView(R.layout.main);
        final EditText name = (EditText) findViewById(R.id.name);
        final EditText sentence = (EditText) findViewById(R.id.sentence);
        final Button send = (Button) findViewById(R.id.send);
        final EditText show = (EditText) findViewById(R.id.show);
        try {
            socket = new Socket("192.168.2.2", 12345);
            this.new ChatThread(socket).start();
        }
        catch (Exception e) {
            e.printStackTrace();
        }
        send.setOnClickListener(new View.OnClickListener() {
            public void onClick(View v) {
                String s1 = name.getText().toString();
                String s2 = sentence.getText().toString();
                String s = s1 + ": " + s2 + "\n";
                if (socket != null) {
                    try {
                        socket.getOutputStream().write(s.getBytes("utf-8"));
                    }
                    catch (Exception e) {
                        e.printStackTrace();
                    }
                    sentence.setText("");
                }
            }
        });
        handler = new Handler() {
```

```
            public void handleMessage(Message msg) {
                super.handleMessage(msg);
                String tmp = show.getText().toString();
                show.setText(tmp + "\n\n" + msg.getData().getString("chat"));
            }
        };
    }
    class ChatThread extends Thread {
        Socket s;
        public ChatThread(Socket client) {
            s = client;
        }
        public void run() {
            try {
                BufferedReader in = new BufferedReader(
                     new InputStreamReader(s.getInputStream(), "utf-8"));
                while (true) {
                    String sentence = in.readLine();
                    if (sentence != null) {
                        Bundle b = new Bundle();
                        b.putString("chat", sentence);
                        Message msg = Message.obtain();
                        msg.setData(b);
                        handler.sendMessage(msg);
                    }
                }
            }
            catch (Exception e) {
                e.printStackTrace();
            }
        }
    }
}
```

上述代码中 Socket 类的构造方法需要两个参数：服务器的 IP 地址和端口号。例子中为 Socket ("192.168.2.2", 12345)。这个构造方法也同时会按照所提供的参数去请求连接服务器，如果连接成功则创建新的 Socket 对象，如果连接不成功，则抛出 UnknownHostException。连接成功后，以该 Socket 对象为参数，启动 ChatThread 线程。

Socket 类的 getOutputStream()方法返回一个输出流，其 write()方法可以向流中写入数据。在本程序中也就是向服务器端发送聊天信息数据。主线程中还需要创建一个 Handler 对象，用于接收子线程发送来的、服务器所发送的新聊天信息，并将这个信息在界面上显示出来。子线程中接收服务器端消息的处理过程和服务器端程序中接收客户端消息的形式完全一样，这里不再赘述。当接收到新的消息，将其封装到一个 Bundle 对象中，再放入 Message 对象中发送给主线程显示。

以上这个聊天的例子只是为了介绍如何使用 ServerSocket 和 Socket 类进行数据通信。例子的程序中有诸多不完善之处，读者可以进一步将其改进，并加入更多的功能，即可做出一个实用的聊天程序。另外，手机中的 Socket 通信也可以实现点对点的通信，而不需要通过额外的服务器来进行转发。这就需要在每个客户端同时编写 ServerSocket 和 Socket 的相关代码，实现消息的发送与接收。感兴趣的用户也可以自己进行测试。

运行本实例，用户一聊天界面如图 8.5 所示，用户二聊天界面如图 8.6 所示。

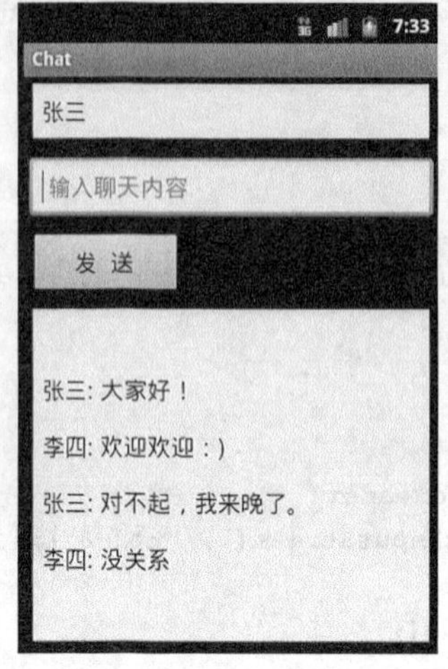

图 8.5 用户一聊天界面

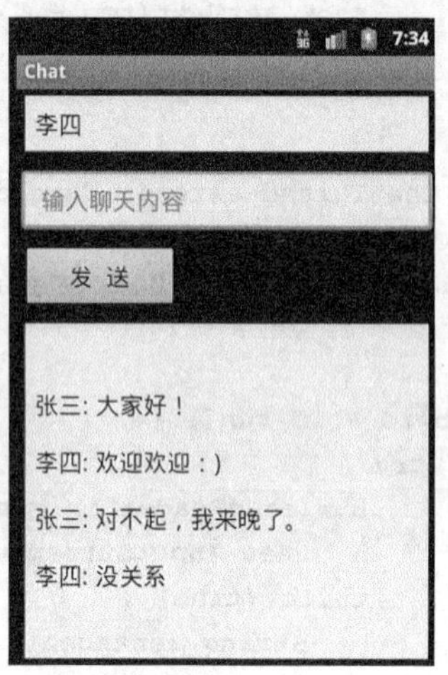

图 8.6 用户二聊天界面

8.3 数据的解析

在进行资源访问获取数据时，通常提供的数据格式有两种最为主流，一种是 JSON 数据，第一种是 XML 数据。其中 JSON 数据是以键值对形式存放的字符串，而 XML 数据则类似于 HTML，是以标签形式提供的文本，下面对这两种格式数据的解析分别予以介绍。

8.3.1 JSON 数据解析

JSON 是目前常用的轻量级网络数据传输格式。Android API 中直接支持对 JSON 数据的处理。JSON 解析中常用的类有 JSONObject、JSONArray、JSONStringer 和 JSONTokener 等。

下面将通过例 8.4 来介绍如何编写应用程序从服务器获取并处理 JSON 数据。本程序中可以使用前面介绍过的 HttpURLConnection 或者 Socket 编程的方式来实现客户端与服务器端的通信。这里介绍另外一种通信方式，通过使用 HttpClient。HttpClient 是 Apache 的开源项目，提供了便捷的访问 Web 服务器的方法，以及管理 session 的能力。Android API 中包含了 HttpClient，一些相关的类包括：

- HttpGet/HttpPost
- HttpResponse
- HttpEntity
- NameValuePair / EntityUtils

利用 HttpClient 进行 Web 访问的一般步骤如下。

- 创建 HttpClient 对象
- 创建 Request 对象
- 调用 execute()方法，向服务器发送请求并获得 HttpResponse 对象

- 解析 Response，获得结果

【例 8.4】　JSON 数据解析应用。

（1）创建一个 Android 工程，名称为 json。在主布局文件中，为默认生成的 TextView 组件设置 id 值为 text。此处代码省略，请用户自行修改。

（2）在主 Activity 程序 JsonActivity.java 中修改该 TextView 的内容，把从服务器读取到的 JSON 数据显示出来，代码如下所示。

```
public class JsonActivity extends Activity {
    @Override
    public void onCreate(Bundle savedInstanceState) {
        super.onCreate(savedInstanceState);
        setContentView(R.layout.main);
        TextView text = (TextView) findViewById(R.id.text);
        try {
            HttpGet request = new HttpGet("http://10.0.2.2:8888/TestServer/Json");
            HttpResponse response = new DefaultHttpClient().execute(request);
            if (response.getStatusLine().getStatusCode() == 200) {
                String msg = EntityUtils.toString(response.getEntity());
                JSONArray array = new JSONArray(msg);
                String s = "";
                for (int i = 0; i < array.length(); i + +) {
                    JSONObject o = (JSONObject) array.get(i);
                    s = s + o.getString("city") + ": " + o.getString("code") + "  ";
                }
                text.setText(s);
            }
        }
        catch (Exception e) {
            e.printStackTrace();
        }
    }
}
```

代码 HttpGet request = new HttpGet("http://10.0.2.2:8888/TestServer/Json")创建一个 HttpClient 的 request 对象，其参数为 URL 地址，指向在测试服务器中编写的 Servlet Json。DefaultHttpClient 类可以创建一个默认的 HttpClient 对象，调用该对象的 execute()方法，以 request 对象为参数，就可以返回一个 HttpResponse 对象。如果 Web 访问正常，该 response 对象中就包含了服务器返回的 JSON 数据。语句 EntityUtils.toString(response.getEntity())提取出了 response 对象中的 JSON 数据，保存在一个字符串中。类 JSONArray 提供了一个构造方法，可以以一个 JSON 字符串为参数，创建一个 JSONArray 对象。接着通过一个循环，调用 get()方法依次取出 JSONArray 对象中的三项数据，将其保存在 JSONObject 对象中。JSONObject 对象的 getString()方法可以通过传递指定名称的参数，获取到实际所需要的数据，本例中为城市的名称和邮政编码。从例子中可以看出通过 Android API 提供的类及其相应的方法，解析和处理 JSON 数据是非常方便的。

确保 Tomcat 服务器已经开启，并且在本工程的 AndroidManifest.xml 文件中添加了相应的权限。运行本实例，结果如图 8.7 所示。界面中所显示的三项数据，就是通过解析服务器返回的 JSON 数据得到的。

图 8.7　JSON 数据处理

8.3.2 XML 数据解析

JSON 适合于处理格式简单的数据。对于格式较为复杂的数据，仍然需要使用 XML 格式，这也是网络程序设计中经常使用的数据传输格式。Android API 中提供了丰富的处理 XML 数据的类。本小节将介绍一些常用的处理 XML 数据的操作方法。XML 数据既可以来自网络，也可以是一个手机上的文件，或者工程的资源文件。从网络服务器中获取 XML 数据所使用的方法可以与前述获取 JSON 数据的方法一样通过 Servlet 来实现，也可以通过数据文件的形式来实现。

本小节的例子中将直接使用服务器中的 XML 数据文件。在 8.1.1 小节中建立的测试服务器 TestServer 的 WebContent 文件夹中新建一个 XML 文件 test.xml，其内容为如下所示。

```
<?xml version = "1.0"?>
<cities>
  <city id = "1">
    <name>beijing</name>
    <code>010</code>
  </city>
  <city id = "2">
    <name>shanghai</name>
    <code>021</code>
  </city>
  <city id = "3">
    <name>xian</name>
    <code>029</code>
  </city>
</cities>
```

Android API 中提供了如下几种处理 XML 数据的方法。

- DOM 方式：基于树的解析与处理。
- SAX 方式：基于流的解析与处理（推方式）。
- XML PULL 方式：基于流的解析与处理（拉方式）。

DOM 方式和 SAX 方式是 Java 中处理 XML 数据最常用的两种方式，Android API 中也提供了相应的支持。XML PULL 方式是 Android API 提供的另外一种有效的处理 XML 数据的方式。本小节中将通过读取和解析同一个 XML 来对比三种处理方式的特点。

1. DOM 方式

XML 数据的组织形式就是树状结构，DOM 方式处理 XML 数据就是先读取所有的 XML 数据，将其构造成一棵树，然后通过一些方法来进行数据的处理。所以这种处理 XML 数据的方式是基于树的方式。

使用 DOM 方式处理 XML 数据的一般步骤如下。

- 建立一个解析器工厂，利用这个工厂来获得一个具体的解析器对象。
- 利用解析器来对 XML 文档进行解析。
- 通过 Documen、NodeList、Node、NamedNodeMap 等对象的相关方法对 XML 数据进行处理。

下面通过例 8.5 来介绍如何使用 DOM 方式来处理 XML 数据。

【例 8.5】 DOM 方式来处理 XML 数据应用。

（1）建立一个 Android 工程，修改主布局文件为线性布局，其中增加一个文本框组件用来显示解析后的 XML 数据。代码如下所示。

```
<?xml version = "1.0" encoding = "utf-8"?>
<LinearLayout xmlns:android = "http://schemas.android.com/apk/res/android"
    android:layout_width = "match_parent"
    android:layout_height = "match_parent"
    android:orientation = "vertical" >
    <EditText android:id = "@+id/text"
        android:layout_width = "match_parent"
        android:layout_height = "match_parent"
        android:gravity = "top"
        android:textSize = "25sp"
        android:editable = "false" />
</LinearLayout>
```

（2）修改主 Activity 文件，代码如下所示。

```
public class DomXmlActivity extends Activity {
    StringBuffer sb = new StringBuffer();
    @Override
    public void onCreate(Bundle savedInstanceState) {
        super.onCreate(savedInstanceState);
        setContentView(R.layout.main);
        EditText text = (EditText) findViewById(R.id.text);
        try {
            DocumentBuilderFactory factory = DocumentBuilderFactory.newInstance();
            DocumentBuilder builder = factory.newDocumentBuilder();
            Document doc = builder.parse("http://10.0.2.2:8888/TestServer/test.xml");
            NodeList nodelist = doc.getElementsByTagName("city");
            for (int i = 0; i < nodelist.getLength(); i + +) {
                Node node = nodelist.item(i);
                sb.append(node.getNodeName() + "-");
                NamedNodeMap map = node.getAttributes();
                Node n = map.item(0);
                sb.append(n.getNodeValue());
                sb.append(node.getTextContent() + "\n");
            }
        }
        catch (Exception e) {
            e.printStackTrace();
        }
    }
}
```

用户通过 DocumentBuilderFactory.newInstance()获取 DocumentBuilderFactory 类的对象 factory。factory 的方法 newDocumentBuilder()可以获得一个解析器 builder，而 builder 的方法 parse ("http://10.0.2.2:8888/TestServer/test.xml")通过参数来指定所需要解析的 XML 数据内容，并进行解析返回一个 Document 对象 doc。该 Document 对象就表示一个 XML 文档的树模型，后续对 XML 文档的操作，都与解析器无关，直接操作该 Document 对象。方法 getElementsByTagName("city")可以将模型中所有节点名称为“city”的节点获取出来并生成一个 NodeList 对象。然后通过 for 循环依次将 NodeList 对象中的各个 Node 的内容获取出来。nodelist.item(i)方法表示获得每个节点，getNodeName()方法表示获得节点名称，getAttributes()方法表示获得节点的属性值，getTextContent()方法表示获得节点的文本内容。以上这些方法获取的数据都追加至 StringBuffer 对象中，最后在界面上显示出来。

运行本实例，结果如图 8.8 所示。

图 8.8　XML 数据处理

2. SAX 方式

尽管 DOM 方式以树形结构来处理 XML 文件思路简单，但是也有其缺点。在处理 DOM 树时，需要读入整个的 XML 文档,然后在内存中创建 DOM 树,生成 DOM 树上的每个 Node 对象。当 XML 文档较小的时候，影响也较小，但是当文档较大时，处理 DOM 就会占用相当资源。

SAX 方式采用了基于流的 XML 数据处理方式，并非事先把所有的 XML 节点都读入内存，再进行数据处理，而是一边读一边对所需要的数据进行处理。与 DOM 比较而言，SAX 是一种比较轻量的方式，适合于处理数据量较大的 XML 文档。SAX 方式处理 XML 数据的一般步骤如下。

- 通过 SAXParserFactory 获取一个解析器工厂实例，然后通过该工厂得到一个解析器对象。
- 创建一个继承自 ContentHandler 类的子类来响应解析事件，根据需要重写其中的相关方法。
- 调用解析器的 parse()方法解析 XML 数据。

下面通过例 8.6 来介绍如何使用 SAX 方式来处理 XML 数据。

【例 8.6】　SAX 方式来处理 XML 数据应用。

（1）建立一个 Android 工程，修改主布局文件同例 8.5。

（2）修改主 Activity 文件，代码如下所示。

```
public class SAXXmlActivity extends Activity {
    StringBuffer sb = new StringBuffer();
    @Override
    public void onCreate(Bundle savedInstanceState) {
        super.onCreate(savedInstanceState);
        setContentView(R.layout.main);
        EditText text = (EditText) findViewById(R.id.text);
        try {
            SAXParser parser = SAXParserFactory.newInstance().newSAXParser();
            parser.parse("http://10.0.2.2:8888/TestServer/test.xml", new SaxHandler());
        }
        catch (Exception e) {
            e.printStackTrace();
        }
        text.setText(sb.toString());
    }
    class SaxHandler extends DefaultHandler {
        int type = 0;
        public void startElement(String uri, String localName, String qName,
                              Attributes attributes) throws SAXException {
          if (qName.equals("city")) {
              sb.append(qName + "-" + attributes.getValue(0)+ "\n");
          }
          else if (qName.equals("name")) {
              type = 1;
```

```
            }
            else if (qName.equals("code")) {
                type = 2;
            }
        }
        public void characters(char[] ch, int start, int length) throws SAXException {
            if(type = =1) {
                sb.append("    " + new String(ch,start,length)+ "\n");
                type = 0;
            }
            if(type = =2) {
                sb.append("    " + new String(ch,start,length)+ "\n\n");
                type = 0;
            }
        }
    }
}
```

上述代码中，SAXParserFactory 的方法 newInstance()可以获取一个解析器工厂实例，然后再调用方法 newSAXParser()得到一个解析器对象。接下来就可以调用方法 parse("http://10.0.2.2:8888/TestServer/test.xml",new SaxHandler())来解析第一个参数所指定的 XML 数据。方法中的第二个参数是一个自定义的 Handler 类，继承自 DefaultHandler。本例是以内部类的形式来组织的，该类中重写了 startElement()和 characters()两个方法。当解析过程中遇到某个节点元素时，startElement()方法会被触发调用，判断如果该元素的名字是“city”，则将其名字和属性值追加至 StringBuffer 对象中。如果该元素的名字是“name”或者“code”，则为一个标记变量 type 设定相应的数值，待后续获取元素内容时进行处理。当解析过程中遇到字符时，characters()方法会被触发调用。如果当前解析到的字符是之前用标记变量 type 标记过的字符，则将这些字符追加至 StringBuffer 对象中。重复这个过程，最后当整个 XML 数据文件全部处理完毕后，将 StringBuffer 对象的值显示在界面上，显示的结果与图 8.8 一致。

此外，Android API 中提供了一种与标准 SAX 解析方式相同的处理 XML 数据的类 android.util.Xml。该类的使用也是要基于一个事先创建好的 ContentHandler 对象，其构造方法与解析处理过程与 SAX 非常类似，此处不再赘述，感兴趣的用户可自行参考 API 文档编写程序进行测试。

3. XML PULL 方式

XML PULL 方式是 Android API 提供的另外一种有效的处理 XML 数据的方式。它和 SAX 方式一样都是基于流的处理方式，但是两者的处理思路略有不同。SAX 方式是基于事件触发方式的，从前面的例子中可以看出，在读取数据流的过程中，解析器在遇到节点开始和结束等时机会触发相应的方法，需要在这些方法中编写相应的处理程序，等待该方法被调用。换句话说，就是等待着把这些数据推送过来。XML PULL 方式采用的是主动去拉取的方式，每当解析器读取到流中的数据，可以主动去判断当前的数据是不是所需要的，如果是就进行处理，如果不是就继续解析后面的数据。

采用 XML PULL 方式处理 XML 数据的一般步骤如下。

- 获取 XmlPullParserFactory 实例。
- 创建 XmlPullParser 解析器。
- InputStream 流作为解析器的输入。

- 通过拉方式解析。
- 关闭流。

下面通过例 8.7 来介绍如何使用 XML PULL 方式来处理 XML 数据。

【例 8.7】 XML PULL 方式来处理 XML 数据应用。

（1）建立一个 Android 工程，修改主布局文件同例 8.5。

（2）修改主 Activity 文件，代码如下所示。

```
public class PullXmlActivity extends Activity {
    StringBuffer sb = new StringBuffer();
    @Override
    public void onCreate(Bundle savedInstanceState) {
        super.onCreate(savedInstanceState);
        setContentView(R.layout.main);
        EditText text = (EditText) findViewById(R.id.text);
        try {
            XmlPullParserFactory factory = XmlPullParserFactory.newInstance();
            XmlPullParser parser = factory.newPullParser();
            InputStream in = new URL("http://10.0.2.2:8888/TestServer/test.xml").openStream();
            parser.setInput(in, "utf-8");
            int eventType = parser.getEventType();
            int type = 0;
            while (eventType != XmlPullParser.END_DOCUMENT) {
                if (eventType == XmlPullParser.START_TAG) {
                    if (parser.getName().equals("city")) {
                        sb.append(parser.getName() + "-"
                                + parser.getAttributeValue(0) + "\n");
                    }
                    else if (parser.getName().equals("name")) {
                        type = 1;
                    }
                    else if (parser.getName().equals("code")) {
                        type = 2;
                    }
                }
                else if (eventType == XmlPullParser.TEXT) {
                    if (type == 1) {
                        sb.append("    " + parser.getText() + "\n");
                    }
                    else if (type == 2) {
                        sb.append("    " + parser.getText() + "\n\n");
                    }
                    type = 0;
                }
                eventType = parser.next();
            }
            in.close();
        }
        catch (Exception e) {
            e.printStackTrace();
        }
        text.setText(sb.toString());
    }
}
```

XmlPullParserFactory 类的 newInstance()方法用于创建一个 factory 对象。调用其 newPullParser()方法可以获得一个 XmlPullParser 的实例，即需要的解析器 parser。使用 new URL("http:

//10.0.2.2:8888/TestServer/test.xml").openStream()获得服务器上测试用 XML 文件的 InputStream 流，并以其为参数，调用 parser 对象的 setInput()方法设定输入流并开始进行数据解析。parser 对象的 getEventType()方法的返回值可以用来判断当前解析出的内容属于哪种类型。常用的类型有：

- XmlPullParser.START_DOCUMENT
- XmlPullParser.END_DOCUMENT
- XmlPullParser.START_TAG
- XmlPullParser.END_TAG
- XmlPullParser.TEXT

本例中仅需要对 XmlPullParser.START_TAG 和 XmlPullParser.TEXT 两种类型进行处理。如果是 XmlPullParser.START_TAG 类型，并且 parser.getName()方法的返回值是“city”，则表示当前的数据是所需要的标签，将标签名字和属性值追加写入 StringBuffer 对象中。如果 parser.getName()方法的返回值是“name”或者“code”，则表示需要获取其具体的文本值，通过一个标记变量 type 记录，待后续处理。如果 parser 对象的 getEventType()方法的返回值是 XmlPullParser.TEXT 类型，并且标记变量 type 的值是之前标记的需要处理的数据，则通过 parser.getText()方法取到这些值并追加写入 StringBuffer 对象中。最后当整个 XML 数据文件全部处理完毕后，将 StringBuffer 对象的值在界面上显示出来。显示的结果与图 8.8 中的一致。

8.4 Web Service 访问

Web Service 即 Web 服务，是一种基于 SOAP 的远程调用标准，该技术能使得运行在不同机器上的不同应用无须借助附加的、专门的第三方软件或硬件，就可相互交换数据或集成。依据 Web Service 规范实施的应用之间，无论所使用的语言、平台或内部协议是什么，都可以相互交换数据。Web Service 是自描述、自包含的可用网络模块，可以执行具体的业务功能。同时 Web Service 基于一些常规的产业标准以及已有的一些技术，诸如 XML 和 HTTP，所以很容易部署，目前已称为业界主流的信息交互方式之一。

Android SDK 中并没有提供解析 Web Service 的库，因此，需要使用第三方的 API 来实现解析。PC 版本的 Web Service 客户端库非常丰富，例如 Axis2、CXF 等，但这些开发包过于庞大，不适于 Android 系统。在 Android 开发中，使用最广泛的 Web Service 客户端包为 Ksoap2，用户可自行下载。将下载的 Ksoap2 包复制到 Android 工程下，并在工程中引用这个 jar 包即可使用。

Web Service 开发通常分为以下 7 个步骤来完成。

（1）实例化 SoapObject 对象，指定 Web Service 的命名空间，以及调用方法名称。

```
SoapObject request =new SoapObject("http://namespace", "methodName");
```

其中，SoapObject 类的第一个参数表示 Web Service 的命名空间，可以从 WSDL 文档中找到 Web Service 的命名空间。第二个参数表示要调用的方法名。

（2）设置调用方法的参数值，如果没有参数，可以省略，设置方法的参数值的代码如下所示。

```
Request.addProperty("param1", "value");
Request.addProperty("param2", "value");
```

要注意的是，addProperty 方法的第一个参数虽然表示调用方法的参数名，但该参数值并不一定与服务端的 Web Service 类中的方法参数名一致，只要设置参数的顺序一致即可。

（3）生成调用 Web Service 方法的 SOAP 请求信息，该信息由 SoapSeria liza tionEnvelope 对象描述。

```
SoapSerializationEnvelope envelope = new
                  SoapSerializationEnvelope(SoapEnvelope.VER11);
envelope.setOutputSoapObject(request);
// Envelope.bodyOut = request;
```

创建 SoapSerializationEnvelope 对象时需要通过 SoapSerializationEnvelope 类的构造方法设置 SOAP 协议的版本号。该版本号需要根据服务端 Web Service 的版本号设置。在创建 SoapSerializationEnvelope 对象后，需要再设置 SOAPSoap SerializationEnvelope 类的 bodyOut 属性，该属性的值就是在第一步创建的 SoapO bject 对象。

（4）创建 HttpTransportsSE 对象。通过 HttpTransportsSE 类的构造方法可以指定 WebService 的 WSDL 文档的 URL。

```
HttpTransportSE ht = new HttpTransportSE("http://IPAddress:port/services/ws?wsdl ");
```

（5）使用 call()方法调用 WebService。

```
ht.call(SOAP_ACTION, envelope);
```

call()方法的第一个参数一般为“命名空间 + 方法名称”，例如“urn:excute”，第二个参数就是在第(3)步创建的 SoapSerializationEnvelope 对象。

（6）使用 getResponse()方法获得 Web Service 方法的返回结果。

```
SoapObject soapObject =(SoapObject) envelope.getResponse();
```

（7）解析返回结果 soapObject。

下面通过例 8.8 来介绍如何使用 Web Service 获取服务器端数据。

【例 8.8】 使用 Web Service 获取服务器端储存的用户信息。

（1）在服务器端建立 Web Service 访问接口，并发布（该内容省略），发布页面如图 8.9 所示。从图中可以看到发布了 7 个 Web Service，分别是 InformWS、AlarmInfoWS、UserInfoWS、BehaviorInfoWS、AdminService、Version 和 PositionInfoWS，每个 Web Service 下还发布了几个不同的方法。

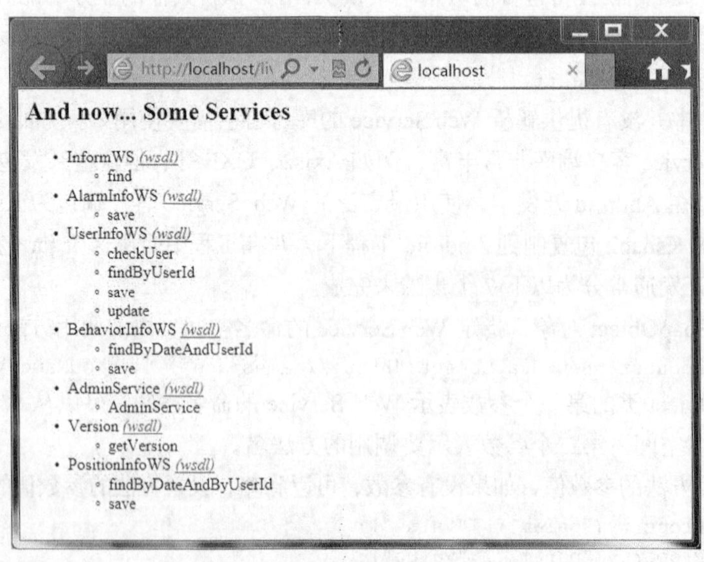

图 8.9 服务器端发布的 Web Service

（2）本例以调用 UserInfoWS 中的 findByUserId()方法来说明如何访问 Web Service。UserInfoWS 中提供的各个方法在 Web 服务部署文件中(WSDL)加以说明，如下代码是 findByUserId

方法的说明片段。从中可以了解该 Web Service 的访问地址为 http://localhost/livingstate/services/UserInfoWS，命名空间地址为 http://ws.state.example.com，并且 findByUserId()方法需要一个 String 类型的参数，返回值为也 String 类型。

```
<?xml version = "1.0" encoding = "UTF-8"?>
<wsdl:definitions targetNamespace = "http://ws.state.example.com" xmlns:apachesoap=
"http://xml.apache.org/xml-soap" xmlns:impl = "http://ws.state.example.com" xmlns:intf=
"http://ws.state.example.com" xmlns:wsdl = "http://schemas.xmlsoap.org/wsdl/" xmlns:wsdlsoap=
"http://schemas.xmlsoap.org/wsdl/soap/" xmlns:xsd = "http://www.w3.org/2001/XMLSchema">
 <wsdl:types>
  <schema elementFormDefault = "qualified"
        targetNamespace = "http://ws.state.example.com"
        xmlns = "http://www.w3.org/2001/XMLSchema">
   <element name = "findByUserId">
    <complexType>
     <sequence>
      <element name = "userId" type = "xsd:string"/>
     </sequence>
    </complexType>
   </element>
   <element name = "findByUserIdResponse">
    <complexType>
     <sequence>
      <element name = "findByUserIdReturn" type = "xsd:string"/>
     </sequence>
    </complexType>
   </element>
   ......
  </schema>
  <wsdl:service name = "UserInfoWSService">
     <wsdl:port binding = "impl:UserInfoWSSoapBinding" name = "UserInfoWS">
       <wsdlsoap:address
               location = "http://localhost/livingstate/services/UserInfoWS"/>
     </wsdl:port>
  </wsdl:service>
</wsdl:definitions>
```

（3）建立一个 Android 工程，将 Ksoap2 包加入工程。修改主布局文件为线性布局，在该布局中依次放置用于提示的文本框、用于输入用户 Id 的输入框、按钮和显示返回数据的列表框，代码如下所示。

```
<LinearLayout xmlns:android = "http://schemas.android.com/apk/res/android"
    android:layout_width = "match_parent"
    android:layout_height = "match_parent"
    android:orientation = "vertical" >
    <TextView
        android:layout_width = "match_parent"
        android:layout_height = "wrap_content"
        android:text = "请输入要查找的用户 Id: " />
    <EditText android:id = "@+id/userId"
        android:layout_width = "match_parent"
        android:layout_height = "wrap_content"
        android:text = "" />
```

```
    <Button
        android:layout_width = "match_parent"
        android:layout_height = "wrap_content"
        android:onClick = "click"
        android:text = "开始访问 web service" />
    <ListView android:id = "@+id/lv"
        android:layout_width = "match_parent"
        android:layout_height = "wrap_content"
        android:text = "@string/hello_world" />
</LinearLayout>
```

（4）修改主 Activity 文件，代码如下所示。

```
public class MainActivity extends Activity {
    //Web Service 地址
    private static final String URL = "http://202.117.132.254:8080/livingstate/services/
UserInfoWS?wsdl";
    //Web Service 命名空间地址
    private static final String NAMESPACE = "http://ws.state.example.com";
    //需调用方法名
    private static final String METHOD_NAME = "findByUserId";
    private static final String SOAP_ACTION = "urn:excute";
    protected void onCreate(Bundle savedInstanceState) {
        super.onCreate(savedInstanceState);
        setContentView(R.layout.activity_main);
    }
    public void click(View view) {
        try {
            invokeWebService();
        }
        catch (Exception e) {
            e.printStackTrace();
        }
    }
    public void invokeWebService() throws IOException, XmlPullParserException {
        ListView lv = (ListView) findViewById(R.id.lv);
        lv.setDividerHeight(1);
        EditText userIdET = (EditText) findViewById(R.id.userId);
        org.ksoap2.serialization.SoapObject rpc = new org.ksoap2.serialization.SoapObject
(NAMESPACE, METHOD_NAME);
        rpc.addProperty("userId", userIdET.getText().toString()); // 传入参数
        HttpTransportSE ht = new HttpTransportSE(URL);
        SoapSerializationEnvelope envelope = new SoapSerializationEnvelope(SoapEnvelope.
VER11);
        envelope.bodyOut = rpc;
        envelope.setOutputSoapObject(rpc);
        ht.call(SOAP_ACTION, envelope);// 发送请求
        SoapObject result = (SoapObject) envelope.bodyIn; // 获取信息
        String[] arr = new String[6];
        // 开始解析数据
        try {
```

```
            for (int i = 0; i < result.getPropertyCount(); i + +) {
                Object obj = result.getProperty(i);
                System.out.println(obj.toString());
                // findByUserId()方法返回值为 JSON 类型,下面解析 JSON
                JSONObject json = new JSONObject(obj.toString());
                arr[0] = "用户 id : " + json.get("userId").toString();
                arr[1] = "用户名 : " + json.get("name").toString();
                arr[2] = "密码 : " + json.get("password").toString();
                arr[3] = "性别 : " + json.get("sex").toString();
                arr[4] = "生日 : " + json.get("birth").toString();
                arr[5] = "地址 : " + json.get("address").toString();
                //将返回值写入适配器
                ArrayAdapter<String> adapter = new ArrayAdapter<String>(this, android.
R.layout.simple_list_item_1, arr);
                //ListView 加载适配器,显示返回数据
                lv.setAdapter(adapter);
            }
        }
        catch (JSONException e) {
            e.printStackTrace();
        }
    }
}
```

（5）在 AndroidManifest.xml 文件中增加网络访问许可。

运行本实例，结果如图 8.10、图 8.11 所示。

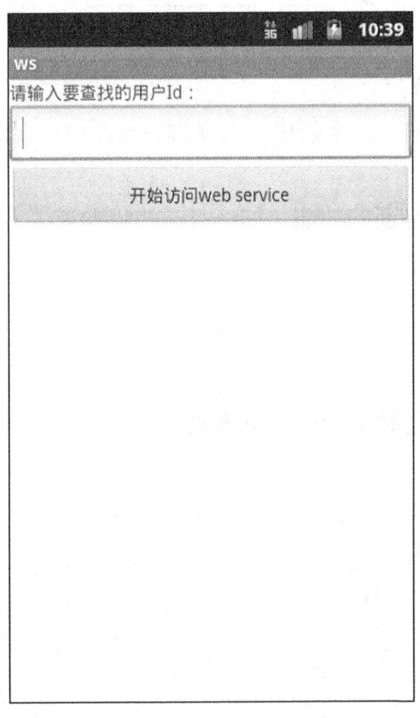

图 8.10　初始页面

图 8.11　访问 bbb 用户结果

8.5 小　结

Android 平台中提供了众多的进行网络通信编程的接口。本章分别介绍了 WebView 组件、HttpURLConnection、Socket 编程、JSON 数据处理、XML 数据处理等内容。WebView 是一个浏览器 UI 组件，它使用了 Webkit 引擎用以显示 Web 页面。利用 HttpURLConnection 类可进行 Web 访问。

Socket 编程使用 ServerSocket 和 Socket 类实现，可以进行 TCP/IP 通信和 UDP 通信。由于服务器端程序要处理多个客户端程序的连接，因此需要使用多线程技术，为每一个客户端启动一个单独的线程来进行处理。在客户端由于接收服务器的信息需要等待，不能直接将这个处理过程写在主线程中，所以也需要通过启动一个独立的线程来实现这个功能，当接收到服务器发送来的消息后，由该线程向 UI 线程发送消息，由 UI 线程将内容显示在界面上。

JSON 数据和 XML 数据是网络编程中常用的数据通信格式。Android API 中提供了对这两种数据格式进行处理的相关类的支持。对 XML 格式数据的处理主要有三种形式：DOM 方式、SAX 方式、XML PULL 方式。这三种方式各有其特点，本章中通过例子对这三种方式进行了对比，实际应用中可根据具体问题的需要选择最适合的方式。

Web Service 作为一个成熟的技术，提供了跨平台、跨语言的信息交互方法，目前是异构网络主流的交互技术。本文使用 Ksoap2 第三方包，介绍了如何在 Android 平台实现 Web Service 的访问。

练　习

1. 基于 WebView 组件编写一个简单的浏览器程序。

2. 利用 HttpURLConnection 类或者 HttpClient 编写一个程序，实现聊天程序中注册新用户和用户登录的功能。

3. 在本章 8.2 中的多人聊天程序的基础上，添加新的功能，如一对一的聊天等。

4. 编写程序解析如下的 JSON 数据。

```
[{"student":"张三","id":"101","score":"77"},
{"student":"李四","id":"102","score":"88"},
{"student":"王五","id":"103","score":"99"}]
```

5. 编写程序解析如下的 XML 数据，尝试使用各种不同的解析方式。

```
<?xml version = "1.0" ?>
<books>
  <book email = "aaa@163.com">
     <name>JAVA 编程思想</name>
     <price>99.0</price>
  </book>
  <book email = "bbb@163.com">
     <name>C + +开发手册</name>
     <price>42.0</price>
```

```
    </book>
    <book email = "ccc@163.com">
        <name>C#高级编程</name>
        <price>110.0</price>
    </book>
    <book email = "ccc@163.com">
        <name>Android 开发从入门到精通</name>
        <price>69.0</price>
    </book>
</books>
```

6. 访问中国天气网提供的 http://flash.weather.com.cn/wmaps/xml/china.xml 接口，实现一个获取全国各省会城市、直辖市的当天天气信息的天气预报应用。

第 9 章 多媒体应用

Android 系统的音频和视频平台，提供了多种常见媒体类型的内置编码和解码功能，因此可以很容易在应用程序中集成音频和视频。另外 Android 框架还包含了对多种摄像头和摄像特性的支持，应用程序可以进行图片和视频的捕获。本文将对 Android 中音视频常用的 API 进行介绍，分析多种方法的使用场合及优缺点，并讨论快速、简便地捕获图像的方法。

9.1 音频与视频的播放

Android 通过 android.media 包提供了对常用的音视频格式的支持,其中音频格式有 MP3、WAV 和 3GP 等，视频格式有 3GP 和 MPEG-4（.mp4）等。

Android 系统中，MediaPlayer 类能够支持播放音频和视频，是播放媒体文件使用最为广泛的类。除了 MediaPlayer 类之外，SoundPool 类也可提供用来播放音频文件，VideoView 类可以用来播放视频等。下面分别予以介绍。

9.1.1 MediaPlayer

MediaPlayer 是设计用来播放音频、视频或流文件的类，可支持播放操作，如停止、开始、暂停和查找等操作，常用方法如表 9.1 所示。

表 9.1 MediaPlayer 常用方法

方法名称	说明
void setDataSource(FileDescriptor fd, long offset, long length)	根据 FileDescriptor 设置多媒体数据来源，并指定开始位置和长度
void setDataSource(FileDescriptor fd)	根据 FileDescriptor 设置多媒体数据来源
void setDataSource(String path)	根据路径设置多媒体数据来源
void setDataSource(Context context, Uri uri)	根据 URI 设置多媒体数据来源
MediaPlayer create(Context context, int resid)	静态方法，通过资源 ID 创建多媒体播放器
MediaPlayer create(Context context, Uri uri)	静态方法，通过 URI 创建多媒体播放器
void pause()	暂停播放
void prepare()	准备播放
void release()	释放 MediaPlayer 对象

续表

方法名称	说　明
void reset()	重置 MediaPlayer 对象
void start()	开始播放或恢复已经暂停的音频播放
void stop()	停止播放
int getCurrentPosition()	得到当前播放位置
int getDuration()	得到文件的时间
boolean isPlaying()	是否正在播放
void seekTo(int msec)	指定播放的位置
void setLooping(boolean looping)	设置是否循环播放
void setVolume(float lVolume, float rVolume)	设置左右声道音量

采用 MediaPlayer 类播放音频，需要创建该类的对象，指定需播放的音频文件（音频文件位置有三种），随后调用它的 start()方法即可。下面将为用户介绍使用 MediaPlayer 播放音频文件的过程。

1. 创建 MediaPlayer 类的对象，并设置音频文件

创建 MediaPlayer 类的对象并设置将播放的音频文件，基本上有两种方式。第一种是使用 new 实例化的方式创建，再通过调用 setDataSource()方法来指定播放文件，第二种是使用 MediaPlayer 的 create()方法来创建。

（1）采用 new 实例化的方式。

采用 MediaPlayer 的无参构造方法创建 MediaPlayer 对象时，还需要指定要播放的文件，这就需要通过调用 setDataSource()方法来完成。在设置音频文件后，MediaPlayer 实际上并未真正加载该音频文件，还需要调用 prepare()方法来真正加载音频文件。使用 new 实例化的方式创建 MediaPlayer 对象，并加载指定音频文件可以使用如下代码。

```
MediaPlayer mp = new MediaPlayer();
try {
    // 调用 setDataSource 方法，指定要加载的音频文件
    mp.prepare();
}
catch(IllegalArgumentException e) {
    e.printStackTrace();
}
catch(SecurityException e) {
    e.printStackTrace();
}
catch(IllegalStateException e) {
    e.printStackTrace();
}
catch(IOException e) {
    e.printStackTrace();
}
```

上文提到过音频资源有三种不同的来源，针对不同来源，使用不同的 setDataSource()方法来加载，使用下面三种不同的方法替换如上代码的注释部分。

a. 音频资源存储在应用程序中的本地资源中

即音频文件存在于该工程的 assets 文件夹中（与 res 平级），如文件名为 mymusic.mp3，加载方式代码如下所示。

```
AssetFileDescriptor fileDescriptor = getAssets().openFd("mymusic.mp3");
mp.setDataSource(fileDescriptor.getFileDescriptor(),
fileDescriptor.getStartOffset(), fileDescriptor.getLength());
```

如不需要指定音频的开始位置和要播放长度，可使用如下代码。

```
AssetFileDescriptor fileDescriptor = getAssets().openFd("mymusic.mp3");
mp.setDataSource(fileDescriptor.getFileDescriptor());
```

b. 音频资源存储在文件系统中

音频存储在文件系统中，即存放在 Android 系统目录或 SD 卡上，可通过指定目录地址来加载，代码如下所示。

```
mp.setDataSource("/sdcard/mymusic.mp3");
```

c. 通过网络连接取得的数据流

音频存储在网络中，需要通过指定 URI 来加载。如要加载的音频文件的 URI 地址为“http://www.baidu.com/mymusic.mp3”，可使用如下代码。

```
mp.setDataSource("http:/www.baidu.com/mymusic.mp3");
```

从网络上提取音频数据，该 Android 工程必须获取访问互联网的许可。在 AndroidManifest.xml 文件中设置互联网许可，代码如下所示。

```
<uses-permission android:name = "android.permission.INTERNET" />
```

（2）使用 MediaPlayer 的 create()方法方式。

MediaPlayer 类的 create()方法是个静态方法，返回新创建的 MediaPlayer 对象，常用的有两种。同样根据音频资源三种不同的来源，使用 create()方法分别以如下方式加载。

a. 音频资源存储在应用程序中的本地资源中

这里音频文件存储于该工程的 res/raw 文件夹下（res 内若没有 raw 文件夹，则新建一个）。如上例，要加载的资源文件为 mymusic.mp3，可以使用代码如下。

```
MediaPlayer mp = MediaPlayer.create(this, R.raw.mymusic);
```

raw 文件夹内的文件名要小写。

b. 音频资源存储在文件系统中

音频存储在文件系统中，假设存放在 SD 卡上，可通过指定 URI 来加载，这里使用了关键字 file，代码如下所示。

```
String urlPath = "file:///sdcard/mymusic.mp3";
MediaPlayer mp = MediaPlayer.create(this, Uri.parse(urlPath));
```

c. 通过网络连接取得的数据流

音频存储在网络中，也需要通过指定 URI 来加载。如加载的音频文件的 URI 地址为“http://www.baidu.com/mymusic.mp3”，可使用如下代码。

```
String urlPath = "http:/www.baidu.com/mymusic.mp3";
MediaPlayer mp = MediaPlayer.create(this, Uri.parse(urlPath));
```

访问网络资源，需设置互联网许可。

2. 播放器的控制

Android 通过控制播放器的状态的方式来控制媒体文件的播放，表 9.1 已列出主要方法，其中以下方法需要加以强调。

- prepare()方法和 prepareAsync()方法分别提供了同步和异步两种方式设置播放器进入 prepare 状态。其中 prepare()方法为阻塞方法，并可阻塞直至媒体播放器准备播放资源，而非阻塞方法 prepareAsync()也可进行实现该功能，但如果播放器用来从流媒体中播放歌曲，并且在播放歌曲之前需要缓冲数据，则应使用非阻塞 prepareAsync()方法。需要注意的是，如果 MediaPlayer 实例是由 create 方法创建的，那么第一次启动播放前不需要再调用 prepare()了，因为 create 方法里已经调用过了。
- seekTo()方法完成定位功能，让播放器从指定的位置开始播放。同时该方法是个异步方法，返回时并不表示定位完成，尤其是播放的网络文件，定位完成时会触发 OnSeekComplete.onSeekComplete()，如果需要可以调用 setOnSeekCompleteListener(OnSeekCompleteListener)设置监听器来处理。
- release()方法可以释放播放器占用的资源，一旦确定不再使用播放器时应当尽早释放资源。
- reset()方法可以使播放器从错误状态中恢复过来，重新回到空闲状态，用于重设 MediaPlayer 播放器。

3. 设置播放器的监听器

MediaPlayer 提供了设置多种监听器的方法，来对播放器的工作状态进行监听，以便及时处理各种情况。如 OnCompletionListener、OnPrepareListener、OnErrorListener、OnBufferingUpdateListener、OnVideoSizeChangedListener、OnInfoListener 和 OnSeekCompleteListener 等。

例如，当播放到音频、视频结束时，可调用 OnCompletionListener 监听器的 onCompletion (MediaPlayer mp)方法，进行循环播放或自动播放列表中下一首歌曲或释放媒体播放器对象。当准备播放媒体时，可调用 OnPrepareListener 监听器的 onPrepared(MediaPlayer mp)方法，开始播放已经预加载好的歌曲。当在异步操作过程中出现错误时（其他错误将在调用方法时抛出异常），将可调用 OnErrorListener 监听器的 onError(MediaPlayer mp, int what, int extra)方法，进行资源回收等工作。

下面将展示一个使用 MediaPlayer 类播放 MP3 音乐的例子，如例 9.1 所示。

【例 9.1】 实现一个带有音量设置和进度设置的音乐播放器。

（1）修改主布局文件，将默认的布局管理器改为 LinearLayout 线性布局，垂直排列（该主布局代码省略），在该布局中添加如下代码。

该布局分为四行，第一行为音乐名称输入框。代码如下所示。

```
<!--- 第一行为音乐名称输入框 ->
<EditText android:id = "@+id/musicName"
   android:layout_width = "match_parent"
   android:layout_height = "40dp" />
```

第二行为一个线性布局，其中水平排列四个按钮，分别为播放、暂停、重播和停止。代码如下所示。

```
<!--- 第二行为播放、暂停、重播和停止按钮 ->
<LinearLayout
   android:layout_width = "match_parent"
   android:layout_height = "wrap_content"
   android:orientation = "horizontal"
   android:gravity = "center_horizontal">
   <Button android:id = "@+id/StartMusic"
      android:layout_width = "80dp"
      android:layout_height = "wrap_content"
      android:background = "@drawable/custom_button"
```

```
        android:text = "播放" />
    <Button android:id = "@+id/pauseMusic"
        android:layout_width = "80dp"
        android:layout_height = "wrap_content"
        android:background = "@drawable/custom_button"
        android:text = "暂停" />
    <Button android:id = "@+id/resetMusic"
        android:layout_width = "80dp"
        android:layout_height = "wrap_content"
        android:background = "@drawable/custom_button"
        android:text = "重播" />
    <Button android:id = "@+id/closeMusic"
        android:layout_width = "80dp"
        android:layout_height = "wrap_content"
        android:background = "@drawable/custom_button"
        android:onClick = "stop"
        android:text = "停止" />
</LinearLayout>
```

第三行也为一个线性布局，水平放置内容为音量的文本框组件，以及表示音量的拖动条组件，和显示当前音量值的文本框组件。代码如下所示。

```
<!--- 第三行为音量文本框和音量拖动条 ->
<LinearLayout
    android:layout_width = "match_parent"
    android:layout_height = "wrap_content"
    android:gravity = "center_horizontal"
    android:orientation = "horizontal" >
    <TextView android:text = "音量"
        android:layout_width = "wrap_content"
        android:layout_height = "wrap_content"
        android:textSize = "17dp" />
    <SeekBar android:id = "@+id/volumeBar"
        android:layout_width = "240dp"
        android:layout_height = "wrap_content"
        android:paddingLeft = "10dp" />
    <TextView android:id = "@+id/curVol"
        android:layout_width = "60dp"
        android:layout_height = "wrap_content"
        android:paddingLeft = "10dp" />
</LinearLayout>
```

最后一行也是一个线性布局，水平排列内容为进度的文本框组件，一个表示音乐进度的拖动条组件，和显示当前进度值的文本框组件。代码如下所示。

```
<!--- 第四行为进度文本框和进度拖动条 ->
<LinearLayout
    android:layout_width = "match_parent"
    android:layout_height = "wrap_content"
    android:gravity = "center_vertical"
    android:orientation = "horizontal" >
    <TextView android:text = "进度"
        android:layout_width = "wrap_content"
        android:layout_height = "wrap_content"
        android:textSize = "17dp" />
    <SeekBar android:id = "@+id/posBar"
        android:layout_width = "240dp"
```

```
        android:layout_height = "wrap_content"
        android:paddingLeft = "10dp" />
    <TextView android:id = "@+id/curPos"
        android:layout_width = "60dp"
        android:layout_height = "wrap_content"
        android:paddingLeft = "10dp" />
</LinearLayout>
```

（2）在主 Activity 类中添加界面组件对应的对象，以及在代码中使用到的成员变量。代码如下所示。

```
private EditText musicName; //音乐名称组件
private Button StartMusic;//播放按钮
private Button pauseMusic;//暂停按钮
private Button resetMusic; //重播按钮
private Button closeMusic;//停止按钮
private SeekBar volumeBar; //音量调节拖动条
private SeekBar posBar;//音量位置拖动条
private TextView curVol; //当前音量
private TextView curPos;//当前播放位置
private MediaPlayer mp; //MediaPlayer 对象
private static String fileName; //文件名
private static int volume = 0;// 当前音量
private static int times = 0; // 音乐总时长
private AudioManager am;// 音频管理类
private File audioFile; //文件名对应的 File 对象
```

（3）在主 Activity 类中添加 setMusicParams()方法，用于设置音量拖动条和播放位置拖动条。

通过 AUDIO_SERVICE 系统服务获取音频管理类，并设置本例调整音量只是针对媒体音乐，再设置音量拖动条最大值为系统音量最大值，以及对音量拖动条增加了 OnSeekBarChangeListener 监听器，当拖动条值改变时，重设当前放音音量，其他不改动的方法留空。

接着设置音乐位置拖动条的最大值为该音频的时长（单位为毫秒），并设置当前位置，以及对位置拖动条增加 OnSeekBarChangeListener 监听器，当拖动条值改变时，先暂停播放，再通过 MediaPlayer 的 seekTo 方法跳到指定位置，接着播放，其他不改动的方法留空。代码如下所示。

```
private void setMusicParams() {
    // 获取音频管理类对象
    am = (AudioManager) getSystemService(AUDIO_SERVICE);
    // 设置当前调整音量只针对媒体音量
    MainActivity.this.setVolumeControlStream(AudioManager.STREAM_MUSIC);
    // 设置音量最大值为系统音量最大值
    volumeBar.setMax(am.getStreamMaxVolume(AudioManager.STREAM_MUSIC));
    volume = am.getStreamVolume(AudioManager.STREAM_MUSIC);// 获取当前音量
    volumeBar.setProgress(volume); // 设置音量条当前音量
    curVol.setText(String.valueOf(volume)); // 设置音量提示框当前音量
    volumeBar.setOnSeekBarChangeListener(new SeekBar.OnSeekBarChangeListener() {
        @Override
        public void onStopTrackingTouch(SeekBar seekBar) {}
        @Override
        public void onStartTrackingTouch(SeekBar seekBar) {}
        @Override
        public void onProgressChanged(SeekBar seekBar, int progress, boolean fUser) {
```

```
            curVol.setText(String.valueOf(progress)); // 设置音量提示框当前音量
            am.setStreamVolume(AudioManager.STREAM_MUSIC, progress,
                    AudioManager.FLAG_PLAY_SOUND); // 设置改变后的音量
        }
    });
    posBar.setMax(times); // 设置音乐位置
    posBar.setOnSeekBarChangeListener(new SeekBar.OnSeekBarChangeListener() {
        @Override
        public void onStopTrackingTouch(SeekBar seekBar) {
            mp.pause();// 暂停播放
            mp.seekTo(seekBar.getProgress());// 设置从指定位置开始播放
            mp.start();// 开始播放
            curPos.setText(seekBar.getProgress() / 1000 + "s");
        }
        @Override
        public void onStartTrackingTouch(SeekBar seekBar) {}
        @Override
        public void onProgressChanged(SeekBar seekBar, int progress, boolean fUser) {}
    });
}
```

（4）修改主 Activity 类中的 onCreate 方法，获取布局中组件的对象，对播放、暂停、重放、停止按钮增加监听器 AudioListener，并实例化 MediaPlayer 播放器。代码如下所示。

```
@Override
public void onCreate(Bundle savedInstanceState)
{
    super.onCreate(savedInstanceState);
    setContentView(R.layout.activity_main);
    musicName = (EditText) findViewById(R.id.musicName);
    StartMusic = (Button) findViewById(R.id.StartMusic);
    pauseMusic = (Button) findViewById(R.id.pauseMusic);
    resetMusic = (Button) findViewById(R.id.resetMusic);
    closeMusic = (Button) findViewById(R.id.closeMusic);
    volumeBar = (SeekBar) findViewById(R.id.volumeBar);
    posBar = (SeekBar) findViewById(R.id.posBar);
    curVol = (TextView) findViewById(R.id.curVol);
    curPos = (TextView) findViewById(R.id.curPos);
    StartMusic.setOnClickListener(new AudioListener());
    pauseMusic.setOnClickListener(new AudioListener());
    resetMusic.setOnClickListener(new AudioListener());
    closeMusic.setOnClickListener(new AudioListener());
    if(fileName != null && !fileName.isEmpty())
        musicName.setText(fileName);
    mp = new MediaPlayer();
}
```

（5）在主 Activity 类中添加播放 play()方法。当未播放时，获取外部存储设备 SD 卡上的音频文件，重设播放器状态，设置播放数据源，并获取音乐的时长，调用 setMusicParams()设置音量拖动条和位置拖动条，最后开始播放音乐。代码如下所示。

```
//播放方法
private void play() throws IOException {
    // 如果播放器正在播放，则不做任何操作
    if(mp.isPlaying())
        return;
    audioFile = new File(Environment.getExternalStorageDirectory(), fileName);
```

```
        mp.reset();// 播放器状态的重设
        mp.setDataSource(audioFile.getAbsolutePath());// 设置播放数据源
        mp.prepare();// 硬件的准备
        // 读取音乐总时长
        if(times == 0)
            times = mp.getDuration();
        setMusicParams(); // 设置音量、进度拖动条参数
        mp.start();// 开始播放
    }
```

（6）在主 Activity 类中添加 onDestroy()方法，用于程序退出时释放与播放器相关的资源。代码如下所示。

```
    @Override
    protected void onDestroy() {
        mp.release();
        super.onDestroy();
    }
```

（7）在主 Activity 类中添加 AudioListener 监听器，在 onClick()方法中处理各按钮单击后的操作。若是播放按钮单击，则调用 play()方法进行播放；若是暂停按钮单击，则先判断是否正在播放，若正在播放，则暂停，并将按钮文字改为继续，如未播放，则开始播放，并将按钮文字改为暂停；若是重放按钮单击，则先判断是否正在播放，若正在播放，则通过 seekTo(0)方法跳到文件开始播放，如未播放，则开始播放；若是停止按钮单击，判断正在播放则停止。代码如下所示。

```
    //自定义监听器
    private class AudioListener implements View.OnClickListener {
        public void onClick(View view) {
            fileName = musicName.getText().toString();
            Button button = (Button) view;
            try {
                //判断各按钮，执行相应操作
                switch(view.getId()) {
                    case R.id.StartMusic:
                        play();
                        break;
                    case R.id.pauseMusic:
                        if(mp.isPlaying()) {
                            mp.pause();
                            button.setText("继续");
                        }
                        else {
                            mp.start();
                            button.setText("暂停");
                        }
                        break;
                    case R.id.resetMusic:
                        if(mp.isPlaying()
                            mp.seekTo(0);
                        else
                            play();
                        break;
                    case R.id.closeMusic:
                        if(mp.isPlaying())
                            mp.stop();
                        break;
```

```
            }
        }
        catch(Exception e) {
            e.printStackTrace();
        }
    }
}
```

运行本实例，显示结果如图 9.1 所示。

本例中，表示音乐进度的拖动条并没有实现随音乐的播放而移动，用户可在此例基础上进行补充。

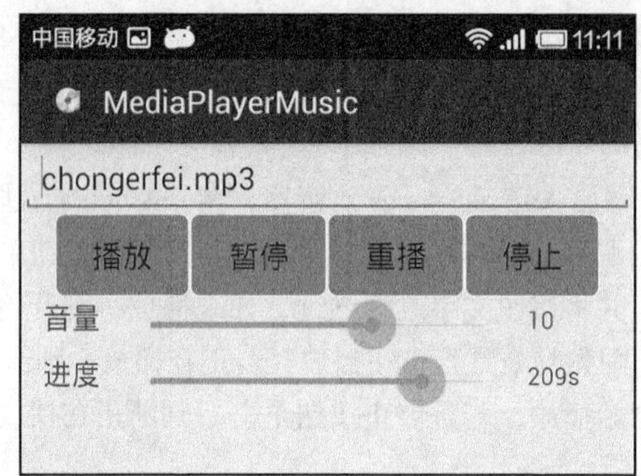

图 9.1　使用 MediaPlayer 实现的播放器

9.1.2　SoundPool

SoundPool 即声音池，主要用于播放一些较短的声音片段。与 MediaPlayer 相比，SoundPool 具有资源占用量低、延迟小、支持多个音频同时播放等优点。游戏中会经常出现短促、密集、延迟程度小的音频，而 SoundPool 的这些特点非常适合此类场景的需要。除此之外，SoundPool 类支持以下功能：

- 设置可同时播放的最高的声音数
- 优先处理声音，达到最大极限时将会减弱优先级别低的声音
- 在完成播放之前暂停和停止声音
- 循环声音
- 更改播放速率
- 设置立体声音量，包含左右声道

SoundPool 的常用方法如表 9.2 所示。

表 9.2　SoundPool 的常用方法

方法名称	说明
SoundPool(int maxStreams, int streamType, int srcQuality)	构造方法，其中 maxStreams 表示允许播放音频的最大数，streamType 表示声音类型，srcQuality 表示声音品质（默认为 0）
int load(Context context, int resId, int priority)	通过资源 id 加载音频文件，priority 表示优先级
int load(String path, int priority)	通过资源路径加载音频文件

续表

方 法 名 称	说　明
int load(AssetFileDescriptor afd, int priority)	用于从 AssetFileDescriptor 对应的文件加载
int load(FileDescriptor fd, long offset, long length, int priority)	用于从 AssetFileDescriptor 对应的文件加载，从 offset 开始，长度为 length
play(int soundID,float leftVolume, float rightVolume,int priority,int loop,float rate)	播放音频，soundID 表示音频 id，leftVolume，rightVolume 分别表示左右声道音量（取值在 0.0 ~ 1.0 之间），priority 表示音频流的优先级（0 最低），loop 表示音乐播放的次数（-1 无限循环，0 播放一次），rate 表示播放速率（取值在 0.5 ~ 2.0，1 代表正常播放）

下面将通过例 9.2 来说明 SoundPool 如何播放音频。

【例 9.2】　实现一个单击不同按钮播放不同提示音的应用。

（1）在工程的 res 下新建 raw 文件夹，放入四个音频文件，分别为 piano.wav、trumpets.wav、siren.wav、bell.wav。

（2）修改主布局文件，将默认的布局管理器改为线性布局，垂直排列，在该布局中添加如下代码。该布局分为两行，第一行放置一个给用户显示提示信息的 TextView 组件，id 值为 hint。代码如下所示。

```
<!-- 第一行放置提示信息 -->
<TextView android:id = "@+id/hint"
    android:layout_width = "match_parent"
    android:layout_height = "d0dp"
    android:textSize = "20sp" />
```

第二行放置一个线性布局，在该布局中放置四个按钮，分别表示号角声、汽笛声、铃声、钢琴声。代码如下所示。

```
<!-- 第二行放置角声、汽笛声、铃声、钢琴声按钮 -->
<LinearLayout
    android:layout_width = "match_parent"
    android:layout_height = "wrap_content"
    android:orientation = "horizontal"
    android:gravity = "center_horizontal" >
    <Button android:id = "@+id/piano"
        android:layout_width = "80dp"
        android:layout_height = "wrap_content"
        android:background = "@drawable/custom_button"
        android:text = "钢琴声" />
    <Button android:id = "@+id/trumpets"
        android:layout_width = "80dp"
        android:layout_height = "wrap_content"
        android:background = "@drawable/custom_button"
        android:text = "号角声" />
    <Button android:id = "@+id/siren"
        android:layout_width = "80dp"
        android:layout_height = "wrap_content"
        android:background = "@drawable/custom_button"
        android:text = "汽笛声" />
    <Button android:id = "@+id/bell"
        android:layout_width = "80dp"
```

```
        android:layout_height = "wrap_content"
        android:background = "@drawable/custom_button"
        android:text = "铃声" />
</LinearLayout>
```

（3）在主 Activity 类中添加在代码中使用到的成员变量。代码如下所示。

```
private int streamVolume; // 音效的音量
private SoundPool soundPool; // 定义 SoundPool 对象
private HashMap<Integer, Integer> soundPoolMap; // 定义 HASH 表
private TextView hint; // 定义提示文本对象
```

（4）在主 Activity 类中添加 init()方法，初始化 SoundPool，并获取音频管理器来设置音量。代码如下所示。

```
private void init() {
    // 初始化 soundPool,允许有 5 个声音同时播放,品质为默认值 0
    soundPool = new SoundPool(5, AudioManager.STREAM_MUSIC, 0);
    soundPoolMap = new HashMap<Integer, Integer>(); // 初始化 HASH 表
    // 获得声音设备和设备最大音量
    AudioManager am = (AudioManager) getSystemService(AUDIO_SERVICE);
    streamVolume = am.getStreamMaxVolume(AudioManager.STREAM_MUSIC);
}
```

（5）在主 Activity 类中添加 loadRes()方法，将音频资源 id 加入到 HashMap 中，并设置播放优先级为 1。代码如下所示。

```
private void loadRes(int no, int raw) {
    // 把资源中的音效加载到指定的 id
    soundPoolMap.put(no, soundPool.load(this, raw, 1));
}
```

（6）在主 Activity 类中添加 play()方法，播放 HashMap 中第 no 个音频，并设置循环方式为 uLoop。代码如下所示。

```
private void play(int no, int uLoop) {
    soundPool.play(soundPoolMap.get(no), streamVolume, streamVolume, 1, uLoop, 1);
}
```

（7）修改主 Activity 类中的 onCreate()方法，在该方法中先调用 init()方法初始化，再将音频资源加入 HashMap 中，获取布局中四个按钮，对每个按钮增加 AudioListener 监听器。代码如下所示。

```
@Override
protected void onCreate(Bundle savedInstanceState) {
    super.onCreate(savedInstanceState);
    setContentView(R.layout.activity_main);
    init(); //初始化 SoundPool 和设置音频音量
    loadRes(1, R.raw.piano); //将音频资源放入 HashMap
    loadRes(2, R.raw.trumpets);
    loadRes(3, R.raw.siren);
    loadRes(4, R.raw.bell);
    hint = (TextView) findViewById(R.id.hint);
    Button piano = (Button) findViewById(R.id.piano);
    Button trumpets = (Button) findViewById(R.id.trumpets);
    Button siren = (Button) findViewById(R.id.siren);
    Button bell = (Button) findViewById(R.id.bell);
    piano.setOnClickListener(new AudioListener()); //增加监听器
```

```
        trumpets.setOnClickListener(new AudioListener());
        siren.setOnClickListener(new AudioListener());
        bell.setOnClickListener(new AudioListener());
    }
```

（8）在主 Activity 类中添加 AudioListener 监听器内部类，在 onClick 方法中，播放各按钮对应的音频。代码如下所示。

```
    //自定义监听器
    private class AudioListener implements View.OnClickListener {
        public void onClick(View view) {
            //判断按钮，播放不同音频
            switch(view.getId()) {
                case R.id.piano:
                    hint.setText("播放钢琴声 1 遍...");
                    play(1, 1);
                    break;
                case R.id.trumpets:
                    hint.setText("播放号角声 2 遍...");
                    play(2, 2);
                    break;
                case R.id.siren:
                    hint.setText("播放汽笛声 3 遍...");
                    play(3, 3);
                    break;
                case R.id.bell:
                    hint.setText("播放铃声 4 遍...");
                    play(4, 4);
                    break;
            }
        }
    }
```

（9）在主 Activity 类中添加 onDestroy()方法，用于程序退出时释放与播放器相关的资源。代码如下所示。

```
    //退出时释放资源
    @Override
    protected void onDestroy() {
        soundPool.release();
        super.onDestroy();
    }
```

运行本实例，单击铃声按钮，结果如图 9.2 所示。

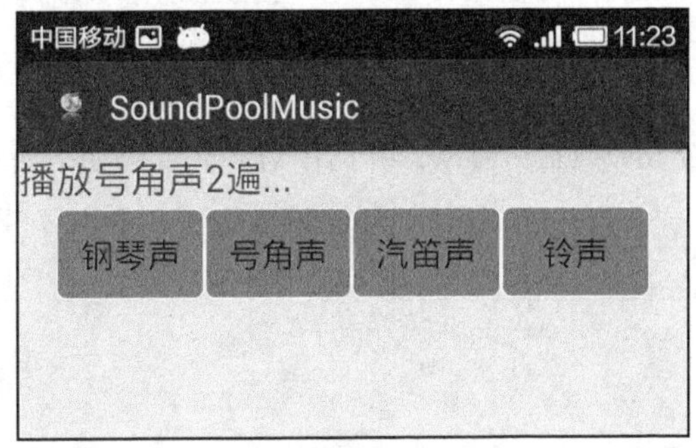

图 9.2　使用 SoundPool 播放音频

SoundPool 存在一些设计上的缺陷，使用时应该注意。

（1）SoundPool 最大只能申请 1 MB 的内存空间，只能使用很短的声音片段，不适合用来播放歌曲或者游戏背景音乐。

（2）SoundPool 虽提供了 pause 和 stop 方法，但调用可能会引起程序终止。

（3）若在初始化中，就调用播放方法 play()进行播放会没有声音，并非因为未执行，而是 SoundPool 需要一定的准备时间。本例将播放方法 play()放到了按钮单击触发中，避免了这个问题。

9.1.3 VideoView

Android 提供了专业化的视图控制组件 VideoView，可从各种来源加载图片，能够计算视频尺寸，还可提供各种显示选项，如缩放比例和着色，也可用来显示 SDCard 中存在的视频文件等。同时在 Android 中，还提供了一个可以与 VideoView 结合使用的 MediaController 组件，用于通过图形控制界面来控制视频的播放。VideoView 组件常用的方法如表 9.3 所示。

表 9.3 VideoView 常用的方法

方法名称	说明
int getBufferPercentage()	得到缓冲的百分比
int getCurrentPosition()	得到当前播放位置
int getDuration()	得到视频文件的时间
void setMediaController (MediaController ctl)	设置播放控制器模式（播放进度条）
void setOnCompletionListener (MediaPlayer.OnCompletionListener l)	当视频文件播放完时触发事件
void setVideoPath (String path)	设置视频源路径
void setVideoURI (Uri uri)	设置视频源地址
void start()	开始播放
void pause()	暂停播放
void suspend()	挂起视频
void resume()	恢复视频
boolean isPlaying()	判断是否正在播放

下面将通过具体例子来说明如何使用 VideoView 播放视频，如例 9.3 所示。

【例 9.3】 实现一个 VideoView 播放视频的应用。

（1）修改主布局文件，将默认的布局改为线性布局，并将背景色改为黑色。在该布局中添加一个 VideoView 组件，设置 id 为 videoView，并居中对齐。代码如下所示。

```
<LinearLayout xmlns:android = "http://schemas.android.com/apk/res/android"
    android:background = "#000000"
    android:layout_width = "match_parent"
    android:layout_height = "match_parent"
    android:orientation = "vertical" >
    <VideoView
        android:id = "@+id/videoView"
        android:layout_width = "wrap_content"
```

```
        android:layout_height = "match_parent"
        android:layout_gravity = "center" />
</LinearLayout>
```

（2）在主 Activity 类中 onCreate()方法最后添加代码，首先获取布局文件中的 VideoView 组件，创建 MediaController 媒体控制器。设置视频路径，这里视频资源是 mp4 文件，位于 SD 卡上，通过 URI 解析地址。对 VideoView 对象添加 OnCompletionListener 监听器，当视频播放完成会调用 onCompletion()方法，输出一个提示。最后使得 VideoView 对象获得焦点，并开始播放。代码如下所示。

```
VideoView videoView = (VideoView) this.findViewById(R.id.videoView);
MediaController mc = new MediaController(this); // 创建 MediaController
videoView.setMediaController(mc); // 设置 MediaController
// 设置视频路径
videoView.setVideoURI(Uri.parse("file:///sdcard/video/myvideo.mp4"));
videoView.setOnCompletionListener(new MediaPlayer.OnCompletionListener()
{
    @Override
    public void onCompletion(MediaPlayer mp)
    {
        Toast.makeText(MainActivity.this, "视频播放完毕!", Toast.LENGTH_LONG).show();
    }
});
videoView.requestFocus(); // 视频播放组件获得焦点
videoView.start(); // 播放
```

运行本实例，显示结果如图 9.3 所示，播放完成效果如图 9.4 所示。

图 9.3　VideoView 播放视频

图 9.4　播放完成效果

播放网络视频方法与播放文件系统视频的方法相同，setVideoURI(Uri uri)方法的参数指向网站即可，例如：

```
videoView.setVideoURI(Uri.parse("http://www.pcvideo.com/myvideo.mp4"));
```

9.1.4　SurfaceView

在 9.1.1 小节中介绍 MediaPlayer 类时，已经讲到 MediaPlayer 除了可以播放音频外，也可以播放视频，只是 MediaPlayer 在播放视频时不能自己提供图像输出界面。在 Android 中提供了 SurfaceView 类来显示图像，SurfaceView 是 View 的子类，内嵌了一个专门用于绘制界面的 Surface，可以控制这个界面的格式、尺寸和位置。SurfaceView 适合 2D 游戏的开发，该类使用双缓冲机制，在新线程中更新画面，所以更新界面速度比普通 view 要快。在播放视频方面，可以配合 MediaPlayer

使用。通过这种方法播放视频大概需要以下四个步骤。

（1）创建 SurfaceView 组件。

创建 SurfaceView 组件可以在布局管理器中完成，也可以使用 Java 代码创建。

（2）创建 MediaPlayer 组件。

与播放音频时创建 MediaPlayer 对象的方法一样，既可以使用无参的构造函数创建，也可以使用静态方法 create()创建。

（3）设置视频输出。

设置视频输出，通过使用 MediaPlayer 的 setDisplay()方法，将视频画面输出到 SurfaceView 上。其中 setDisplay()方法语法格式如下。

```
setDisplay(SurfaceHolder sh);
```

SurfaceHolder 是一个接口，其作用类似于 Surface 界面的监听器。上述语法中参数 sh 就是 SurfaceHolder 的一个对象，可以通过 SurfaceView 对象的 getHolder()方法获得。

（4）控制视频播放。

通过使用 MediaPlayer 的 play()、pause()和 stop()等方法，可以控制视频的播放、暂停、停止等。

下面将通过一个简单的例子来说明如何使用 MediaPlayer 和 SurfaceView 播放视频，如例 9.4 所示。

【例 9.4】 实现一个使用 MediaPlayer 和 SurfaceView 播放视频的应用。

（1）新建 Android 工程，在工程的 res 的 drawable 对应的目录下分别拷入表示播放、暂停、停止的图片。

（2）修改主布局文件，将默认的布局改为线性布局，布局方向为垂直，并将背景色设置为黑色。在该布局中添加如下代码。

该布局分为三行，第一行添加 SurfaceView 组件，将 android:keepScreenOn 属性值设置为 true，表示播放视频时打开屏幕。代码如下所示。

```
<!-- 第一行放置 SurfaceView 组件 -->
<SurfaceView android:id = "@+id/surface"
   android:layout_width = "700px"
   android:layout_height = "530px"
   android:layout_gravity = "top|center_horizontal"
   android:keepScreenOn = "true" />
```

在该布局第二行添加一个 SeekBar 组件，表示视频进度。代码如下所示。

```
<!-- 第二行放置 SeekBar 组件，表示视频进度 -->
<SeekBar android:id = "@+id/posBar"
   android:layout_width = "700px"
   android:layout_height = "wrap_content"
   android:layout_gravity = "center_horizontal" />
```

在该布局中第三行添加一个线性布局，水平放置三个 ImageView 组件，分别为播放、暂停和停止按钮。代码如下所示。

```
<!-- 第三行放置播放、暂停和停止按钮组件-->
<LinearLayout
   android:layout_width = "match_parent"
   android:layout_height = "wrap_content"
   android:gravity = "center_horizontal"
   android:orientation = "horizontal" >
   <ImageView android:id = "@+id/play"
      android:layout_width = "80px"
      android:layout_height = "80px"
```

```
        android:layout_margin = "5px"
        android:src = "@drawable/play"
        android:text = "播放" />
    <ImageView android:id = "@+id/pause"
        android:layout_width = "80px"
        android:layout_height = "80px"
        android:layout_margin = "5px"
        android:src = "@drawable/pause"
        android:text = "暂停" />
    <ImageView android:id = "@+id/stop"
        android:layout_width = "80px"
        android:layout_height = "80px"
        android:layout_margin = "5px"
        android:src = "@drawable/stop"
        android:text = "停止" />
</LinearLayout>
```

（3）在主 Activity 类中添加界面组件对应的对象，以及在代码中使用到的成员变量和表示路径的常量。代码如下所示。

```
private final String VIDEO_PATH = "/sdcard/video/myvideo.mp4"; //视频路径
private MediaPlayer player;
private SurfaceView surface; //SurfaceView组件对象
private SurfaceHolder surfaceHolder; //SurfaceHolder对象
private ImageView play, pause, stop; //播放控制组件
private SeekBar posBar; //进度拖动条组件
private int times = 0; //视频时长
```

（4）在主 Activity 类中添加 init()方法，实例化 MediaPlayer 对象，加载视频文件，读取视频时长，并对视频进度拖动条设置 OnSeekBarChangeListener 监听器，当拖动进度时，从拖动后的位置开始播放。代码如下所示。

```
private void init() {
    player = new MediaPlayer(); //实例化MediaPlayer
    try {
        player.setDataSource(VIDEO_PATH); //设置视频路径
        player.prepare();
    }
    catch(Exception e) {
        e.printStackTrace();
    }
    if(times == 0)
        times = player.getDuration();  // 读取视频总时长
    posBar.setMax(times); // 设置视频当前位置
    posBar.setOnSeekBarChangeListener(new SeekBar.OnSeekBarChangeListener() {
        @Override
        public void onStopTrackingTouch(SeekBar seekBar) {
            player.pause();// 暂停播放
            player.seekTo(seekBar.getProgress());// 设置从指定位置开始播放
            player.start();// 开始播放
        }
        @Override
        public void onStartTrackingTouch(SeekBar seekBar){}
        @Override
```

```
        public void onProgressChanged(SeekBar seekBar, int progress, boolean fUser){}
    });
}
```

（5）在主 Activity 类中修改 onCreate 方法。先设置屏幕全屏，随后获取布局组件的对象，调用 init()方法进行初始化，对 play、pause、stop 对象添加 OnClickListener 监听器，分别设置当单击播放、暂停、停止时的对应操作。代码如下所示。

```
@Override
protected void onCreate(Bundle savedInstanceState) {
    requestWindowFeature(Window.FEATURE_NO_TITLE); //隐藏标题栏
    this.getWindow().setFlags(WindowManager.LayoutParams.FLAG_FULLSCREEN,
WindowManager.LayoutParams.FLAG_FULLSCREEN);  //隐藏营运商图标、电量等
    super.onCreate(savedInstanceState);
    setContentView(R.layout.activity_main);
    play = (ImageView) findViewById(R.id.play);
    pause = (ImageView) findViewById(R.id.pause);
    stop = (ImageView) findViewById(R.id.stop);
    surface = (SurfaceView) findViewById(R.id.surface);
    posBar = (SeekBar) findViewById(R.id.posBar);
    init(); //初始化
    //播放按钮增加监听器
    play.setOnClickListener(new View.OnClickListener() {
        @Override
        public void onClick(View v) {
            player.reset();
            surfaceHolder = surface.getHolder();
            // 设置视频显示在 SurfaceView 上
            player.setDisplay(surfaceHolder);
            try {
                player.setDataSource(VIDEO_PATH);
                player.prepare();
            }
            catch(Exception e) {
                e.printStackTrace();
            }
            player.start();
        }
    });
    //暂停按钮增加监听器
    pause.setOnClickListener(new View.OnClickListener() {
        @Override
        public void onClick(View v) {
            if(player.isPlaying())
                player.pause();
            else
                player.start();
        }
    });
    //停止按钮增加监听器
    stop.setOnClickListener(new View.OnClickListener() {
        @Override
        public void onClick(View v) {
            if(player.isPlaying())
            {
                player.stop();
```

```
            }
        }
    });
}
```

（6）在主 Activity 类中增加资源回收时的处理方法，代码如下所示。

```
@Override
protected void onDestroy() {
    if(player.isPlaying()) {
        player.stop();
    }
    player.release();
    super.onDestroy();
}
```

运行本实例，显示结果如图 9.5 所示。

图 9.5　使用 MediaPlayer 和 SurfaceView 播放视频

注意

本例中，表示视频进度的拖动条也没有实现随播放而移动，用户可在此例基础上进行补充。

9.2　摄像头的使用

Android 系统中包含了对摄像头的支持，通过摄像头意图 Intent 或 Camera 类的 API 都可以进行图像和视频的捕获。本节先介绍摄像头意图的使用，这种方式快速、简便，随后介绍一种更高级的可为用户创建自定义摄像功能的方法。

9.2.1　摄像头意图 Intent

Android 中可以不用 Camera 类进行摄像头的控制，而是使用系统提供的相关意图来调用内置的摄像头应用程序。摄像头意图会请求通过内置摄像应用来捕获图像或视频，并把控制权返回给应用程序。

使用摄像头意图的方法，一般可分为以下几个步骤。

（1）创建一个摄像头意图。

```
Intent intent = new Intent(参数);
```

创建一个意图如上所示，其中参数可为以下两种类型。

- MediaStore.ACTION_IMAGE_CAPTURE：向内置摄像头程序请求图像的意图动作类型。
- MediaStore.ACTION_VIDEO_CAPTURE：向内置摄像头程序请求视频的意图动作类型。

（2）启动摄像头意图。

用 startActivityForResult()方法执行摄像头 Intent，启动完毕后摄像头应用的界面就会出现在屏幕上，用户可以进行拍照或摄像。

（3）接收意图结果。

在应用程序中设置 onActivityResult()方法，用于接收从摄像头 intent 返回的数据。当用户拍摄完毕后（或者取消操作），系统会调用此方法。

下面通过一个简单的例子来说明拍照意图的使用，如例 9.5 所示。

【例 9.5】 实现一个使用摄像头意图 Intent 进行拍照以及查看的应用。

（1）修改主布局文件，将默认的布局改为帧布局，并在该布局中添加一个按钮用于拍照，一个 ImageView 组件用于显示拍照后的照片。代码如下所示。

```
<FrameLayout xmlns:android = "http://schemas.android.com/apk/res/android"
    android:layout_width = "match_parent"
    android:layout_height = "match_parent" >
    <Button android:id = "@+id/start"
        android:layout_width = "match_parent"
        android:layout_height = "wrap_content"
        android:layout_gravity = "bottom"
        android:text = "拍照" />
    <ImageView android:id = "@+id/pic"
        android:layout_width = "match_parent"
        android:layout_height = "wrap_content"
        android:layout_gravity = "top" />
</FrameLayout>
```

（2）在主 Activity 类中添加请求编码常量，照片的保存位置和 ImageView 组件对象等变量。代码如下所示。

```
//定义请求信息编号为 200
private static final int REQUEST_SUCCESS_CODE = 200;
private String imagePath;
private ImageView image;
```

（3）修改主 Activity 类的 onCreate()方法。在该方法中获取拍照按钮和显示照片的视图组件，并对按钮增加 OnClickListener 监听器。监听器中首先设置照片名为 SD 卡中 mypic.jpg，接着建立摄像头意图用于拍照，最后通过 startActivityForResult()方法启动摄像头意图。代码如下所示。

```
@Override
protected void onCreate(Bundle savedInstanceState) {
    super.onCreate(savedInstanceState);
    setContentView(R.layout.activity_main);
    Button b = (Button) findViewById(R.id.start);
    image = (ImageView) findViewById(R.id.pic);
    b.setOnClickListener(new View.OnClickListener() {
        @Override
        public void onClick(View v) {
            // 获取 SD 卡上的文件路径
            imagePath = Environment.getExternalStorageDirectory().getAbsolutePath()
```

```
                + "/mypic.jpg";
        File imageFile = new File(imagePath);
        // 将字符串表示的路径转为 URI 表示
        Uri imageFileUri = Uri.fromFile(imageFile);
        // 创建使用摄像头意图
        Intent intent = new Intent(MediaStore.ACTION_IMAGE_CAPTURE);
        // 给意图中添加文件路径作为参数
        intent.putExtra(MediaStore.EXTRA_OUTPUT, imageFileUri);
        // 启动意图
        startActivityForResult(intent, REQUEST_SUCCESS_CODE);
    }
  });
}
```

（4）在接收意图的 onActivityResult()方法中，通过 BitmapFactory 将照片编码为位图，然后通过 ImageView 组件显示在屏幕上。代码如下所示。

```
@Override
protected void onActivityResult(int requestCode, int resultCode, Intent intent) {
    if(requestCode == REQUEST_SUCCESS_CODE) {
        if(resultCode == RESULT_OK) {
            // 将照片编码为位图
            Bitmap bmp = BitmapFactory.decodeFile(imagePath);
            // 布局中 ImageView 组件加载该照片
            image.setImageBitmap(bmp);
        }
        else
            Toast.makeText(this, "放弃拍照", Toast.LENGTH_LONG).show();
    }
    super.onActivityResult(requestCode, resultCode, intent);
}
```

运行本实例，打开应用界面，如图 9.6 所示；单击拍照按钮调用摄像头界面，如图 9.7 所示；拍照完毕后回到主页显示结果，如图 9.8 所示。

图 9.6　主界面

图 9.7　调用拍照界面

图 9.8　拍照后主界面

9.2.2 Camera 类

在 Android 中还可以使用 android.hardware 包中的 Camera 类进行摄像头的控制，Camera 类没有构造函数，但可通过提供的 open()方法打开摄像头。Camera 类常用的方法如表 9.4 所示。

表 9.4　Camera 常用的方法

方法名称	说明
void autoFocus(Camera.AutoFocusCallback cb)	设置自动对焦
Camera.Parameters getParameters()	得到相机参数
Camera open()	启动相机服务
void release()	释放相机服务
void setParameters(Camera.Parameters params)	设置参数
void setPreviewDisplay(SurfaceHolder holder)	设置预览
void startPreview()	开始预览
void stopPreview()	停止预览
void takePicture(Camera.ShutterCallback shutter, Camera.PictureCallback raw, Camera.PictureCallback postview, Camera.PictureCallback jpeg)	拍照

在开始开发摄像头的应用之前，应该确保已经在 manifest 中正确声明了对摄像头的使用及其他相关的资源的权限。可以在 AndroidManifest.xml 文件中添加所需权限，其中包括：

（1）Camera 权限。

应用程序必须请求摄像头的使用权限，语法如下。

```
<uses-permission android:name = "android.permission.CAMERA" />
```

如果是通过意图来使用摄像头的，应用程序则不必请求该权限。

（2）Camera Feature。

应用程序必须同时声明对 camera feature 的使用，语法如下。

```
<uses-feature android:name = "android.hardware.camera" />
```

该项声明会使 AndroidMarket 在没有摄像头或不支持指定 feature 的设备上禁止安装该应用程序。如果应用程序对摄像头或摄像头 feature 的使用，不是必需的，则应在 AndroidManifest.xml 中指定 android:required 属性为 false，语法如下。

```
<uses-feature android:name = "android.hardware.camera" android:required = "false" />
```

（3）存储权限。

如果应用程序要把图像或视频保存到外部存储（SD 卡）上，则还须在 AndroidManifest.xml 中加入写入权限，语法如下。

```
<uses-permission android:name = "android.permission.WRITE_EXTERNAL_STORAGE" />
```

（4）录音权限。

若应用还需录音则必须请求音频捕获权限，语法如下。

```
<uses-permission android:name = "android.permission.RECORD_AUDIO" />
```

在应用中，用户可能需要在自定义外观中使用摄像头，或者需要提供特殊的功能。相比使用 intent 而言，建立定制摄像的界面需要编写的代码会更多，但是却能结合自己的需要控制更多的功能。

创建一个定制的摄像界面，一般可分为以下几个步骤。

（1）检测并访问摄像头。

创建代码以检测摄像头是否存在，并请求访问。

（2）创建预览类。

创建继承自 SurfaceView 并实现 SurfaceHolder 接口的摄像预览类，该类能预览摄像的实时图像。

（3）设置监听器。

为用户组件添加相应的监听器，使其能响应用户捕获图像或视频。

（4）捕获并保存文件。

建立捕获图片或视频并保存到输出文件的代码。

（5）释放摄像头。

摄像头使用完毕后，应用程序必须正确地将其释放，便于其他程序的使用。

下面通过一个具体的实例来说明控制摄像头拍照的具体过程，如例 9.6 所示。

【例 9.6】　实现一个使用摄像头类完成的照相机应用。

（1）首先在 AndroidManifest.xml 文件中添加摄像头、SD 卡存储等权限。代码如下所示。

```
<uses-permission android:name = "android.permission.CAMERA" />
<uses-permission android:name = "android.permission.MOUNT_UNMOUNT_FILESYSTEMS" />
<uses-permission android:name = "android.permission.WRITE_EXTERNAL_STORAGE" />
<uses-feature android:name = "android.hardware.camera" />
<uses-feature android:name = "android.hardware.camera.autofocus" />
```

（2）其次，在 AndroidManifest.xml 文件中的对应 activity 标签内，设置移动设备屏幕方向为横屏。代码如下所示。

```
<activity android:name = "com.example.cameraclass.MainActivity"
    android:screenOrientation = "landscape"
    android:label = "@string/app_name" >
```

（3）修改主布局文件，将默认的布局管理器改为 FrameLayout 帧布局，背景设为黑色（该主布局代码省略），在主布局中添加如下代码。

主布局将屏幕分成三列，第一列设置垂直线性布局，左对齐，其内组件垂直居中。布局中分别放置表示关闭闪光灯、打开闪光灯、自动闪光灯的 ImageView 组件。

```
<!-- 第一列放置关闭闪光灯、打开闪光灯、自动闪光灯图片 -->
<LinearLayout
    android:layout_width = "wrap_content"
    android:layout_height = "match_parent"
    android:layout_gravity = "left"
    android:gravity = "center"
    android:orientation = "vertical" >
    <ImageView
        android:layout_width = "125px"
        android:layout_height = "125px"
        android:layout_margin = "15px"
        android:src = "@drawable/closeflash" />
    <ImageView
        android:layout_width = "125px"
        android:layout_height = "125px"
```

```
        android:layout_margin = "15px"
        android:src = "@drawable/openflash" />
    <ImageView
        android:layout_width = "125px"
        android:layout_height = "125px"
        android:layout_margin = "15px"
        android:src = "@drawable/autoflash" />
</LinearLayout>
```

第二列为帧布局，在 Java 代码中，会在此布局中放置一个 SurfaceView 组件，而该 SurfaceView 组件用于显示摄像头的预览画面。代码如下所示。

```
<!-- 第二列放置预览画面组件 -->
<FrameLayout android:id = "@+id/camera_preview"
    android:layout_width = "wrap_content"
    android:layout_height = "match_parent"
    android:layout_gravity = "center" />
```

第三列为一个线性布局，右对齐，其内组件垂直居中。布局中分别放置表示前后摄像头切换、拍照、设置的 ImageView 组件。代码如下所示。

```
<LinearLayout
    android:layout_width = "wrap_content"
    android:layout_height = "match_parent"
    android:layout_gravity = "right"
    android:gravity = "center"
    android:orientation = "vertical" >
    <ImageView
        android:layout_width = "125px"
        android:layout_height = "125px"
        android:layout_margin = "15px"
        android:src = "@drawable/turn" />
    <ImageView
        android:id = "@+id/takePhoto"
        android:layout_width = "125px"
        android:layout_height = "125px"
        android:layout_margin = "15px"
        android:src = "@drawable/take" />
    <ImageView
        android:layout_width = "125px"
        android:layout_height = "125px"
        android:layout_margin = "15px"
        android:src = "@drawable/set" />
</LinearLayout>
```

（4）在主 Activity 类中添加 Camera 摄像头对象，CameraPreview 预览类对象和类 PictureCallback 对象。其中 PictureCallback 类中实现拍照动作，首先停止预览，通过文件输出流将拍摄的照片保存到 SD 卡上，最后再次开启预览。代码如下所示。

```
private Camera camera;
private CameraPreview cPreview;
private PictureCallback picCallBack = new PictureCallback() {
    @Override
    public void onPictureTaken(byte[] data, Camera camera) {
        camera.stopPreview(); // 停止预览
        File pictureFile = getOutputMediaFile();
```

```
        if(pictureFile == null) {
            Log.d("提示", "不能建立照片文件!");
            return;
        }
        try {
            FileOutputStream fos = new FileOutputStream(pictureFile);
            fos.write(data);
            fos.flush();
            fos.close();
            camera.startPreview(); // 保存完毕，开始预览
        } catch(FileNotFoundException e) {
            Log.d("提示", "未发现文件");
        } catch(IOException e) {
            Log.d("提示", "访问文件出错");
        }
    }
};
```

（5）在主 Activity 类中添加检测摄像头是否存在方法。代码如下所示。

```
private boolean checkCameraHardware(Context context) {
    if(context.getPackageManager().hasSystemFeature(
      PackageManager.FEATURE_CAMERA))
        return true; // 摄像头存在
    else
        return false; // 摄像头不存在
}
```

（6）在主 Activity 类中添加安全获取 Camera 对象的方法。代码如下所示。

```
// 安全获取 Camera 对象
public static Camera getCameraInstance() {
    Camera c = null;
    try {
        c = Camera.open(); // 试图获取 Camera 实例
    } catch(Exception e) {
        Log.d("提示", "摄像头不可用（正被占用或不存在）";
    }
    return c;
}
```

（7）在主 Activity 类中添加创建照片文件名的方法。该方法首先检查 SD 卡是否存在，然后通过 Environment 类的静态方法获取系统图片路径，最后设置照片文件名构成形式为“IMG_年月日_时分秒.jpg”。代码如下所示。

```
// 为保存图片或视频创建 File
private File getOutputMediaFile() {
    // 检查是否存在 SD 卡
   if(!Environment.getExternalStorageState().equals(Environment.MEDIA_MOUNTED)){
        Toast.makeText(this, "SD 卡不存在", Toast.LENGTH_LONG).show();
    }
    File mediaStorageDir = new File(Environment.getExternalStoragePublicDirectory(
                    Environment.DIRECTORY_PICTURES).getAbsolutePath());
    // 创建媒体文件名
```

```
    String timeStamp = new SimpleDateFormat("yyyyMMdd_HHmmss").format(
                    new Date());
    File mediaFile = new File(mediaStorageDir.getPath() + File.separator + "IMG_"
                    + timeStamp + ".jpg");
    return mediaFile;
}
```

（8）在主 Activity 类中添加 onDestroy()方法，用于程序退出时释放摄像头资源。代码如下所示。

```
@Override
protected void onDestroy() {
    camera.stopPreview();
    camera.release();
    super.onDestroy();
}
```

（9）在主 Activity 类中修改 onCreate 方法。首先设置应用全屏；然后检查摄像头是否存在，若存在，则创建 Camera 实例；再获取布局文件中第二列帧布局对象，设置布局大小和位置；创建 CameraPreview 实例，并将该实例表示的预览画面放入第二列布局中；最后调用摄像头类的 takePicture()方法进行拍照。代码如下所示。

```
@Override
protected void onCreate(Bundle savedInstanceState) {
    // 设置全屏
    requestWindowFeature(Window.FEATURE_NO_TITLE);
    this.getWindow().setFlags(WindowManager.LayoutParams.FLAG_FULLSCREEN, WindowManager.
LayoutParams.FLAG_FULLSCREEN);
    super.onCreate(savedInstanceState);
    setContentView(R.layout.activity_main);
    if(!checkCameraHardware(this)) {
        Toast.makeText(this, "无摄像头,不能使用本应用",
                    Toast.LENGTH_LONG).show();
        return;
    }
    camera = getCameraInstance(); // 创建 Camera 实例
    if(camera == null) {
        Toast.makeText(MainActivity.this, "摄像头对象没有建立",
                    Toast.LENGTH_LONG).show();
        return;
    }
    // 设置 CameraPreview 组件布局参数
    FrameLayout.LayoutParams lparams = new FrameLayout.LayoutParams(1000, 720);
    lparams.gravity = Gravity.CENTER;
    cPreview = new CameraPreview(this, camera);
    cPreview.setLayoutParams(lparams);
    FrameLayout layout = (FrameLayout) findViewById(R.id.camera_preview);
    layout.addView(cPreview);
    ImageView takePhoto = (ImageView) findViewById(R.id.takePhoto);
    takePhoto.setOnClickListener(new View.OnClickListener() {
        @Override
        public void onClick(View v) {
            // 从摄像头获取图片
```

```
                camera.takePicture(null, null, picCallBack);
                Toast.makeText(MainActivity.this, "照片保存!",
                        Toast.LENGTH_SHORT).show();
            }
        }
    );
}
```

（10）新建 CameraPreview 类，代码如下所示。该类继承 SurfaceView 类，实现 SurfaceHolder.Callback 接口。

```
public class CameraPreview extends SurfaceView implements SurfaceHolder.Callback {
    // 待添加的内容
}
```

（11）在 CameraPreview 类中，定义 SurfaceHolder 对象和摄像头对象。在其构造方法中获取 SurfaceHolder 对象，设置回调方法。代码如下所示。

```
private SurfaceHolder holder;
private Camera camera;
// 构造函数，建立 SurfaceHolder 对象
public CameraPreview(Context context, Camera camera) {
    super(context);
    if(camera == null) {
        Toast.makeText(context, "摄像头对象没有建立", Toast.LENGTH_LONG).show();
        return;
    }
    this.camera = camera;
    // 定义一个 SurfaceHolder.Callback，创建和销毁底层 surface 时能够获得通知
    holder = getHolder();
    holder.addCallback(this);
}
```

（12）在 CameraPreview 类中，添加创建方法，并设置开始预览。代码如下所示。

```
// surface 已被创建，把预览画面的位置通知摄像头
public void surfaceCreated(SurfaceHolder holder) {
    try {
        if(holder == null) {
            System.out.println("holder = =null");
            return;
        }
        // 设置 surfaceHolder 自己不维护缓冲
        holder.setType(SurfaceHolder.SURFACE_TYPE_PUSH_BUFFERS);
        camera.setPreviewDisplay(holder);
        camera.startPreview();
        camera.autoFocus(null);
    } catch(IOException e) {
        Log.d("提示", "不能进行预览");
    }
}
```

（13）在 CameraPreview 类中，添加预览方向改变时触发方法，先停止预览，再设置预览显示，最后再打开预览。代码如下所示。

```
// 预览方向改变时触发方法
```

```
public void surfaceChanged(SurfaceHolder holder, int format, int w, int h) {
    if(holder.getSurface() == null) { // 预览 surface 不存在
        return;
    }
    try { // 更改时停止预览
        camera.stopPreview();
        camera.setPreviewDisplay(holder);
        camera.startPreview();
    } catch(Exception e) {
        Log.d("提示", "停止不存在的预览");
    }
}
//空实现
public void surfaceDestroyed(SurfaceHolder holder) {}
```

运行本实例，显示结果如图 9.9 所示。

图 9.9　使用 Camera 类实现照相机

本例中只完成了拍照功能，其他闪光灯、前后摄像头切换和设置功能都未完成，请感兴趣的用户在此例基础上自行补充。

9.3　小　结

本章主要介绍了在 Android 中如何播放音频和视频，以及如何控制相机进行拍照等内容。MediaPlayer 既可以播放音频，也可以播放视频。播放音频时每次只能播放一个，适用于播放长音乐和背景音乐，播放视频时由于没提供输出界面，所以需要与 SurfaceView 配合使用。SoundPool 适合同时播放多个短小的音频，例如消息提示，游戏音效等。VideoView 和 MediaController 组件能够实现比较完善的视频播放功能。

最后讲解了两种摄像头的使用方法，一种是通过摄像头意图，编写代码简单、快速；另一种是使用 Camera 类，该方法提供了一种更高级的，可为用户创建自定义摄像的功能，使用起来更加灵活。

练　　习

1. 使用 MediaPlayer 顺序播放 SD 卡的 music 目录下的所有音频。
2. 应用 SoundPool 同时播放 SD 卡的 sound 目录下的所有音频。
3. 使用 VideoView 播放 SD 卡上的一个视频文件，并使用 MediaController 进行控制。
4. 使用 MediaPlayer 和 SurfaceView 播放 SD 卡上的一个视频文件。
5. 分别使用摄像头意图和 Camera 类设计一个照相机应用。

实验一 Android 开发环境搭建

【目的】

1. 完成 Android 开发环境的安装配置
2. 掌握 Android SDK 的目录结构
3. 掌握 Android 模拟器上传文件
4. 掌握 Android 模拟器的常用操作命令
5. 掌握 Android 模拟器安装软件

【理论】

Android 开发语言是 Java 语言，其开发环境包含 Java 开发包 JDK（建议使用 JDK1.6 或以上），开发工具推荐为 Eclipse（建议使用 3.5 版本或以上），以及 Android 开发包 Android Development Kit（简称 ADT），和其他组件如模拟器，即 Android Virtual Device（简称 AVD）。

Android SDK 还提供了一些常用命令来提高编程、调试效率，命令位于 Android SDK 中的 tools 和 platform-tools 文件夹下。

【内容】

1. 安装 Android 开发环境
2. 使用 Android 常用命令安装和删除应用程序

【步骤】

1. 安装 Android 开发环境

（1）安装 JAVA JDK。

下载网址：http://www.oracle.com/technetwork/cn/java/javase/downloads。

安装，并配置环境变量 JAVA_HOME、PATH、CLASSPATH。

（2）安装 Eclipse。

下载网址：http://www.eclipse.org/downloads。

解压即可。

（3）Eclipse 中文包的安装（可省略）。

中文包网址：http://www.eclipse.org/babel/downloads.php。

打开 Eclipse 开发工具，在“Help”菜单中单击“install new software”，打开“install”窗口，

单击“add”按钮，显示一个仓库对话框，在“Name”中输入语言包名称，在“Location”中粘贴 Babel 语言包的地址，选择其中的中文包项“Babel Language Packs in Chinest(Simplified)”安装即可，如图 1 所示。

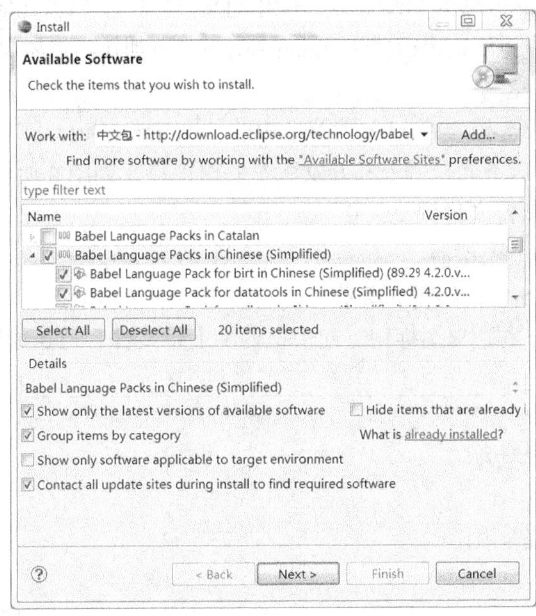

图 1　安装中文包插件

（4）安装 ADT。

在 Eclipse 开发工具中单击“Help”菜单的“install new software”，打开“install”窗口，单击“Add...”按钮，在弹出对话框中 Name 处输入“ADT”，Location 处输入“https://dl-ssl.google.com/android/eclipse”，选择相应项安装，如图 2 所示。

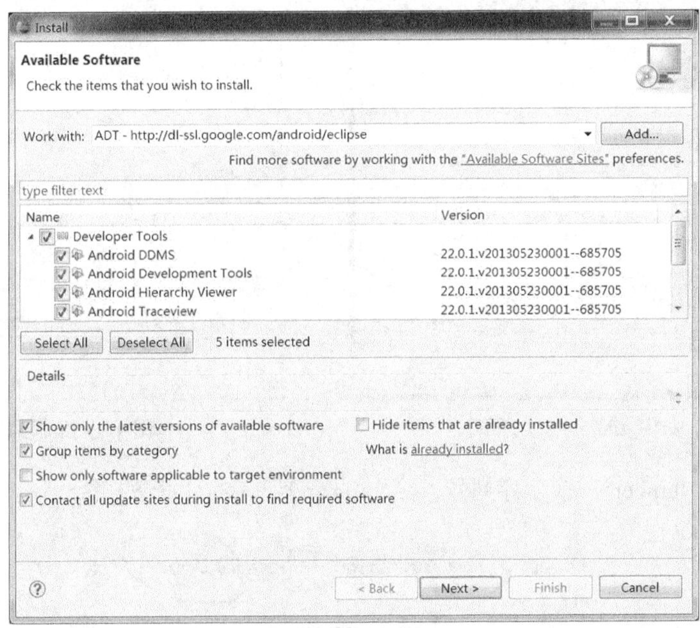

图 2　安装 ADT 插件

（5）安装 Android SDK。

下载网址：http://developer.android.com/sdk/index.html#download。

解压到相应目录，在 Eclipse 中配置 Android Preferences 中的“SDK Location”。并单击 Eclipse 的“Window”菜单中的“Android SDK Manager”，打开 SDK 管理器，安装相应版本的 SDK 文档、SDK 平台、实例和基于不同硬件系统的模拟器映像插件等，如图 3 所示，最后设置 SDK 环境变量。

（6）创建模拟器 AVD。

在 Eclipse 开发工具中选择“Window”菜单中的“Android Virtual Device Manager”打开模拟器管理窗口，创建几个针对不同分辨率和开发版本的模拟器，如图 4 所示。

（7）Android 模拟器上传文件到 SD 卡。

- 打开 Eclipse 和一个模拟器。
- 单击右上角的 DDMS，如图 5 所示。
- 单击 File Explorer 右边图标用于向模拟器上上传文件，通过如图 6 所示的 File Explorer 查看是否安装成功。

（8）掌握 adb 操作命令。

- 查询模拟器或设备实例

```
adb devices
```

- 设定模拟器或设备实例

```
adb -s <serialNumber> <command>
```

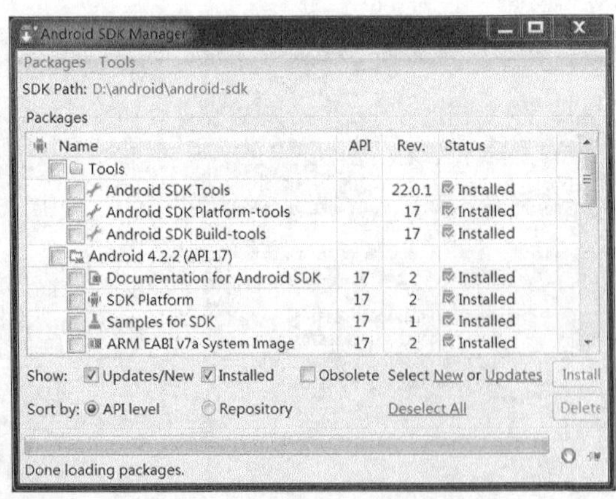

图 3　安装 Android SDK 插件

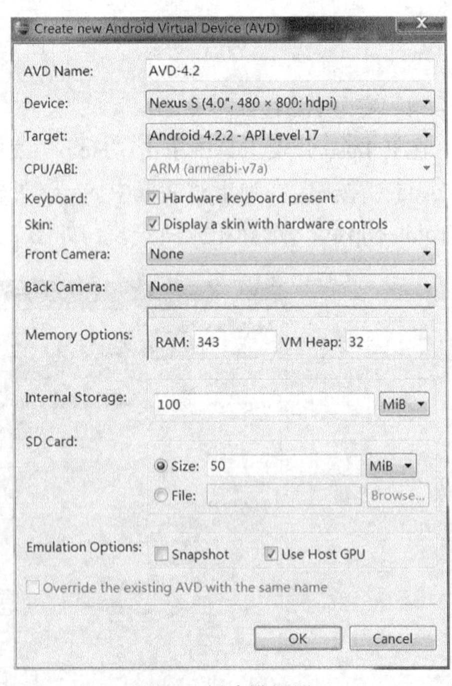

图 4　创建模拟器

其中“<serialNumber>”表示序列号，“<command>”表示执行的命令。

图 5　DDMS

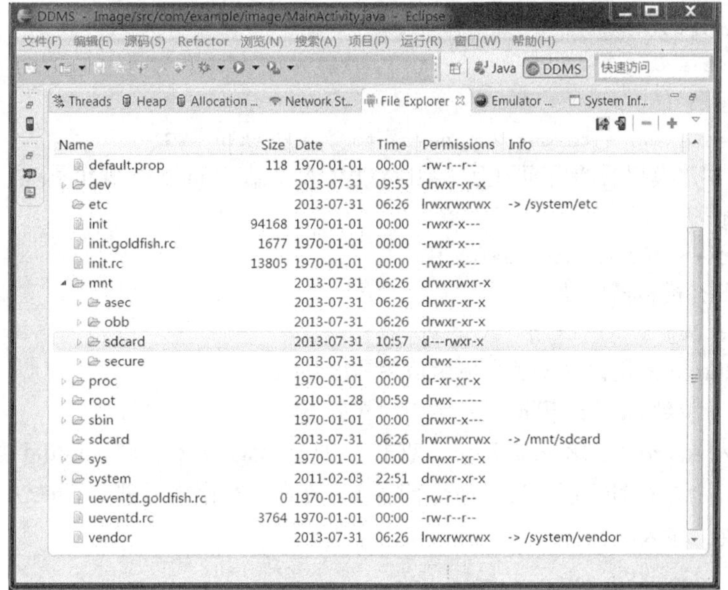

图 6　File Explorer 窗口

例如需要在 emulator-5554 上安装搜狗输入法 apk 应用时，命令格式如下。

```
adb -s emulator-5554 install sogouInput4.7.5.apk
```

如果存在多个模拟器或实例正在运行，则必须指定设备。

- 文件导出

将模拟器或设备上的文件导出到计算机，命令格式如下，参数含义同上。

```
adb pull <remote> <local>
```

- 文件导入

将计算机本地文件导入到模拟器或设备，命令格式如下，其中“<local>”表示计算机上的文件位置，“<remote>”表示模拟器或设备实例上的文件位置。

```
adb push <local> <remote>
```

- 打开 shell

```
adb shell
```

Android 平台底层采用 Linux 内核，可以使用 shell 来进行操作。

- 安装应用程序

```
adb install <path_to_apk_file>
```

如果只有一个设备在运行，可以通过该命令将计算机上的 apk 应用安装到模拟器或设备上，如果多个设备在运行，需指定设备实例安装。

（9）掌握 android 操作命令。

- 列出模拟器类型

```
android list targets
```

- 创建模拟器

```
android create avd -n <avd_name> -t <target_id> [-<option> <value>] ...
```

--target 后面跟创建模拟器的 id，--name 后面跟模拟器的名字。

- 列出已创建的模拟器

```
android list avd
```

- 切换模拟器样式

```
android create avd --target 2 --name cupcake --skin QVGA
```

切换样式在创建模拟器的后面加上“--skin QVGA”，在 Windows 操作系统下按 F7。

- 删除模拟器

```
android delete avd -n <avd_name>
```

- 指定用什么模拟器启动

```
emulator -debug avd-config -avd cupcake
```

2. 使用 Android 常用命令安装和删除应用程序

（1）将 apk 安装到 android 模拟器，操作步骤如下。

- 首先进入 Android SDK 安装目录，双击 AVD Manager.exe 启动 Android 模拟器。
- 然后打开命令行对话框，进入命令行模式。在命令行模式下进入 Android SDK 安装目录下的 tools 文件夹，输入：

```
adb install c:\sogou4.7.5.apk
```

注意

sogou4.7.5.apk 应用位于 C 盘根目录下，同时要求只启动了一个模拟器，结果如图 7、图 8 所示。

图 7　在模拟器上安装应用

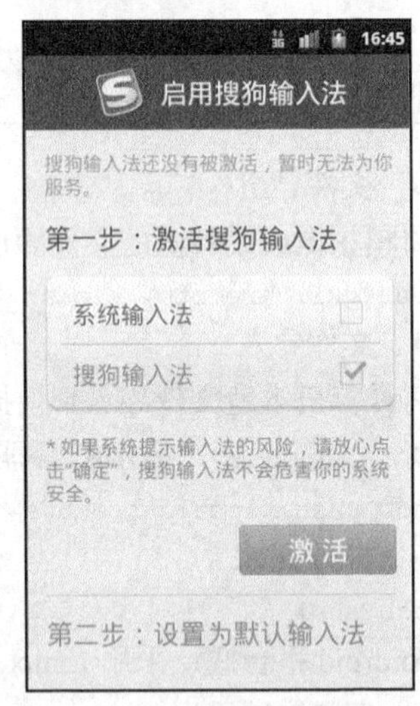

图 8　安装成功配置应用

（2）卸载模拟器中的 apk 文件，操作步骤如下。

- 如上方式首先启动 Android 模拟器。
- 然后打开命令行对话框，进入命令行模式。在命令行模式下进入 Andoid SDK 安装目录下的 tools 文件夹，依次输入：

```
abd shell
cd data
cd app
ls
rm com.sohu.inputmethod.sogou-2.apk
```

注意

（1）其中“cd data”和“cd app”分别为进入 data 和 app 目录，应用程序一般安装在系统的“data\app”路径下。

（2）在不清楚文件完全名称时，可以用 ls 命令先显示出来，结果如图 9 所示。上例搜狗输入法安装成功后的文件名为 com.sohu.inputmethod.sogou-2.apk。

（3）使用 rm 命令删除要卸载的 apk 包，然后重新启动模拟器，会发现搜狗输入法已删除。

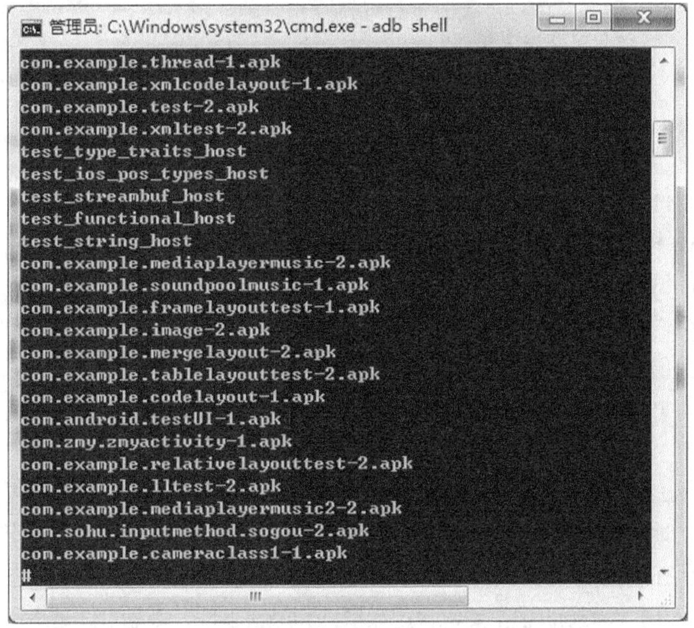

图 9　shell 中 list 命令查看应用名

实验二
界面设计：基本组件

【目的】

1. 掌握常用组件在布局文件中的设置
2. 掌握在 Java 程序中获取组件值
3. 掌握对组件值的验证
4. 掌握基本组件常用的监听器，和事件处理
5. 掌握将组件值提交到下一个 Activity 活动的方法

【理论】

Android 通过 widget 包提供了各种 UI 组件，包括按钮、列表框、单选按钮、复选按钮、进度条、图片、日期选择器、时间选择器等基本组件，以及进度条、拖动条、评分条、选项卡、图像切换器等高级组件。各个组件具有一些通用的属性，但需要注意的是各组件特有属性的应用。图形界面应用都是通过事件来实现人机交互，Android 中主要包含键盘事件和触摸事件，键盘事件是指用户使用物理键盘过程中产生的事件，如按下、弹起等；而触摸事件是指用户触摸屏幕产生的事件，如按下、弹起、滑动、双击等。同时 Android 中还提供了消息提示框、状态栏上通知、对话框等用户提示方式，需要用户掌握。

【内容】

1. 补充完全下例空缺处，完成如图 1 ~ 图 5 所示的注册信息界面。

2. 根据本例，再设计一个问卷调查 UI 界面，并实现在 Java 程序中获取数据、验证数据，以及提交数据到下一个 Activity 页面的功能。

图 1 注册界面　　图 2 部门列表框　　图 3 单击确定检查　　图 4 提交成功　　图 5 接收页面

【步骤】

1. 设计 UI 界面

（1）将主布局修改为线性布局 LinearLayout，垂直排列，代码如下所示。

```
<LinearLayout xmlns:android="http://schemas.android.com/apk/res/android"
    android:layout_width="match_parent"
    android:layout_height="match_parent"
    android:orientation="vertical" >
    <!-- 待添加布局组件 -->
</LinearLayout>
```

（2）在主布局中添加用户名文本框和输入框，请完成如下代码中的提示部分。

```
<TextView android:text = "用户名"
     请在此位置添加用户名文本框其他属性 />
<EditText android:id = "@+id/name"
     请在此位置添加用户名编辑框其他属性 />
```

（3）在主布局中添加密码文本框和输入框，请完成如下代码中的提示部分。

```
<TextView android:text="密码"
      请在此位置添加密码文本框其他属性 />
<EditText android:id="@+id/password"
      android:inputType="textPassword"
      请在此位置添加密码编辑框其他属性 />
```

（4）在主布局中添加性别文本框和复选框，请完成如下代码中的提示部分。

```
<TextView android:text="性别"
     请在此位置添加性别文本框其他属性 />
<RadioGroup android:id="@+id/sex"
     android:orientation="horizontal"
     请在此位置添加单选按钮组其他属性 >
     <RadioButton android:id="@id/man"
         android:text="男"
         请在此位置添加单选按钮其他属性 />
     <RadioButton android:id="@+id/woman"
         android:text="女"
         请在此位置添加单选按钮其他属性 />
</RadioGroup>
```

（5）在主布局中添加联系电话文本框和输入框，设置输入类型为 android:inputType = "text|phone"，请完成如下代码中的提示部分。

```
<TextView
     android:text="联系电话"
     请在此位置添加联系电话文本框其他属性 />
<EditText android:id="@+id/tel"
     android:inputType="text|phone"
     请在此位置添加联系电话编辑框其他属性 />
```

（6）在主布局中添加部门文本框和列表框，设置列表框数据来源为 depts.xml 文件，请完成如下代码中的提示部分。

```
<TextView
     android:text="部门"
     请在此位置添加部门文本框其他属性 />
```

```
<Spinner android:id="@+id/dept"
    android:entries="@array/dept"
    请在此位置添加部门编辑框其他属性 />
```

（7）在主布局中添加爱好文本框和一个线性布局，该布局中放置四个 CheckBox 组件，分别表示书籍、运动、音乐和电影，请完成如下代码中的提示部分。

```
<TextView android:text="爱好"
    请在此位置添加爱好文本框其他属性 />
</TextView>
<LinearLayout
   请在此位置添加线性布局其他属性 />
   <CheckBox android:id="@+id/book"
      android:text="书籍"
      请在此位置添加书籍复选按钮其他属性 />
   <CheckBox android:id="@+id/sport"
      android:text="运动"
      请在此位置添加运动复选按钮其他属性 />
   <CheckBox android:id="@+id/music"
      android:text="音乐"
      请在此位置添加音乐复选按钮其他属性 />
   <CheckBox android:id="@+id/movie"
      android:text="电影"
      请在此位置添加电影复选按钮其他属性 />
</LinearLayout>
```

（8）在主布局中最后添加一个确定按钮，设置 android:onClick 属性值为 myclick，即当单击确定按钮时，会调用 myclick 方法，请完成如下代码中的提示部分。

```
<Button android:id="@+id/ok"
   android:layout_gravity="center_horizontal"
   android:text="确定"
   android:onClick="myclick"
   请在此位置添加确定按钮其他属性 />
</Button>
```

如果界面组件过多使得当前界面元素无法完全显示时，可以将界面主布局设置在组件 ScrollView 组件中，将 xmlns:android 属性放到 ScrollView 内，代码如下所示。

```
<ScrollView xmlns:android="http://schemas.android.com/apk/res/android"
   android:layout_width="match_parent"
   android:layout_height="wrap_content" >
   <LinearLayout
      android:layout_width="match_parent"
      android:layout_height="match_parent"
      android:orientation="vertical" >
      <!--布局中各组件 -->
   </LinearLayout>
</ScrollView>
```

2. 表示部门的 Spinner 组件，其数据来源文件 depts.xml 位于 res/values 目录下，代码如下所示。

```
<resources>
<string-array name = "dept">
```

```
<item>人力资源部</item>
<item>销售部</item>
<item>财务部</item>
<item>开发部</item>
</string-array>
</resources>
```

3. 设计后台程序

（1）在主 Activity 文件中，定义布局中的各组件对象和一个存放爱好中各复选框对象的 favs 动态数组。

```
private ArrayList<CheckBox> favs;
private EditText userName;
//定义布局中各组件对象，此处省略
```

（2）在 onCreate()方法中，获取各组件。

```
userName=(EditText) findViewById(R.id.name);
//获取个组件对象，此处省略
favs=new ArrayList<CheckBox>();
//将各项爱好对象存入数组
favs.add(book);
favs.add(sport);
favs.add(music);
favs.add(movie);
```

（3）获取性别方法。

```
public String getSex() {
    RadioButton radioButton=(RadioButton) findViewById(
                                        sex.getCheckedRadioButtonId());
    return radioButton.getText().toString();
}
```

（4）获取爱好方法，爱好以逗号分隔。

```
public String getFavorite() {
    String favo="";
    for(CheckBox cb : favs) {
        if(cb.isChecked()) {
            favo+=cb.getText().toString();
            favo+=",";
        }
    }
    if(!"".equals(favo))
        favo=favo.substring(0, favo.length() - 1);
    else
        favo="您未选择爱好！";
    return favo;
}
```

（5）当检查通过时，输出注册信息，提交到下一个 Activity 页面。

```
public void myclick(View view) {
    if(check()) {
        StringBuilder sb=new StringBuilder();
        sb.append("用户名：" + userName.getText().toString() + "\n");
        sb.append("性别：" + getSex() + "\n");
        sb.append("电话：" + tel.getText().toString() + "\n");
```

```
        sb.append("部门: " + dept.getSelectedItem().toString() + "\n");
        sb.append("爱好: " + getFavorite() + "\n");
        Toast.makeText(this, sb.toString(), Toast.LENGTH_LONG).show();
        Intent intent=new Intent(); //将注册信息提交到 ResultActivity 页面
        intent.setClass(this, ResultActivity.class);
        intent.putExtra("info", sb.toString());
        this.startActivity(intent);
    }
}
```

（6）创建一个 result_activity.xml 布局文件，放置一个文本框组件，并创建 ResultActivity 类，修改 onCreate()方法如下，显示上页传来的数据。

```
protected void onCreate(Bundle savedInstanceState) {
    super.onCreate(savedInstanceState);
    requestWindowFeature(Window.FEATURE_NO_TITLE);
    setContentView(R.layout.result_activity);
    TextView result=(TextView) findViewById(R.id.result);
    result.setText("从前一个页面传过来的内容如下: \n\n"
            + this.getIntent().getStringExtra("info"));
}
```

请用户补充完此例，使得运行结果如图 1～图 5 所示。并依照此例，完成一个问卷调查 UI 界面设计，详情见内容部分要求。

实验三
界面设计：布局管理器

【目的】

1. 了解四种布局管理器的区别和各自特有的属性
2. 掌握四种布局管理器的应用场合和用法
3. 灵活使用四种布局管理器的嵌套实现各种复杂布局
4. 掌握复用 XML 布局文件的方法
5. 掌握代码控制 UI 界面的方法

【理论】

布局管理器能够管理 Android 应用中用户 UI 的各种视图组件，Android 提供了四种常用布局管理器，即线性布局 LinearLayout、表格布局 TableLayout、帧布局 FrameLayout 和相对布局 RelativeLayout，另外在设计 UI 布局时，布局管理器能够嵌套，对于相同的布局可以通过<include>和<merge>标签进行布局文件的复用。为了提高布局的灵活性，Android 还提供了使用代码控制 UI 界面，以及代码和 XML 联合控制 UI 界面的方法。

【内容】

请按照图 1 设计的样式，完成如图 2 的 Android 应用 UI 的开发（图片不限）。

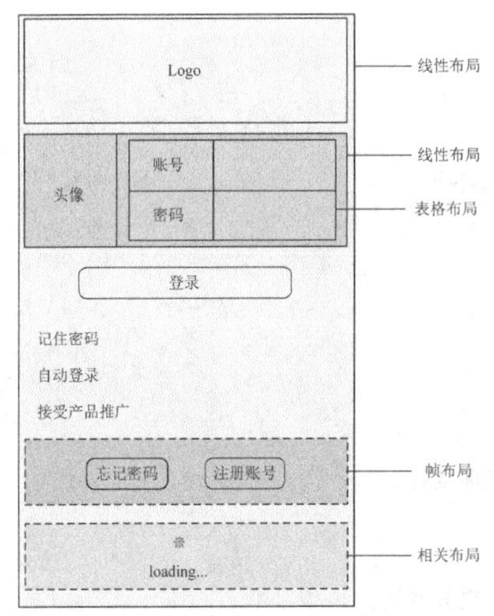

图 1　UI 布局样式

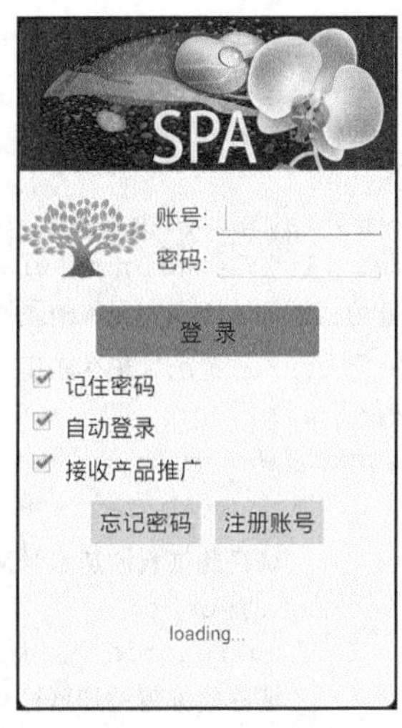

图 2　布局完成效果

【步骤】

1. 用 Java 代码设置全屏

打开工程src目录下的主Activity文件，在onCreate方法中的执行语句super.onCreate(savedInstanceState)之前，添加如下两条语句：

```
requestWindowFeature(Window.FEATURE_NO_TITLE); //隐藏标题栏
this.getWindow().setFlags(WindowManager.LayoutParams.FLAG_FULLSCREEN,
WindowManager.LayoutParams.FLAG_FULLSCREEN); //隐藏营运商图标、电量等
```

2. 按照图 1 设计的样式，完成一个 Android 应用 UI 的开发

（1）将主布局修改为线性布局，如下代码所示，请在 LinearLayout 标签内添加背景色和布局方向。

```
<LinearLayout xmlns:android="http://schemas.android.com/apk/res/android"
    android:layout_width="match_parent"
    android:layout_height="match_parent"
    请在此位置添加背景色和布局方向 >
    <!-- 待添加布局元素 -->
</LinearLayout>
```

（2）在主布局中添加一个显示 logo 的 ImageView 组件，可根据实际屏幕大小设置图片宽高，请完成如下代码中的提示部分。

```
<ImageView
        android:layout_width="720px"
        android:layout_height="281px"
        请在此位置添加 logo 图片其他属性 />
```

（3）在主布局中继续添加一个线性布局，该布局分为两列，分别放置头像和表格布局，表格布局也分为两行，放置用户名和密码，请完成如下代码中的提示部分。

```
<LinearLayout
    android:layout_width="match_parent"
    android:layout_height="wrap_content"
    请在此位置添加该线性布局其他属性 >
    <!-- 第一列，显示头像 -->
    <ImageView
        android:src="@drawable/head"
        请在此位置添加图片其他属性 />
    <!-- 第二列，放置表格布局 -->
    <TableLayout
        android:layout_width="wrap_content"
        android:layout_height="wrap_content"
        请在此位置添加该表格布局其他属性 >
        <!-- 第一行，放置账号文本框和文本输入框 -->
        <TableRow>
            <TextView
                    android:text="账号:"
                    请在此位置添加账号文本框其他属性 />
                <EditText
                    android:id="@+id/userName"
                    请在此位置添加账号文本输入框其他属性 />
        </TableRow>
```

```
        <!-- 第二行，放置密码文本框和文本输入框 -->
        <TableRow>
            <TextView
                android:text="密码:"
                请在此位置添加密码文本框其他属性  />
            <EditText
                android:id="@+id/userPass"
                请在此位置添加密码输入框其他属性  />
        </TableRow>
    </TableLayout>
</LinearLayout>
```

（4）在主布局中继续添加一个登录按钮，请完成如下代码中的提示部分。

```
<Button
    android:text="登  录"
    android:onClick="login"
    请在此位置添加登录按钮其他属性  />
```

（5）在主布局中继续添加三个已选中的复选框，分别为记住密码、自动登录和接收产品推广，请完成如下代码中的提示部分。

```
<CheckBox
    android:text="记住密码"
    请在此位置添加记住密码其他属性  />
<CheckBox
    android:text="自动登录"
    请在此位置添加自动登录其他属性  />
<CheckBox
    android:text="接收产品推广"
    请在此位置添加接收产品推广其他属性  />
```

（6）在主布局中继续添加一个布局管理器（查看实验要求），该布局管理中放置忘记密码和注册账号两个按钮，请完成如下代码中的提示部分。

```
<请在此位置添加布局管理器名和其他属性 >
    <Button
        android:text="忘记密码"
        android:onClick="forgetPass"
        请在此位置添加忘记密码其他属性  />
    <Button
        android:text="注册账号"
        android:onClick="register"
        请在此位置添加注册账号其他属性  />
</布局管理器名>
```

（7）在主布局中继续添加一个布局管理器（查看实验要求），该布局管理中放置进度条和表示加载中的文本框，请完成如下代码中的提示部分。

```
<请在此位置添加布局管理器名和其他属性
    android:layout_width="match_parent"
    android:layout_height="match_parent" >
    <ProgressBar
        style="?android:attr/progressBarStyleSmall"
        请在此位置添加进度条其他属性  />
```

```
    <TextView
        android:id="@+id/loading"
        请在此位置添加加载中文本框其他属性  />
</布局管理器名>
```

（8）在主 Activity 文件中，添加用于登录的 login 方法、用于处理忘记密码的 forgetPass 方法、用于打开注册界面 register 方法，请完成如下代码中的提示部分。

```
public void login(View view) {
    // 请添加登录处理
}
public void forgetPass(View view) {
    // 请添加打开忘记密码界面
}
public void register(View view) {
    // 请添加打开注册界面
}
```

补充完毕上述注释区域，运行实例，能正常显示界面和处理登录、忘记密码、注册按钮的单击事件。

思考：

（1）不使用题目要求的布局管理器嵌套层次，能否采用其他嵌套方式，实现该题目要求。

（2）能否通过<include>和<merge>标签进行布局文件的复用。

实验四 多线程应用

【目的】

1. 掌握多线程的基本概念
2. 掌握 Handler 类的基本用法
3. 掌握 AsyncTask 类的基本用法
4. 掌握 Timer 类的基本用法
5. 掌握多线程综合应用

【理论】

多线程经常用于游戏、多媒体程序、有网络连接的程序、进行耗时计算等的程序中。Android 使用 Thread 类创建线程，并提供用于发送消息和处理消息的类 Handler。在后台工作比较简单，只需要向 UI 线程传递一些简单数据的环境下，Android 还提供了 AsyncTask 类，AsyncTask 是 Android 提供的一个轻量级的基于多线程的进行后台异步工作处理的类。另外，Timer 定时器也是常用的实现多线程程序的方式。

【内容】

启动应用先显示一个图片“正在获取网络图片”，开启新线程从网络下载图片，下载完毕替换“正在获取网络图片”；单击“查看更多”按钮，在打开新界面中使用多线程从本地资源中随机选取图片进行切换显示。完成后的效果如图 1 ~ 图 4 所示。

图 1　下载图片前

图 2　下载图片后

图 3　图片切换 1

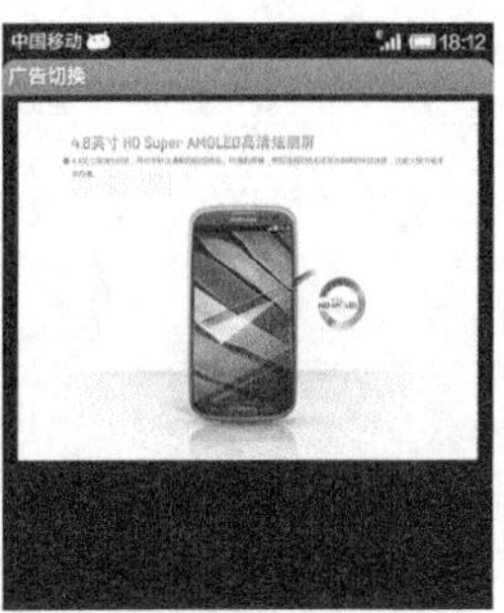

图 4　图片切换 2

【步骤】

（1）创建 Android 项目，修改布局文件如下所示，请将下列空出代码部分补全。

```
<LinearLayout xmlns:android="http://schemas.android.com/apk/res/android"
    android:layout_width="match_parent"
    android:layout_height="match_parent"
    请在此位置添加背景色和布局方向 >
    <TextView
        android:layout_width="match_parent"
        android:layout_height="wrap_content"
        请在此位置添加产品介绍文本框其他属性
        android:text="产品介绍" />
    <ImageView android:id="@+id/iv"
        android:layout_width="wrap_content"
        android:layout_height="wrap_content"
        请在此位置添加图片组件其他属性，和"正在获取网络图片"对应图片 />
    <Button android:id="@+id/button"
        请在此位置添加按钮属性和 onClick 属性
        android:text="查看更多" />
</LinearLayout>
```

（2）主 Activity 类 MainActivity.java，代码如下所示。

```
public class MainActivity extends Activity {
    private ImageView iv = null;
    @Override
    protected void onCreate(Bundle savedInstanceState) {
        super.onCreate(savedInstanceState);
        setContentView(R.layout.main);
        iv = (ImageView) findViewById(R.id.iv);
        //开启一个线程从网络加载图片
        new Thread(new Runnable() {
            public void run() {
                //可修改为其他网路图片地址
                final Bitmap bm = getPic("http://img0.pconline.com.cn/pconline/1303/11/
3209483_25_500.jpg");
                try {
                    Thread.sleep(2000); //睡眠 2 秒
                }
                catch (Exception e) {
                    e.printStackTrace();
                }
                //再开启一个线程设置从网络下载的图片
                iv.post(new Runnable() {
                    public void run() {
                        //请在此位置设置下载的图片显示
                    }
                });
            }
        }
        ).start();
    }
    //从网络下载图片方法
```

```
    private Bitmap getPic(String path) {
        Bitmap bm = null;
        try {
            URL url = new URL(path);
            URLConnection conn = url.openConnection();
            conn.connect();
            InputStream is = conn.getInputStream();
            bm = BitmapFactory.decodeStream(is);
        } catch (Exception e) {
            e.printStackTrace();
        }
        return bm;
    }
    //启动第二个界面
    public void myclick(View view) {
        Intent intent = new Intent();
        intent.setClass(this, MyThread.class);
        startActivity(intent);
    }
}
```

（3）修改布局文件 activity_thread.xml，请将下列空出代码部分补全。

```
<LinearLayout xmlns:android="http://schemas.android.com/apk/res/android"
    android:layout_width="match_parent"
    android:layout_height="match_parent"
    请在此位置添加线性布局相关属性 >
    <ImageView android:id="@+id/iv1"
        android:layout_width="wrap_content"
        android:layout_height="wrap_content"
        android:adjustViewBounds="true"
        请在此位置设置图片适应屏幕大小相关属性  />
</LinearLayout>
```

（4）修改 Activity 类文件 MyThread.java，代码如下所示。

```
public class MyThread extends Activity implements Runnable {
    private ImageView iv=null;
    private Handler handler=null;
    private int[] path= { R.drawable.img1, R.drawable.img2, R.drawable.img3,
                R.drawable.img4, R.drawable.img5, R.drawable.img6  };
    private Thread t;
    @SuppressLint("HandlerLeak")
    @Override
    protected void onCreate(Bundle savedInstanceState) {
        super.onCreate(savedInstanceState);
        setContentView(R.layout.activity_thread);
        iv=(ImageView) findViewById(R.id.iv1);
        t=new Thread(this);
        t.start();
        handler=new Handler() {
            @Override
            public void handleMessage(Message msg) {
                if(msg.what == 0x111)
                    iv.setImageResource(path[msg.arg1]);
                super.handleMessage(msg);
            }
        };
```

```
    }
    @Override
    public void run() {
        int index=0;
        while(!Thread.currentThread().isInterrupted()) {
            index=new Random().nextInt(path.length);
            Message msg=handler.obtainMessage();
            msg.arg1=index;
            msg.what=0x111;
            handler.sendMessage(msg);
            try {
                Thread.sleep(1000);
            }
            catch(Exception e) {
                e.printStackTrace();
            }
        }
    }
    @Override
    protected void onDestroy() {
        super.onDestroy();
        //请在此位置关闭线程
    }
}
```

（5）在 AndroidManifest.xml 文件中增加网络访问许可，代码如下所示。

```
<uses-permission android:name = "android.permission.INTERNET"/>
```

补充完毕本例，运行如图 1 ~ 图 4 所示的效果。再参照本例使用 AsyncTask 类下载图片，并使用 Timer 定时器定时加载本地图片资源。

实验五 基于文件的日程安排

【目的】

1. 掌握首选项方式的存储和读取
2. 掌握位于内存上的数据文件的存储和读取
3. 掌握位于 SD 卡上的数据文件的存储和读取

【理论】

用户首选项信息是以键值对的形式存放在应用中的 XML 文件中，用于保存程序配置信息等少量数据。Android 提供了 SharedPreferences 类及相关的一系列方法来操作和处理这些数据信息。数据文件也可以保存在机身内存或者 SD 卡中，Android 数据文件的读写操作采用了 FileInputStream 类和 FileOutputStream 类，以及相关的一系列方法。

【内容】

实现基于文件的日程安排应用，运行结果如图 1～图 3 所示。通过单击增加图标打开新增活动界面，通过单击保存按钮将活动内容写入到文件，并返回前页。

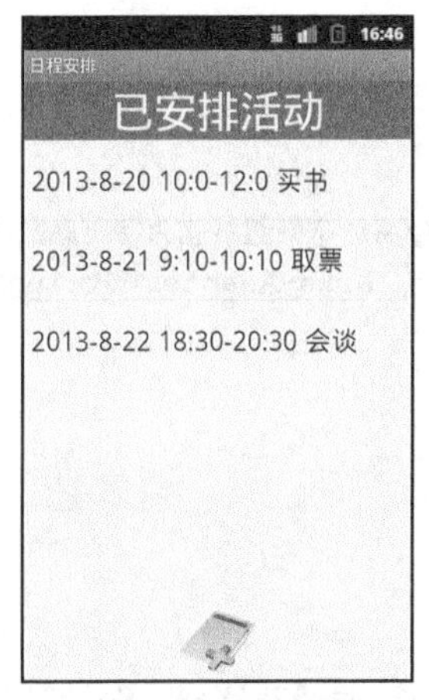

图 1 已安排活动页

图 2 设置新活动时间

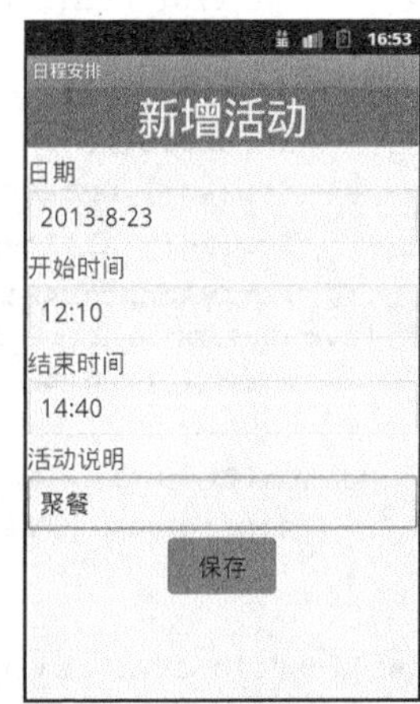

图 3 新增活动

【步骤】

（1）修改主布局如下所示，请将下列空出代码部分补全。

```
<RelativeLayout xmlns:android = "http://schemas.android.com/apk/res/android"
    请在此位置添加主布局其他属性 >
    <TextView android:id = "@+id/title"
        请在此位置添加文本框其他属性
        android:text = "通讯录" />
    <ListView android:id = "@+id/listView"
        请在此位置添加列表框其他属性 />
    <ImageView
        请在此位置添加图片组件其他属性
        android:onClick = "add" />
</RelativeLayout>
```

（2）修改主 Activity 文件，通过 getItemList()方法从文件 item 中读取数据显示到 ListView 组件中。单击新增打开增加活动界面，新增活动界面关闭返回时在 onActivityResult()方法中再次调用 getItemList()方法获取最新数据并显示，代码如下所示。

```
public class MainActivity extends Activity {
    private ListView itemList;
    @Override
    protected void onCreate(Bundle savedInstanceState) {
        super.onCreate(savedInstanceState);
        setContentView(R.layout.main);
        itemList = (ListView) findViewById(R.id.itemList);
        getItemList();
    }
    private void getItemList() {
        ArrayList<String> list = new ArrayList<String>();
        try {
            FileInputStream in = openFileInput("item");
            BufferedReader br = new BufferedReader(new InputStreamReader(in));
            String line = "";
            while((line = br.readLine()) != null) {
                list.add(line);
            }
            br.close();
            in.close();
            String[] contents= {};
            String[] allItem = list.toArray(contents); //将动态数组转化为字符串数组
            ArrayAdapter<String>      adapter = new      ArrayAdapter<String>(this,
android.R.layout.simple_list_item_1, allItem);
            itemList.setAdapter(adapter);
        }
        catch(Exception e) {
            e.printStackTrace();
        }
    }
    public void add(View view) {
        Intent intent = new Intent(MainActivity.this, AddItemActivity.class);
        startActivityForResult(intent, 0x111); // 打开新界面
    }
```

```
    @Override
    protected void onActivityResult(int requestCode, int resultCode, Intent data) {
        super.onActivityResult(requestCode, resultCode, data);
        if(requestCode == 0x111 && resultCode == 0x111) {
            getItemList();
        }
    }
}
```

（3）创建增加活动界面布局文件 additem.xml，请将下列空出代码部分补全。

```
<LinearLayout xmlns:android = "http://schemas.android.com/apk/res/android"
    请在此位置添加主布局其他属性 >
    <TextView
        android:text = "新增活动"
        请在此位置添加新增活动文本框其他属性 />
    <TextView
        android:text = "日期"
        请在此位置添加日期文本框其他属性 />
    <EditText android:id = "@+id/date"
        android:onClick = "chooseDate"
        请在此位置添加日期编辑框其他属性 />
    <TextView
        android:text = "开始时间"
        请在此位置添加开始时间文本框其他属性 />
    <EditText android:id = "@+id/startTime"
        android:onClick = "chooseStartTime"
        请在此位置添加开始时间编辑框其他属性 />
    <TextView
        android:text = "结束时间"
        请在此位置添加结束时间文本框其他属性 />
    <EditText android:id = "@+id/endTime"
        android:onClick = "chooseEndTime"
        请在此位置添加结束时间编辑框其他属性 />
    <TextView
        android:text = "活动说明"
        请在此位置添加活动说明文本框其他属性 />
    <EditText android:id = "@+id/item"
        请在此位置添加活动说明编辑框其他属性 />
    <Button android:id = "@+id/save"
        android:onClick = "save"
        android:text = "   保存   "
        请在此位置添加保存按钮其他属性 />
</LinearLayout>
```

（4）创建增加活动界面 Activity 文件 AddItemActivity.java，输入新增活动的日期、开始时间、结束时间、活动名称，保存到文件 item 中，代码如下所示。

```
public class AddItemActivity extends Activity {
    private EditText date, startTime, endTime, item;
    @Override
    protected void onCreate(Bundle savedInstanceState) {
        super.onCreate(savedInstanceState);
        setContentView(R.layout.additem);
```

```
        date = (EditText) findViewById(R.id.date);
        startTime = (EditText) findViewById(R.id.startTime);
        endTime = (EditText) findViewById(R.id.endTime);
        item = (EditText) findViewById(R.id.item);
    }
    // 通过日期对话框组件选择日期
    public void chooseDate(View view) {
        DatePickerDialog datepd = new DatePickerDialog(this, new DatePickerDialog.OnDateSetListener() {
            @Override
            public void onDateSet(DatePicker dp, int year, int month, int day) {
                date.setText(year + "-" + (month + 1) + "-" + day); // 写入到日期框
            }
        }, 2013, 8, 20); // 默认为 2013-8-20 日
        datepd.setMessage("请选择日期");
        datepd.show();
    }
    // 通过时间对话框组件选择开始时间
    public void chooseStartTime(View view) {
        TimePickerDialog timepd = new TimePickerDialog(this, new TimePickerDialog.OnTimeSetListener() {
            @Override
            public void onTimeSet(TimePicker view, int hourOfDay, int minute) {
                startTime.setText(hourOfDay + ":" + minute);
            }
        }, 10, 0, true);  //设置时间对话框，并设置默认时间为 24 小时制 10 点整
        timepd.setMessage("请选择开始时间");
        timepd.show();
    }
    // 通过时间对话框组件选择结束时间
    public void chooseEndTime(View view) {
        TimePickerDialog timepd = new TimePickerDialog(this, new
TimePickerDialog.OnTimeSetListener() {
            @Override
            public void onTimeSet(TimePicker view, int hourOfDay, int minute) {
                endTime.setText(hourOfDay + ":" + minute);
            }
        }, 12, 0, true);
        timepd.setMessage("请选择结束时间");
        timepd.show();
    }
    // 保存活动信息
    public void save(View view) {
        StringBuilder result = new StringBuilder();
        result.append(date.getText().toString() + " ");
        result.append(startTime.getText().toString() + "-");
        result.append(endTime.getText().toString() + " ");
        result.append(item.getText().toString());
        try {
            //以追加方式保存活动信息到 item 文件中
            FileOutputStream out = openFileOutput("item", MODE_APPEND);
            PrintStream ps = new PrintStream(out);
            ps.println(result.toString());
            ps.close();
            out.close();
```

```
            Toast.makeText(this, "保存完毕!", Toast.LENGTH_LONG).show();
            Intent intent = getIntent();
            setResult(0x111, intent);
            finish(); //关闭当前界面，返回到前页
        }
        catch(Exception e) {
            e.printStackTrace();
        }
    }
}
```

（5）在 AndroidManifest.xml 文件中添加新增活动界面的< activity>标记，代码如下所示。

```
<activity
      android:name = "com.example.schedule.AddItemActivity"
      android:label = "@string/app_name" >
</activity>
```

运行本实例，完成结果如图 1 ~ 图 3 所示，另请增加修改和删除功能。

实验六
基于 SQLite 的通信录

【目的】

1. 掌握 SQLiteOpenHelper 类结构
2. 掌握基于 SQLite 数据库的应用开发过程
3. 掌握 Content Provider 发布数据的方法
4. 掌握 Content Resolver 获取数据的方法

【理论】

Android 系统中集成了 SQLite 数据库，并且为数据库的操作提供了相关的类和方法，便于没有数据库开发经验的开发者编写程序。另外，Android 平台中利用 Content Provider 机制来实现跨应用程序数据共享。一个应用程序可以通过 Content Provider 来发布自己的数据，其他的应用程序可以通过 Content Resolver 来获取共享数据。

【内容】

实现基于 SQLite 数据库的通信录应用，运行结果如图 1 ~ 图 3 所示。通过单击增加图标打开添加通信录界面，通过单击通信录中的各条信息可删除选中项。

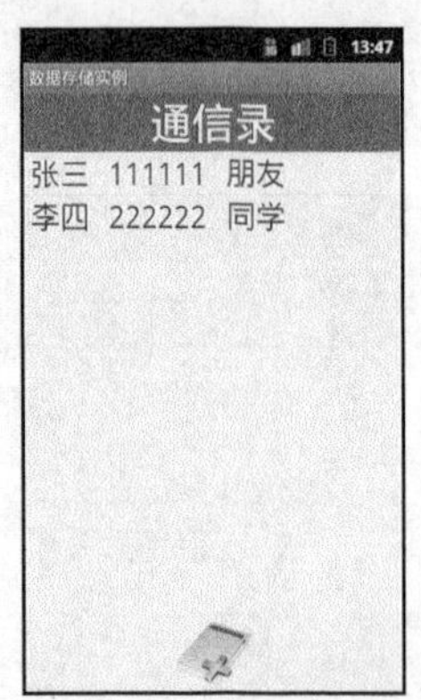

图 1　通信录列表页

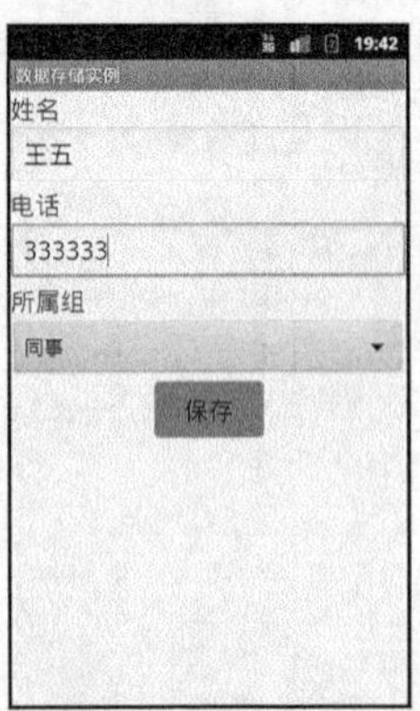

图 2　添加联系人

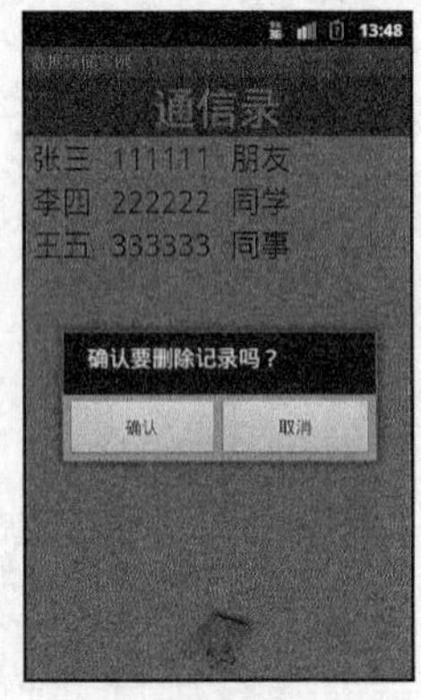

图 3　删除联系人

【步骤】

（1）建立数据库操作类 DatabaseHelper，代码如下所示。

```
public class DatabaseHelper extends SQLiteOpenHelper {
    private static final String DB_NAME = "MyRelation.db";
    private static final String TABLE_NAME = "relation";
    private static final String CREATE_TABLE = "create table relation(_id integer
primary key autoincrement,name text,tel text,groupName text);"; //字段_id为主键，自动增长
    private SQLiteDatabase db; // 创建SQLiteDatabase实例
    DatabaseHelper(Context context) { //构造方法
        super(context, DB_NAME, null, 2);
    }
    public void insert(ContentValues values) { //插入
        SQLiteDatabase db = getWritableDatabase();
        db.insert(TABLE_NAME, null, values);
        db.close();
    }
    public void del(int id) { //删除
        if(db == null)
            db = getWritableDatabase();
        db.delete(TABLE_NAME, "_id = ?", new String[]{ String.valueOf(id) });
    }
    public Cursor query() { //查询
        SQLiteDatabase db = getWritableDatabase();
        Cursor cursor = db.query(TABLE_NAME, null, null, null, null, null, null);
        return cursor;
    }
    public void close() { // 关闭数据库
        if(db != null) db.close();
    }
    public void onCreate(SQLiteDatabase db) { //重写onCreate() 方法
        this.db = db;
        db.execSQL(CREATE_TABLE);
    }
    public void onUpgrade(SQLiteDatabase db, int oldVersion, int newVersion) {}
}
```

（2）修改主布局如下，请将下列空出代码部分补全。

```
<RelativeLayout xmlns:android = "http://schemas.android.com/apk/res/android"
    请在此位置添加主布局其他属性 >
    <TextView android:id = "@+id/title"
        请在此位置添加文本框其他属性
        android:text = "通信录" />
    <ListView android:id = "@+id/listView"
        请在此位置添加列表框其他属性 />
    <ImageView
        请在此位置添加图片组件其他属性
        android:onClick = "add" />
</RelativeLayout>
```

（3）修改主 Activity 文件 MainActivity.java，代码如下所示。

```
public class MainActivity extends Activity {
    private ListView listView;
    @Override
    protected void onCreate(Bundle savedInstanceState) {
        super.onCreate(savedInstanceState);
```

```
        setContentView(R.layout.main);
        listView = (ListView) findViewById(R.id.listView);
        getRelationFromDB();
    }
    private void getRelationFromDB() { //从数据加载联系人信息到列表视图
        final DatabaseHelper dbHelper = new DatabaseHelper(this);
        Cursor cursor = dbHelper.query();
        String[] from= { "_id", "name", "tel", "groupName" }; //数据库字段数组
        int[] to= { R.id._id, R.id.name, R.id.tel, R.id.group }; //显示布局 id 数组
        SimpleCursorAdapter scadapter = new SimpleCursorAdapter(this, R.layout.relationlist,
cursor, from, to);
        listView.setAdapter(scadapter);
        listView.setOnItemClickListener(new AdapterView.OnItemClickListener() {
            public void onItemClick(AdapterView<?> adapter, View view, int position, long id) {
                final long temp = id;
                AlertDialog.Builder adBuilder = new AlertDialog.Builder(MainActivity.this);
                adBuilder.setMessage("确认要删除记录吗? ").setPositiveButton("确认", new
DialogInterface.OnClickListener() {
                    public void onClick(DialogInterface dialog, int which) {
                        dbHelper.del((int) temp);
                        Cursor cursor = dbHelper.query();
                        String[] from = { "_id", "name", "tel", "groupName" };
                        int[] to= { R.id._id, R.id.name, R.id.tel, R.id.group };
                        SimpleCursorAdapter scadapter = new SimpleCursorAdapter(get
ApplicationContext(), R.layout.relationlist, cursor, from, to);
                        MainActivity.this.listView.setAdapter(scadapter);
                    }
                }).setNegativeButton("取消", new DialogInterface.OnClickListener() {
                    public void onClick(DialogInterface dialog, int which) {}
                });
                AlertDialog aleraDialog = adBuilder.create();
                aleraDialog.show();
            }
        });
        dbHelper.close();
    }
    public void add(View view) { //增加按钮单击
        Intent intent = new Intent(MainActivity.this, AddRelationActivity.class);
        startActivityForResult(intent, 0x111); //打开新界面
    }
    @Override
    protected void onActivityResult(int requestCode, int resultCode, Intent data) {
        super.onActivityResult(requestCode, resultCode, data);
        if(requestCode == 0x111 && resultCode == 0x111) {
            getRelationFromDB(); //增加新联系人后返回刷新列表框
        }
    }
}
```

(4) 创建列表视图布局文件 relationlist.xml，请将下列空出代码部分补全。

```
<?xml version = "1.0" encoding = "utf-8"?>
<LinearLayout xmlns:android = "http://schemas.android.com/apk/res/android"
    请在此位置添加主布局其他属性 >
    <!-- 对应主键字段,自动增长,不显示 -->
```

```
    <TextView android:id = "@+id/_id"
        请在此位置设置文本框其他属性 />
    <TextView android:id = "@+id/name"
        请在此位置设置显示姓名的文本框其他属性 />
    <TextView android:id = "@+id/tel"
        请在此位置设置显示电话的文本框其他属性 />
    <TextView android:id = "@+id/group"
        请在此位置设置显示联系人群组的文本框其他属性 />
</LinearLayout>
```

（5）创建增加联系人界面布局文件 addrelation.xml，请将下列空出代码部分补全。

```
<LinearLayout xmlns:android = "http://schemas.android.com/apk/res/android"
    请在此位置添加主布局其他属性 >
    <TextView android:text = "姓名"
        请在此位置添加姓名文本框其他属性 />
    <EditText android:id = "@+id/addName"
        请在此位置添加姓名编辑框其他属性 />
    <TextView android:text = "电话"
        请在此位置添加电话文本框其他属性 />
    <EditText android:id = "@+id/addTel"
        请在此位置添加电话的编辑框其他属性 />
    <TextView android:text = "所属组"
        请在此位置添加所属组文本框其他属性 />
    <Spinner android:id = "@+id/addGroup"
        请在此位置添加所属组下拉列表其他属性  />
    <Button android:id = "@+id/save"
        android:text = "   保存   "
        请在此位置添加保存按钮的其他属性 />
</LinearLayout>
```

（6）创建增加联系人界面主 Activity 文件 AddRelationActivity.java，代码如下所示。

```
public class AddRelationActivity extends Activity  {
    private EditText addName, addTel;
    private Spinner addGroup;
    @Override
    protected void onCreate(Bundle savedInstanceState) {
        super.onCreate(savedInstanceState);
        setContentView(R.layout.addrelation);
        addName = (EditText) findViewById(R.id.addName);
        addTel = (EditText) findViewById(R.id.addTel);
        addGroup = (Spinner) findViewById(R.id.addGroup);
    }
    public void save(View view) {
        final ContentValues values = new ContentValues();
        values.put("name", addName.getText().toString());
        values.put("tel", addTel.getText().toString());
        values.put("groupName", addGroup.getSelectedItem().toString());
        final DatabaseHelper dbHelper = new DatabaseHelper(getApplicationContext());
        final AlertDialog.Builder adBuilder = new AlertDialog.Builder(this);
        adBuilder.setMessage(" 确 认 保 存 记 录 吗 ？ ").setPositiveButton(" 确 认 ",  new
DialogInterface.OnClickListener() {
            public void onClick(DialogInterface dialog, int which) {
```

```
                dbHelper.insert(values); //插入数据
                Intent intent = getIntent();
                setResult(0x111, intent);
                AddRelationActivity.this.finish(); //关闭当前界面，返回首页
            }
        }).setNegativeButton("取消", new DialogInterface.OnClickListener() {
            public void onClick(DialogInterface dialog, int which) {}
        });
        AlertDialog aleraDialog = adBuilder.create();
        aleraDialog.show(); //显示对话框
    }
}
```

（7）在 AndroidManifest.xml 文件中添加 AddRelationActivity 界面应用。

```
<activity
    android:name = "com.example.relationperson.AddRelationActivity"
    android:label = "@string/app_name" >
</activity>
```

运行本实例，完成结果如图 1 ~ 图 3 所示，另请增加批量删除功能。

实验七 天气预报应用

【目的】

1. 理解网络通信基础概念
2. 掌握 HTTP 通信原理与方法
3. 掌握 Socket 通信原理与方法

【理论】

在 Android SDK 内提供了多种网络连接处理方式。Android 系统通过引入了开源项目 HttpClient，为已经熟悉 HttpClient 的开发人员提供了更加便捷的网络访问方法。Android 除了提供了访问 HTTP 服务的基本功能，还提供了 HTTP 请求队列管理，以及 HTTP 连接池管理，提高并发请求情况下的处理效率，以及提供了网络状态监视等接口。Socket 称作“套接字”，用于描述 IP 地址和端口。Android 支持应用程序通过“套接字”向网络发送请求或者应答网络请求。另外，Android 还对移动终端中的蓝牙和 Wi-Fi 通信提供了强大的支持能力。

【内容】

实现一个天气预报的应用，如图 1 ~ 图 3 所示。

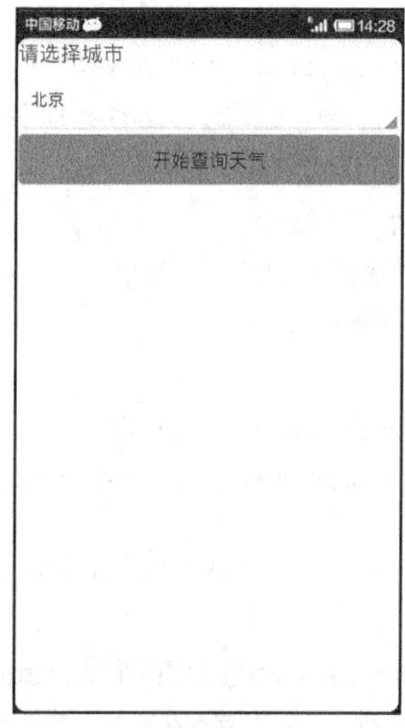

图 1 天气预报首页

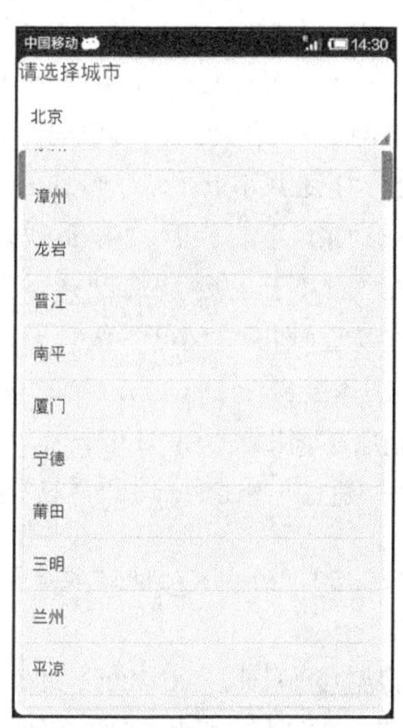

图 2 选择城市

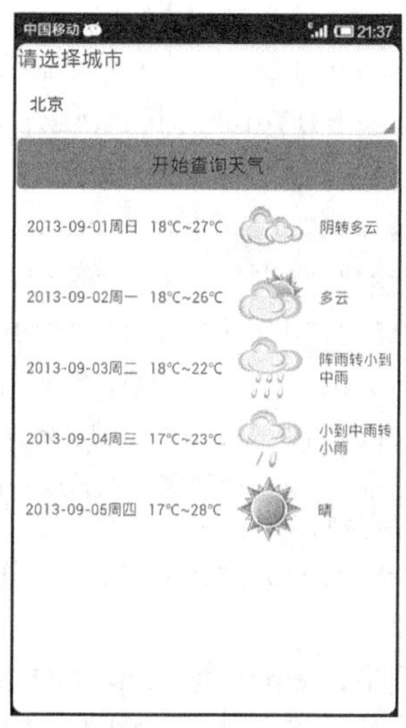

图 3 城市天气预报

【步骤】

Android 系统下获取天气预报除了使用 Google 的 API 之外，还可以使用 http://webservce.webxml.com.cn/WebServices/WeatherWS.asmx 接口，该接口数据来自中央气象局。Google 的 API 不稳定，而中央气象局接口中的免费版每天又有访问次数的限制，使用起来存在一些不便。WebXml.com.cn 提供的接口，可访问 400 多个城市天气预报 Web 服务，包含 2～300 个以上的中国城市和 100 个以上的国外城市天气预报数据，数据每 2.5 小时左右自动更新一次，非常方便。最后就是中国天气网提供的接口，类型丰富，使用最为广泛。

其中，中国天气网提供的 XML 接口如下：

- http://flash.weather.com.cn/wmaps/xml/china.xml

该接口可返回全国各省会城市、直辖市的当天天气信息，以 XML 文件形式提供，需配合 XML 解析器使用。

中国天气网提供的 JSON 接口为：

- http://www.weather.com.cn/data/sk/101010100.html
- http://www.weather.com.cn/data/cityinfo/101010100.html

其中 101010100 为北京的编码，返回北京当天天气，信息项略有差异。

- http://m.weather.com.cn/data/101010100.html

该接口能返回指定城市 6 天内的天气信息，返回结果如下所示。


```
{"weatherinfo":{"city":"北京","city_en":"beijing","date_y":"2013年8月23日","date":"",
"week":"星期五","fchh":"11","cityid":"101110101","temp1":"32℃～24℃",
"temp2":"32℃～25℃","temp3":"28℃～23℃","temp4":"27℃～22℃","temp5":"31℃～23℃",
"temp6":"31℃～22℃","tempF1":"89.6℉～75.2  ","tempF2":"89.6℉～77  ",
"tempF3":"82.4℉～73.4  ","tempF4":"80.6℉～71.6  ","tempF5":"87.8℉～73.4  ",
"tempF6":"87.8℉～71.6  ","weather1":"多云转阴","weather2":"阴转多云",
"weather3":"多云","weather4":"阴","weather5":"小雨","weather6":"阴转多云","img1":"1",
"img2":"2","img3":"2","img4":"1","img5":"1","img6":"99","img7":"2","img8":"99","img9":"7",
"img10":"99","img11":"2","img12":"1","img_single":"1","img_title1":"多云",
"img_title2":"阴","img_title3":"阴","img_title4":"多云","img_title5":"多云",
"img_title6":"多云","img_title7":"阴","img_title8":"阴","img_title9":"小雨",
"img_title10":"小雨","img_title11":"阴","img_title12":"多云","img_title_single":"多云",
"wind1":"东风小于3级","wind2":"东北风小于3级","wind3":"东北风小于3级",
"wind4":"东风小于3级","wind5":"东风小于3级","wind6":"东风小于3级",
"fx1":"东风","fx2":"东风","fl1":"小于3级","fl2":"小于3级","fl3":"小于3级",
"fl4":"小于3级","fl5":"小于3级","fl6":"小于3级","index":"炎热",
"index_d":"天气炎热，建议着清凉夏季服装。","index48":"炎热",
"index48_d":"天气炎热，建议着清凉夏季服装。","index_uv":"弱","index48_uv":"弱",
"index_xc":"适宜","index_tr":"适宜","index_co":"较不舒适","st1":"32","st2":"22",
"st3":"33",
"st4":"25","st5":"27","st6":"23","index_cl":"较适宜","index_ls":"适宜","index_ag":"
易发"}}
```


其中，temp1 到 temp6 为六天的温度范围，weather1 到 weather6 为六天的天气情况，img1 到 img12 是六天早晚天气图片的编号等。通过解析相应的键值，即可获取所需要的信息。下面使用这个接口实现一个天气预报的应用，步骤如下。

（1）根据图 1～图 3 所示的布局，修改布局文件如下所示，请将下列空出的代码部分补全。

```
<LinearLayout
    android:background = "#FFFFFFFF"
    android:orientation = "vertical"
    请在此位置添加主布局其他属性  >
    <!-- 第一行为提示本文-->
    <TextView
        请在此位置添加本文框其他属性
        android:text = "请选择城市"
        android:textSize = "20sp" />
    <!-- 第二行为城市下拉列表框-->
    <Spinner android:id = "@+id/citySpinner"
        android:layout_width = "match_parent"
        android:layout_height = "60dp" />
    <!-- 第三行为查询按钮-->
    <Button
        请在此位置添加按钮其他属性
        android:id = "@+id/btn"
        android:onClick = "search"
        android:text = "开始查询天气" />
    <!-- 第四行为天气预报第一天布局-->
    <LinearLayout
        请在此位置添加线性布局属性 >
        <TextView android:id = "@+id/date1"
            请在此位置添加日期文本框其他属性 />
        <TextView android:id = "@+id/temp1"
            请在此位置添加温度文本框其他属性 />
        <ImageView android:id = "@+id/img1"
            请在此位置添加天气图片组件其他属性 />
        <TextView android:id = "@+id/weather1"
            请在此位置添加天气文本框其他属性 />
    </LinearLayout>
    <!-- 请仿照第一天天气预报布局，完成下面第二天到第五天布局 -->
</LinearLayout>
```

（2）在主 Activity 文件中建立城市名称和城市编码对照表，所有城市的编号信息请登录中国天气网查询。

```
private String[][] citycode= {
    { "北京", "天津", "上海", "石家庄", "郑州", "合肥", ...... },
    { "101010100", "101030100", "101020100", "101090101", "101180101",
     "101220101", ...... }
};
```

（3）将中国天气网提供的天气图片（共 32 个）复制进入工程的 res 对应的 drawable 目录下，并在主 Activity 文件中建立对应关系。

```
private int[] ico= {
    R.drawable.b_0, R.drawable.b_1, R.drawable.b_2, R.drawable.b_3, R.drawable.b_4,
    R.drawable.b_5, R.drawable.b_6, ......, R.drawable.b_31
};
```

（4）在主 Activity 文件中建立布局组件对象。

```
private Button btn; //查询按钮
private Spinner citySpinner; //下拉列表框
private int selectedPos; //选中城市的编号，通过该编号可取出城市编码
private TextView date1, date2, date3, date4, date5; //日期
private TextView weather1, weather2, weather3, weather4, weather5; //天气
private TextView temp1, temp2, temp3, temp4, temp5; //温度
private ImageView img1, img2, img3, img4, img5; //天气图片
private String result = ""; //查询天气返回结果
```

（5）在主 Activity 文件中，建立 HTTP 访问中国天气网的方法，通过访问 http://m.weather.com.cn/data/城市编码.html 文件获取天气。

```
private boolean getWeather() {
    boolean isSuccess = false;
    String URL = "http://m.weather.com.cn/data/";
    URL url = null;
    try {
        URL = URL + citycode[1][selectedPos] + ".html";
        url = new URL(URL);
    } catch(MalformedURLException e) {
        e.printStackTrace();
        return isSuccess;
    }
    if(url != null) {
        try {
            // 使用 HttpURLConnection 打开连接
            HttpURLConnection urlConn = (HttpURLConnection) url.openConnection();
            // 得到读取的内容
            InputStreamReader in = new InputStreamReader(urlConn.getInputStream());
            // 为输出创建 BufferedReader
            BufferedReader buffer = new BufferedReader(in);
            result = buffer.readLine();
            in.close();
            urlConn.disconnect();
            isSuccess = true;
        } catch(Exception e) {
            e.printStackTrace();
        }
    }
    return isSuccess;
}
```

（6）在主 Activity 文件中，建立查询按钮单击后的处理方法。

```
public void search(View view) {
    Calendar cal = Calendar.getInstance();  //获取当前日期
    //设置日期输出格式
    SimpleDateFormat sdf = new SimpleDateFormat("yyyy-MM-ddEEE");
    if(getWeather()) {  // 获取天气成功
        if(!result.equals("")) { //天气返回结果不空
            try {  //解析 JSON 格式的返回结果
                JSONObject root = new JSONObject(result);
```

```
            JSONObject json = new JSONObject(root.getString("weatherinfo"));
            date1.setText(sdf.format(cal.getTime()));  //输出日期
            cal.add(Calendar.DAY_OF_MONTH, 1);  //第二天
            date2.setText(sdf.format(cal.getTime()));
            cal.add(Calendar.DAY_OF_MONTH, 1);  //第三天
            date3.setText(sdf.format(cal.getTime()));
            cal.add(Calendar.DAY_OF_MONTH, 1);  //第四天
            date4.setText(sdf.format(cal.getTime()));
            cal.add(Calendar.DAY_OF_MONTH, 1);  //第五天
            date5.setText(sdf.format(cal.getTime()));
            weather1.setText(json.getString("weather1"));  //输出天气
            weather2.setText(json.getString("weather2"));
            weather3.setText(json.getString("weather3"));
            weather4.setText(json.getString("weather4"));
            weather5.setText(json.getString("weather5"));
            temp1.setText(json.getString("temp1"));  //输出温度
            temp2.setText(json.getString("temp2"));
            temp3.setText(json.getString("temp3"));
            temp4.setText(json.getString("temp4"));
            temp5.setText(json.getString("temp5"));
            img1.setImageResource(ico[json.getInt("img1")]);  //输出天气图片
            img2.setImageResource(ico[json.getInt("img3")]);
            img3.setImageResource(ico[json.getInt("img5")]);
            img4.setImageResource(ico[json.getInt("img7")]);
            img5.setImageResource(ico[json.getInt("img9")]);
        } catch(Exception e) {
            e.printStackTrace();
        }
    }
  } else
    Toast.makeText(this, "对不起，未获取到天气信息!",
            Toast.LENGTH_LONG).show();
}
```

（7）在主 Activity 文件中，修改 onCreate()方法。

```
protected void onCreate(Bundle savedInstanceState) {
    //隐藏标题栏
    requestWindowFeature(Window.FEATURE_NO_TITLE);
    super.onCreate(savedInstanceState);
    setContentView(R.layout.activity_main);
    //设置访问许可
    if(android.os.Build.VERSION.SDK_INT > 9) {
        StrictMode.ThreadPolicy policy = new
                StrictMode.ThreadPolicy.Builder().permitAll().build();
        StrictMode.setThreadPolicy(policy);
    }
    btn = (Button) findViewById(R.id.btn);
    citySpinner = (Spinner) findViewById(R.id.citySpinner);
    //将城市名对应的数组 citycode[0]加入适配器,放入下拉列表框,让用户选择
    ArrayAdapter<String> adapter = new ArrayAdapter<String>(this,
                    android.R.layout.simple_spinner_item, citycode[0]);
    adapter.setDropDownViewResource(
                    android.R.layout.simple_spinner_dropdown_item);
```

```
        citySpinner.setAdapter(adapter);
        citySpinner.setOnItemSelectedListener(new OnItemSelectedListener() {
            @Override
            public void onItemSelected(AdapterView<?> parent, View view, int pos, long id) {
                selectedPos = pos;
                //当选择城市完成后立刻查询天气
                search(btn);
            }
            @Override
            public void onNothingSelected(AdapterView<?> arg0) {}
            });
        date1 = (TextView) findViewById(R.id.date1);
        weather1 = (TextView) findViewById(R.id.weather1);
        temp1 = (TextView) findViewById(R.id.temp1);
        img1 = (ImageView) findViewById(R.id.img1);
        //此处省略获取其他四个日期、天气、温度、图片组件对象
    }
```

（8）在 AndroidManifest.xml 文件中，增加访问网络许可。

```
<uses-permission android:name = "android.permission.INTERNET" />
```

运行本实例，显示结果如图 1 ~ 图 3 所示。请参考本例，自行实现上文介绍的通过天气网站提供的其他接口完成天气预报应用。

另请查询资料，在网络访问前先判断移动网络或 Wi-Fi 网络是否可用。

实验八 音乐播放器及相机拍摄

【目的】

1. 掌握使用 MediaPlayer 播放音频
2. 掌握使用 SoundPool 播放音频
3. 掌握使用 VideoView 播放视频
4. 掌握使用 MediaPlayer 和 SurfaceView 播放视频
5. 掌握通过摄像头意图拍照
6. 掌握通过 Camera 类拍照

【理论】

Android 系统对常见的媒体类型提供了多种内置编解码功能，较好地支持了在程序中集成音频和视频的应用。其中 MediaPlayer 既可以播放音频，也可以播放视频，但播放音频时每次只能播放一个，适用于播放长音乐和背景音乐，播放视频时由于没提供输出界面，所以需要与 SurfaceView 配合使用；SoundPool 适合同时播放多个短小的音频，如消息提示、游戏音效等；MediaController 配合 VideoView 组件能够实现比较完善的视频播放功能。

Android 系统还提供了两种摄像头的使用方法，一种是通过摄像头意图，编写代码简单、快速；另一种是使用 Camera 类，提供了一种更高级的可为用户创建自定义摄像功能的方法。

【内容】

1. 完成一个音乐播放器的开发，如图 1 所示。
2. 完成照相机应用，如图 2 所示。

图 1　音乐播放器界面

图 2　照相机应用界面

【步骤】

1. 完成一个音乐播放器的开发

（1）修改主布局如下所示，请将下列空出代码部分补全。

```
<LinearLayout xmlns:android = "http://schemas.android.com/apk/res/android"
    xmlns:tools = "http://schemas.android.com/tools"
    请在此位置添加主布局其他属性  >
    <!-- 第一行放置四个播放控制按钮 -->
    <LinearLayout
        请在此位置添加该布局属性  >
        请在此位置添加上一首、播放、下一首、停止四个按钮
    </LinearLayout>
    <!-- 第二行放置音量相关组件 -->
    <LinearLayout
        请在此位置添加该布局属性  >
        请在此位置添加音量文本、音量拖动条、音量大小文本
    </LinearLayout>
    <!-- 第三行放置进度相关组件 -->
    <LinearLayout
        请在此位置添加该布局属性  >
        请在此位置添加进度文本、进度拖动条、进度大小文本
    </LinearLayout>
    <!-- 第四行放置表示音乐名的列表视图组件 -->
    <ListView
        请在此位置添加列表视图属性  />
</LinearLayout>
```

（2）在主 Activity 类添加代码中使用到的部分成员变量。

```
private ListView musicView; //布局中列表框组件
private static final String MUSIC_PATH = new String("/sdcard/音乐"); //音乐存放路径
private ArrayList<String> musicFileList = new ArrayList<String>(); //播放文件名列表
private ArrayList<String> musicDirList = new ArrayList<String>();//播放文件路径列表
```

（3）在主 Activity 类的 onCreate()方法中添加如下代码所示。

```
musicView = (ListView) findViewById(R.id.musicView);
setMusicList(); //获取/sdcard/音乐目录下所有 mp3
```

（4）在主 Activity 类添加设置文件列表的方法。

```
private void setMusicList() {
    musicFileList.clear(); //清除文件名列表
    musicDirList.clear(); //清除文件路径列表
    getAllMusic(new File(MUSIC_PATH)); //设置文件名和文件路径列表
    ArrayAdapter<String> musicAdapter = new ArrayAdapter<String>(this,
    android.R.layout.simple_list_item_1, musicFileList);
    musicView.setAdapter(musicAdapter); //将文件名置入列表组件
}
```

（5）在主 Activity 类添加一个递归获取“/sdcard/音乐”目录下所有 mp3 文件的方法，分别将文件名和绝对路径存入不同列表。

```
private void getAllMusic(File dir) {
```

```
    File[] files = dir.listFiles();
    for(File file : files) {
        if(file.isDirectory())
            getAllMusic(file);
        else
            if(file.getAbsolutePath().toLowerCase().endsWith(".mp3")) {
                musicFileList.add(file.getName());
                musicDirList.add(file.getAbsolutePath());
                Log.i(file.getName(), file.getAbsolutePath());
            }
    }
}
```

运行本实例，结果如图 1 所示，现请用户添加播放功能，完成播放上一首、下一首、播放/暂停、停止功能。可进一步实现单击列表框 mp3 名称进行播放。

2. 完成照相机应用

（1）修改主布局文件，将下列空出代码部分补全。

```
<FrameLayout xmlns:android = "http://schemas.android.com/apk/res/android"
    请在此位置添加主布局其他属性  >
    <!-- 第一行放置表格布局 -->
    <TableLayout
        请在此位置添加该布局属性  >
        <TableRow>
            请在此位置添加员工编号文本框和输入框组件
        </TableRow>
        <TableRow>
            请在此位置添加员工姓名文本框和输入框组件
        </TableRow>
    </TableLayout>
    <!-- 第二行放置显示界面组件 -->
    <SurfaceView
        请在此位置添加显示界面组件属性  />
    <!-- 第三行放置拍照组件 -->
    <ImageView
        请在此位置添加拍照组件属性  />
</FrameLayout>
```

（2）在主 Activity 类中添加代码中使用到的部分成员变量。

```
private Camera camera;
private SurfaceView surface;
private CameraPreview cPreview;
```

（3）在主 Activity 类的 onCreate 方法中添加如下代码，来隐藏营运商图标、电量栏。

```
this.getWindow().setFlags(WindowManager.LayoutParams.FLAG_FULLSCREEN,
    WindowManager.LayoutParams.FLAG_FULLSCREEN);//隐藏营运商图标、电量栏
```

（4）在主 Activity 类添加安全获取 Camera 对象实例的方法。

```
// 安全获取 Camera 对象实例的方法
public static Camera getCameraInstance() {
    Camera c = null;
    try {
        c = Camera.open(); //获取 Camera 实例
        c.setDisplayOrientation(90); //如遇到翻转 90° 情况，可使用该方法
```

```
    }
    catch(Exception e) {
        Log.d("提示", "摄像头不可用(正被占用或不存在)");
    }
    return c;
}
```

(5)在主 Activity 类添加释放资源方法。

```
@Override
protected void onDestroy() {
    camera.stopPreview();
    camera.release();
    super.onDestroy();
}
```

本例实现一个输入员工编号和姓名，再进行照相保存员工基本信息的功能，现已将界面布局设计完成，请用户添加拍照、保存功能，使之运行结果如图 2 所示。

参考文献

[1] 姚尚朗，靳岩．Android 开发入门与实战．北京：人民邮电出版社，2013.

[2] 李刚．疯狂 Android 讲义．北京：电子工业出版社，2013.

[3] 软件开发技术联盟．Android 开发实战．北京：清华大学出版社，2013.

[4] 吴亚峰，于复兴，杜化美．Android 应用案例开发大全．北京：人民邮电出版社，2013.

[5] Wallace Jackson．Android 应用开发入门．北京：人民邮电出版社，2013.

[6] 达尔文，姚军．Android 应用开发攻略．北京：机械工业出版社，2013.

[7] Satya Komatineni，Dave MacLean，曾少宁，杨越．精通 Android．北京：人民邮电出版社，2013.

[8] 王雅宁．轻松学 Android 开发．北京：电子工业出版社，2013.

[9] 李宁．Android 开发权威指南．北京：人民邮电出版社，2013.

[10] 林少丹．移动终端应用开发技术——Android 实战．北京：机械工业出版社，2013.

[11] 徐诚．零点起飞学 Android 开发．北京：清华大学出版社，2013.

[12] 范怀宇．Android 开发精要．北京：机械工业出版社，2012.

[13] 汪永松．Android 平台开发之旅．北京：机械工业出版社，2012.

[14] 扶松柏．Android 开发从入门到精通．北京：兵器工业出版社， 2012.

[15] 明日科技．Android 从入门到精通．北京：清华大学出版社，2012.

[16] 李鸥．实战 Android 应用开发．北京：清华大学出版社，2012.

[17] 王世江，余志龙，陈昱勋，郑名杰．Android SDK 开发范例大全．北京：人民邮电出版社，2011.